Conflict, Social Capital and Managing Natural Resources

A West African Case Study

Conflict, Social Capital and Managing Natural Resources

A West African Case Study

Edited by

Keith M. Moore

Sustainable Agriculture and Natural Resource Management (SANREM)

Collaborative Research Support Program (CRSP) West Africa Site

Office of International Research, Education, and Development Virginia Polytechnic Institute and State University

CABI Publishing

CABI Publishing is a division of CAB International

CABI Publishing
CAB International
Wallingford
Oxfordshire OX10 8DE
UK

Tel: +44 (0)1491 832111
Fax: +44 (0)1491 833508
E-mail: cabi@cabi.org
Website: www.cabi-publishing.org

CABI Publishing
875 Massachusetts Avenue
7th Floor
Cambridge, MA 02139
USA

Tel: +1 617 395 4056
Fax: +1 617 354 6875
E-mail: cabi-nao@cabi.org

A catalogue record for this book is available from the British Library, London, UK.

Library of Congress Cataloguing-in-Publication Data
Conflict, social capital, and managing natural resources : a West African case study / edited by Keith M. Moore.
p. cm.
"Sustainable Agriculture and Natural Resource Management (SANREM), Collaborative Research Support Program (CRSP)."
Includes bibliographical references and index.
ISBN 0-85199-948-4 (alk. paper)
1. Natural resources–Africa, West–Management. 2. Land tenure–Environmental aspects–Africa, West. 3. Grazing–Environmental aspects–Africa, West. 4. Soil conservation–Africa, West. 5. Sustainable agriculture–Africa, West. 6. Infrastructure (Economics)–Africa, West. 7. Conflict management–Africa, West. 8. Sustainable development–Africa, West. I. Moore, Keith M. II. SANREM.

HC1000.Z65C66 2004
333.7′0966–dc22 2004012409

ISBN 0 85199 948 4

Typeset in Melior by Servis Filmsetting Ltd, Manchester
Printed and bound in the UK by Cromwell Press, Trowbridge

Contents

Contributors

Abaye, O., *Department of Crop, Soil and Environmental Science, Virginia Tech, Blacksburg, Virginia 24061-0404, USA. E-mail: cotton@vt.edu*

Ampagoomian, C., *Tulane School of Public Health and Tropical Medicine Tulane University, New Orleans, Louisiana, USA. E-mail: catampagoomian@yahoo.com*

Badini, O., *International Programs/Development Cooperation, Washington State University, Pullman, Washington 99164-5121, USA. E-mail: obadini@wsu.edu*

Ballo, A., *ESPGRN Team Leader, Centre Régional de la Recherche Agronomique, Institut d'Economie Rurale, BP 205, Mopti, Mali, Africa. E-mail: Adama.Ballo@ier.ml*

Bertelsen, M., *College of Agricultural and Life Sciences, Virginia Tech, Blacksburg, Virginia 24061-0334, USA. E-mail: bertel@vt.edu*

Bingham, S., *Allan Savory Center for Holistic Management, 394 Vanderbilt Road, Asheville, NC 28803, USA. E-mail: sbingham@igc.org*

Brewster, C., *Office of International Research, Education and Development, Virginia Tech, 840 University City Boulevard, Blacksburg, Virginia 24061-0378, USA. E-mail: brewster@vt.edu*

Cissé, S., *Centre Régional de la Recherche Agronomique, Institut d'Economie Rurale, B.P. 205, Mopti, Mali, Africa. E-mail: Salmana.Cisse@ier.ml*

Crane, T.A., *Department of Anthropology, University of Georgia, Athens, Georgia 30602, USA. E-mail: tcrane@uga.edu*

Dioni, L., *Centre Régional de la Recherche Agronomique-Sotuba, Institut d'Economie Rurale, BP 258, Bamako, Mali, Africa. E-mail: Lassana.Dioni@ier.ml*

Earl, J., *Preventing Harm Minnesota, Minneapolis, Minnesota 55407, USA. E-mail: JuliaBill@compuserve.com*

El Hadj, M., *Department of Crop and Soil Environmental Science, Virginia Tech, Blacksburg, Virginia 24061-0404, USA. E-mail: mhadj@vt.edu*

Goebel, J., *Goebel & Associates, PO Box 2503, Redmond, Oregon 97756, USA. E-mail: gobel@aboutlistening.com*

Kaboré, D., *Institut de l'environnement et des recherches agricoles, Burkina Faso, 04 BP 7192, Ouagadougou, Africa. E-mail: kaboredaniel@hotmail.com*

Keïta, M., *Centre Régional de la Recherche Agronomique, Institut d'Economie Rurale, B.P. 205, Mopti, Mali, Africa. E-mail: Moussa.Keita@ier.ml*

Kodio, A., *Centre Régional de la Recherche Agronomique, Institut d'Economie Rurale, B.P. 205, Mopti, Mali, Africa. E-mail: Amandou.Kodio@ier.ml*

Moore, K.M., *Office of International Research, Education and Development, Virginia Tech, Blacksburg, Virginia 24061-0378, USA. E-mail: keithm@vt.edu*

Ouattara, A.K., *Centre Régional de la Recherche Agronomique, Institut d'Economie Rural, BP 205, Mopti, Mali, Africa. E-mail: AbdoulKarim.ouattara@ier.mali*

Sengupta, N., *Center for Environmental Applications of Remote Sensing, Virginia Tech, Blacksburg, Virginia 24061-0324, USA.*

Touré, A., *Local Governance Advisor, CARE/Tombouctou, Mali, Africa. E-mail: Abtour2002@yahoo.fr*

Touré, M.S.M., *Centre Régional de la Recherche Agronomique, Institut d'Ecomonie Rurale, BP 205, Mopti, Mali, Africa. E-mail: Mohamed.Toure@ier.ml*

Traoré, B., *Centre Régional de la Recherche Agronomique, Institut d'Economie Rurale, BP 205, Mopti, Mali, Africa. E-mail: Boureima.Traore@ier.ml*

Wyeth, P., *International Programs-Development Cooperation, Washington State University, Pullman, Washington 99164-5121, USA. E-mail: pwyeth@wsu.edu*

Wynne, R.H., *Center for Environmental Applications of Remote Sensing, Virginia Tech, Blacksburg, Virginia 24061-0324, USA. E-mail: wynne@vt.edu*

Preface

This book is the product of a 6-year programme involving the population and territory of one decentralized local government unit in West Africa. On one hand, it recounts the efforts of research and development agents engaged in developing and transferring the appropriate social and biophysical technologies and decision support tools for sustainable development. On the other hand, it describes the experience and conditions of the local population seeking a pathway from poverty and food insecurity to a healthy economy and environment.

The book's contributors are part of a global programme, the Sustainable Agriculture and Natural Resource Management (SANREM) Collaborative Research Support Program (CRSP) initiated in 1993 with sites in Ecuador, the Philippines and Mali. The SANREM CRSP is financed jointly by the United States Agency for International Development (USAID) and participating institutions. Through SANREM, US universities partner with host country institutions for the purpose of building and testing interdisciplinary tools and methodologies to assist agricultural and natural resource management (NRM) decision makers.

In the USA, SANREM West Africa is led by Virginia Tech in partnership with Washington State University, the University of Georgia, and trainers from Tufts University and the Center for Holistic Management. Despite SANREM's academic roots, it is also highly participatory and seeks to ensure relevance and encourage adoption of improved NRM processes by developing the tools and methodologies collaboratively with local institutions and partners. This inclusive approach provides the added benefit of being able to build on all SANREM partners' experience. In Mali, SANREM's principal partners are led by the *Institut d'Economie Rurale* (IER), CARE-Mali and Groupe de Recherche Action pour la Développement (GRAD) as well as the men and women, farmers, pastoralists and fisherfolk of the Commune of Madiama.

SANREM began Phase II (1998–2004) of its activities in West Africa with the objective of advancing NRM in the Sahel by addressing what was increasingly being perceived as the major stumbling block to progress in community-based NRM: violent conflict over access to and use of natural resources. We immediately realized that as partners in a research programme we had no comparative advantage in conflict resolution *per se*. However, we did feel we had something to contribute through the development and testing of a model institutional framework by which communities could learn to work together to improve management of both their natural resources and the conflicts embedded in their exploitation. This community-based NRM would be both good science and good development.

Acknowledgements

A collaborative effort such as this involves a large supporting cast who have built the SANREM CRSP West Africa site programme from its base in Madiama through IER and their partners on the local and national scene in Mali, to Virginia Tech, Washington State, Tufts, and the University of Georgia, and finally, to the Land Resources Management Team in the Office of Natural Resource Management at USAID/EGAT/NRM/LRM in Washington, DC. The SANREM CRSP West Africa team would like to thank USAID for supporting the research and development activities described in this book. We are especially grateful to Bob Hedlund, SANREM CRSP Project Manager, and Christine Bergmark, who preceded him. We would also like to thank the SANREM Management Entity at the University of Georgia: Carlos Perez and the late Bob Hart, Program Directors, Carla Roncoli and Constance Neely, Associate Program Directors, and Tammy Cotton and Gina Thomas, Administrative Assistants for their support and assistance. This is the team that assured the effective implementation of this effort over the past 6 years.

Applied researchers cannot function without the assistance of cooperative partners who may or may not benefit from the null hypothesis. The faith and cooperative support of the residents of the Commune of Madiama has been a source of inspiration and renewal for our work. We especially would like to thank Commune Mayor Konaté and his staff who have always been available and encouraging in our activities. Our gratitude is also extended to the chiefs of the ten villages composing the commune, their councillors and those who shared in our research activities whether in the household or in the field throughout the seasons. In addition, we deeply appreciate the support and advice of Macké Cissé of the Sub-Prefecture of Sofara, as well as that of Amadou Hadj, President of the Nérékoro Herders Association and M. Askofaré of the Mopti Office Riz. We also want to thank the Institut d'Economie Rurale (IER) technicians Ibrahima Diallo, Issa Kané and Albedja Touré, who are based in Madiama. Of course, we owe a deep debt of gratitude to the new leaders of Madiama on the Natural Resource Management Advisory Committee (NRMAC) who have been our guides and assistants. Among those we would particularly like to thank are President Issa Sao, Secretaries Moussa Sao and Abdoulaye Cissé, Herder Representative Amadou Cissé, Hunters Representative Malicki Camara, and Women's Association Representative Kadidia Konaté.

This multi-institutional programme could not have functioned without the coordination and support of our Malian partners.

We wish to express our gratitude to Director Generals Alpha Maïga and Bino Témé and Scientific Director Bouréma Dembélé of IER for facilitating our work. We deeply appreciate the leadership provided by SANREM CRSP-Mali National Coordinators, Lassine Diarra and Salmana Cissé (in particular for his wisdom and humour) and the hospitality of Centre Régionale de la Recherche Agronomique (CRRA)/Mopti Center Director, Amadou Kodio. Thanks are owed to Christie Collins and Garth Van't Hul at CARE/Mali and Idrissa Maïga and Gouro Diallo of Group de Recherche Action pour la Développement (GRAD) for backstopping the field team in Djenné and Madiama. All of us owe a deep debt of gratitude to Abdoulaye Touré (CARE/Djenné) who as NRMAC Facilitator kept the wheels turning through all seasons. Thanks are also due to Chéry Traoré (GRAD) for picking up the reins in the last year of the project.

We would also like to thank S.K. De Datta, SANREM CRSP Principal Investigator and Director, Office of International Research, Education, and Development, Virginia Tech, and the late Jan Noel, Director of International Programs, Washington State University, for lightening the load of this collaborative research endeavour.

I would personally like to express my deep appreciation for the colleagueship of the authors, practitioners, researchers, technicians and friends who have brought this work to its present state. Thank you all. In particular, I would like to extend my gratitude to my predecessor as Program Manager, Mike Bertelsen, a constant source of inspiration, Oumarou Badini for late night discussions, Charlene Brewster for calming advice and council, Catherine Ampagoomian for research assistance, John Lipovsky for clerical support, and copy editors Sara Thorne-Thomsen and Kirk Neal for making this all readable.

Finally, this book was made possible through support provided by the G/EGAD/AFS Global, United States Agency for International Development under terms of Grant No. PCE-A-00-98-000-19-00 to Virginia Polytechnic Institute and State University (Virginia Tech). The opinions expressed herein are those of the authors and do not necessarily reflect the views of the United States Agency for International Development.

Keith M. Moore

1 Conflict and Agropastoral Development in the Sahel

Keith M. Moore[1], Michael Bertelsen[2], Salmana Cissé[3] and Amadou Kodio[3]

[1]Office of International Research, Education and Development, Virginia Tech, Blacksburg, VA 24061-0378, USA; [2]College of Agricultural and Life Sciences, Virginia Tech, Blacksburg, VA 24061-0334, USA; [3]Centre Régional de la Recherche Agronomique, Institut d'Economie Rurale, B.P. 205, Mopti, Mali

---

The Inland Delta and hinterlands of the Niger River, like many natural resource systems throughout the world, are transitioning to more intensified agriculture and animal husbandry production. These dominant sectors serve as the engines of sustainable economic development that provide food security and alleviate poverty. Although open range opportunistic grazing management by transhumant herders has been a way of sustaining life for centuries in this region, increasing population pressure, changing political structures, declining and erratic rainfall, and degrading natural resources have forced both agricultural and pastoral communities to transform their production systems and the social relations on which they are based. Unfortunately, this transformation has brought about increasingly violent conflicts over natural resources.

After the introduction of animal traction effectively increased cultivated land, subsequent technological fixes have largely failed to increase agricultural or livestock productivity and have even resulted in higher rates of natural resource degradation. The reason for such poor results is generally perceived as stemming from the weak involvement of the local population in the conception, design, implementation and evaluation of these interventions. This, in turn, led to new community-based approaches such as *gestion de terroir*[1] and the development of pastoral organizations. These sectoral approaches despite their more participatory nature have suffered from three interrelated weaknesses: the limited scope of intervention with respect to the natural resource base to be managed; inherent biases towards sedentary populations; and the lack of mechanisms to cope with natural resource-based conflicts at the local level. Indeed, until recently, non-governmental organizations (NGOs) implementing local development programmes have systematically avoided all situations where conflict was manifested.

Historically, the State has been either unwilling or unable to effectively reverse the trend towards increased conflict and related natural resource degradation. In fact, the State has often incited conflict by introducing projects, programmes, and policies favouring one group of resource users over another. Throughout the Sahelian region of West Africa there is a growing consensus that central governments are poorly placed to make many of the decisions critical to their citizens' welfare. The Interstate Committee for the Fight against Desertification in the Sahel (CILSS)-sponsored Nouakchott Conference (1984) was the first time West African governments had formally recognized the need for local involvement in development projects. This conference was followed by two other CILSS-sponsored conferences, one in Ségou (1989) that further stressed the need for decentralized natural resource management (NRM) governance, and the other in Praia (1994) that highlighted the relationships between decentralization and land tenure. Subsequently, West African states embarked on a policy of decentralization, devolving responsibility for governance to local administrative structures with far-reaching implications for governance in the region. At the same time, multi-national and bilateral donor agencies (The World Bank, United States Agency for International Development (USAID), etc.) and NGOs have promoted local civil society organizations (CSOs) as a core element in their new strategies for rural development and NRM.

The advent of decentralization has provided a renewed impetus mobilizing rural civil society in the transition to more intensified production systems and transforming the vicious cycle of poverty into a virtuous circle of poverty alleviation. Recent research on social capital has shown that participation in development activities is a more nuanced phenomenon than once imagined (Dasgupta and Serageldin, 2000; Isham *et al.*, 2002; and Grootaert and van Bastelaer, 2002). In order to confront the over-determined conflict between production system actors as they transition to more intensified and sustainable livelihood systems, the interconnections between groups sharing common resources need to be addressed. Bonding, bridging and linking relationships that build trust and common interests and the networks that support them within, between and beyond communities (collectively referred to as social capital) need to be examined, strengthened and, where appropriate, created. Although participation is both necessary and appropriate, what matters most is the quality of participatory relationships and how those relations are interconnected (Pretty and Buck, 2002).

NRM is one of the powers being decentralized to the local level across West Africa, shifting governance and consensus building to those best equipped to address these issues. However, this has placed a significant burden on emerging local NRM decision-making institutions. Local governments are poorly equipped and trained to deal with NRM issues. In addition, members of the Rural Councils are fully occupied in becoming acquainted with the large number of additional duties and responsibilities they have assumed under decentralization. A consistent and effective approach to building and reinforcing the social and human capital necessary to facilitate local NRM decision making is needed.

This book details an approach to social and human capital development for improved NRM decision-making through the Sustainable Agriculture and Natural Resource Management (SANREM) Collaborative Research Support Program (CRSP)-West Africa programme. Within these chapters we describe the development of networks among the local population, NGOs and local and foreign researchers, and we explore the biophysical technologies and decision support tools for facilitating improved and locally driven, natural resource decision making. In this chapter we review efforts that have been made to accelerate agricultural and pastoral development in the Sahel since independence. We further analyse the origins and impacts of natural resource-based conflicts. We discuss the four pillars on which SANREM work is based (participation, interdisciplinarity, multi-stakeholder involvement, landscape/lifescape scales) and explore the SANREM approach to improving natural resource and conflict management at the commune level. The chapter concludes with an overview of the chapters in this volume.

Development Assistance Strategies and Agropastoralism

Development assistance actors

Leadership and support for development assistance to agropastoral populations have traditionally included multinational banks, bi-lateral donor agencies, international agricultural research centres (IARCs), national ministries, research institutes, and locally based NGOs. The most significant of these have been the national ministries and research institutes whose legacy of control over long-term programme implementation often dates from the colonial period. These organizations have provided sectoral guidance and set national priorities. The role of multinational and bi-lateral agencies has been to set the trends in development project design, building on the lessons learned from successive waves of pastoral and agricultural development projects.

Various government ministries have promoted agricultural and pastoral policies, programmes, and projects by sector. These ministries have rarely coordinated their work. Often the ministry for agriculture is separate from that of livestock (or subsumed as a separate component). Recently evoked environmental issues have been the domain of a third ministry. If, as Sanders *et al.* (1996) suggest, government policies have neglected agriculture, they have even further neglected livestock. Overall government development policies have favoured the intensification of agricultural production at the expense of transhumance. This bias extends to the almost complete omission of pastoralist concerns in the development of famine/drought early warning systems (Sommer, 1998).

After independence in the 1960s, the World Bank, IARCs, and other donor agencies supported this sectoral approach. Agricultural research and development focused on the development and introduction of new crop varieties. Productivity gains were expected to result from genetic enhancement of export crops, rather than from improved agronomic practices (Lynam and Blackie, 1994). The introduction of new cultivars has been a major thrust of the IARCs (following the Green Revolution successes) and National Agriculture Research Services (NARSs) have also followed this path (Sanders *et al.*, 1996). Extension activities have focused on the introduction of inorganic fertilizer and new cultivars. Despite some successes, the adoption of new varieties has not caught on with African farmers and consequently population growth has outdistanced agricultural growth (Lynam and Blackie, 1994). While American range science began to influence livestock research during this period (Pratt *et al.*, 1997), increased livestock productivity has not occurred either. Livestock development and extension activities have stressed little more than providing veterinary services and finding new sources of water.

The Sahelian drought of 1968–74 brought about many changes in the way agropastoral development issues were addressed. In 1973, the Permanent Inter-State Committee for Drought Control in the Sahel (CILSS) was established with the objective of finding permanent solutions to food self-sufficiency and socioeconomic well-being in the Sahel through the coordinated efforts of nine countries (Burkina Faso, Cape Verde, Chad, The Gambia, Guinea Bissau, Mali, Mauritania, Niger and Senegal). Their first actions involved emergency aid by coordinating the international relief efforts of the time. CILSS has moved on to mobilizing development resources, conducting research through the Institute of the Sahel (INSAH) and national research institutes, disseminating information, and promoting awareness throughout the region and internationally on drought and desertification (Djalbord Diard, 1992).

Projects and programmes

To alleviate the effects of periodic drought and promote development, projects and programmes using different types of strategies have been implemented to assist agropastoral groups. These include drought relief actions, targeted interventions, systematic or programmatic interventions, and participatory interventions.

Drought relief

Drought relief actions consist of providing food to people under current drought conditions. Pastoralists who lost their livestock or farmers who lost their crops during the drought benefit from such actions. Food is distributed to needy households. Such interventions are frequent in all Sahelian countries. However, recurrent drought relief has done little more than further debilitate existing production systems.

Targeted interventions

Apart from the above punctual drought relief actions, specific interventions have often been implemented for pastoral development. The objectives of such interventions have been: (i) to increase herd size (or re-establish/restock herd), (ii) to increase milk yield, (iii) to maintain appropriate herd structure, and (iv) to develop disease resistance by selective breeding (Niamir, 1991). Vaccines, the eradication of tsetse flies, dams and boreholes have contributed to increased livestock production. Unfortunately drought relief actions, livestock salvage operations, and targeted interventions have been conducted on an ad hoc basis, and not as a part of a coordinated, coherent long-term strategy. As a result, they are relatively costly and have not been sustainable (Okai, 1987).

Systematic and programmatic interventions

Systematic and programmatic interventions are intended to develop the whole system of production (agriculture or livestock) instead of targeting actions on only restricted and isolated system components. Irrigation projects originally determined the model for this type of development intervention. While major new irrigation systems are not currently being developed, older systems have been improved or expanded, the number of micro-systems increased, and agricultural production expanded into wetland areas where herders traditionally bring their herds during the dry season. Both pastures and sources of drinking water are diminishing rapidly due to this trend. Governments and NGOs have supported the intensification of agriculture and livestock production. Both require fixed investments in land and, therefore, the permanent settlement of the producing population(s). Lack of success has occurred for many reasons, including an increased concentration of animals contributing to environmental degradation and increased conflicts between herders and farmers.

Commercial ranching, group ranching and permanent settlements designed to increase meat and milk production were the primary emphasis of pastoral development projects during this period. Although some World Bank projects addressed land tenure and policy issues, priority was given to capital investments, centralized planning, and reduced stocking rates (Pratt *et al.*, 1997). National and bi-laterally funded projects followed this general model. For the most part, these efforts have been characterized as either remarkable for their universal lack of success (Horowitz and Little, 1987) or simply a failure of farmers and herders to adopt the proposed technologies (McIntire *et al.*, 1992). Preston and Leng (1994) suggest that animal science made little or no contribution to increasing livestock productivity. Pratt *et al.* note the primary causes of failure as a lack of understanding of the traditional pastoral systems; an assumption that these systems were market-driven; an imposition of rigid organizational forms, tenure rights systems, and inappropriate incentive frameworks; and institutional weaknesses of the implementing agencies.

Livestock market development has been generally coupled with permanent settlements. The aim has been to facilitate livestock exports and taxation of pastoralists. Since livestock products represent 20–30% of Malian exports, there is good reason to believe that developing and formalizing such markets is worthwhile. Agreements between Communaute Economique du Betail et de la Viande (CEBV) countries theoretically delineate corridors for livestock inter-state movement, mostly from Mali, Burkina Faso and Niger to coastal countries (Côte d'Ivoire, Benin and Togo) in order to ensure international transhumance patterns that minimize conflicts between farmers and herders. Unfortunately, these agreements are less and less respected by farmers who expand food crop production to meet the demand of the growing population. Furthermore, the efficiency of corridors is questioned even at the national level.

In Burkina Faso, the government, in collaboration with the World Bank, established four ranches (a pastoral management zone) on land for which full title was not established. In doing so, use rights were left undefined for project beneficiaries. The majority of colonists in these ranches were farmer–herders (80%) and the rest, Peul herders. The agricultural communities grew cereals, cotton, citrus, bananas and vegetables. The bananas and vegetables were cultivated along the streams and in wetland areas. In fact, the cultivators left little or no room in their land use plans for animal corridors, access to water points, or dry season pasture (Sanon, 1996). In a similar project in Sideradougou (Bary, 1996), conflict resulting from the grazing of a herd in a cultivated field left a Peul herder dead, and later led to the death of the Karboro father of the assailant. Both Peul and Karboro communities accused government forces (which defused the immediate confrontation) of siding with their opponents.

Consequently, sector-based strategies have come into disrepute throughout the Sahel. Current strategies apply integrated approaches focusing on natural resource management, a prioritization of food production, flexibility and mobility in herd management, and an emphasis on local institution building (particularly among pastoralists). These approaches shift agricultural research priorities from breeding to improved soil management technologies (Lynam and Blackie, 1994) in order to increase agricultural and livestock productivity. Farmers and herders should be provided with a menu of technological options from which they may select or adapt to their production circumstances. Intensification is already beginning to occur, as Sanders *et al.* (1996) note, with respect to the use of *zaï* and bunds combined with manure. They recommend going beyond simple labour-intensive technological innovations to increased investment by farmers in combination with capital intensive technologies including inorganic fertilizer.

NGO contributions to agropastoral development typically began with emergency relief activities. While this led to a top-down approach, with increasing regularity NGOs have become involved in more routine development activities (Brown, 1993). These activities have focused on small scale interventions targeting villagers and local problems. The NGOs have begun to recognize the importance of natural resource tenure issues and associated conflicts, but they have tended to avoid involvement in these issues. Their lack of partnerships and contact with government services, the small scale of activities, the absence of baseline and monitoring data, and low level of technical expertise have all diminished NGOs' impact (Toulmin and Moorehead, 1993). Nevertheless, NGOs are beginning to support pastoral systems that attempt to enable pastoralists to exploit spatial and temporal availability of rangeland resources.

Participatory interventions

Although most interventions are systematic or implemented in a programmatic framework, they fail to increase agricultural or livestock production and result in even greater natural resource degradation. The reason for such bad results is the weak involvement of the local population, as they were not involved in project conception, design, implementation or evaluation. Such failures called for new community-based approaches such as *gestion de terroir* and the development of pastoral organizations in the Sahelian countries.

Gestion de terroir is a multi-sector and global strategy aiming to establish a new socioeconomic and ecological equilibrium in order to achieve food self-sufficiency and to preserve/regenerate the productive potential of natural resources (Rochette, 1989). *Gestion de terroir* involves the local people in natural resources planning and gives them full responsibility for NRM within the limits of their *terroir*. Land tenure must be secure enough in this unit in order to ensure a sustainable use of resources. All resource users in the *terroir* should be included in the management unit (e.g. ethnic, age and gender groupings).

This approach was developed in the early 1980s and first initiated in some Sahelian countries like Burkina Faso and Senegal. In Burkina Faso, the Programme National de Gestion des Terroirs (PNGT), financed by the World Bank, has been conducted for a decade in regions of the country where NRM is critical. The subsequent World Bank Projet de Gestion des Ressources Naturelles (PGRN) has disseminated the methodology throughout the Sahelian countries in order to implement local level NRM. In Senegal, the USAID-sponsored Community-Based Natural Resource Management (CBNRM) Project also worked with a similar approach but applied it at the multi-village level. Despite these attempts, many problems have been noted, including: (i) lack of fundamental research-based data that could help projects or NGOs in implementation; (ii) village *terroirs* imprecisely defined and limited in scope; (iii) the agricultural area has nearly always been the unit of analysis instead of livestock; and (iv) avoidance of conflict situations (Lowenberg-DeBoer *et al.*, 1994; Painter *et al.*, 1994; Benjaminsen, 1997).

Pastoral organizations

An emerging general consensus on the importance of livestock to the Sahelian economies led decision-makers to consider pastoral development in their policies. In particular, pastoral organizations supported by The World Bank, the African Development Bank, IFAD, France, and Norway were developed in Mali, Senegal, Mauritania and Niger as a new form of community-based NRM (Shanmugaratnam *et al.*, 1992). Pastoral organizations (POs) cover all types of institutional arrangements that regulate individual and collective actions by pastoralists to safeguard and promote their economic, social, cultural and political interests. They are legally organized by the government and dependent on the state for technical assistance, supplies and financial resources (Vedeld, 1992). PO functions include acquiring secure land tenure, resource management, provision of services, communication of information, external relations and the building and maintenance of community coherence and morale. Key factors that determine the viability of POs include food security, water security, land security, herd ownership, credit, veterinary services, marketing, economic self-sufficiency and literacy (Shanmugaratnam *et al.*, 1992). Implementation problems have lessened PO impacts. In particular, drought has led pastoralists to disperse as migrants in search of food. In addition, the identification of beneficiaries has not been clear in communities and there has been a noted incapacity of POs to implement project plans. Shanmugaratnam *et al.* provide a complete list of the problems.

Increasingly, international donor agencies and national NGOs are recognizing the importance of conflict resolution as a major component of natural resource management research and extension efforts. This has resulted from their increased understanding of pastoral management systems and their problems shifting project focus to institution building, and decentralized and flexible decision making processes (Pratt *et al.*, 1997). For their part, the NARSs only indirectly address issues of land tenure and conflict in agropastoral systems. NARS major contributions continue to be the development of resource-intensifying technologies designed to increase and diversify incomes, thereby increasing food security and reducing the likelihood of resource conflict.

NRM-related Conflicts

The connection between food insecurity, NRM and conflict is well documented in the literature. Messer *et al.* (1998) point out that food insecurity and natural resource scarcities are major causes and consequences of conflict. The authors cite data indicating that conflict in countries of sub-Saharan Africa has been associated with per capita food production declines of more than 12% per year. In the Sahel, the most severe conflict has generally involved NRM practices associated with agricultural and pastoral production systems.

The heightened competition between livestock and agricultural production through the conversion of pastureland into cropland and the expansion of cultivation into irrigated agriculture along water points has restricted access to water and to dry-season pasture of *bourgou* (*Echinochloa stagnina*). Additionally, by cultivating livestock corridors (enriched by animal manure), sedentary farmers have disturbed transhumance patterns and herders have been obliged to change their strategies in order to adapt to the new landscape. Maintaining control of local resources in the face of non-local users or migrant 'strangers' can be a powerful force mobilizing a sedentary agricultural community. For herders, herd mobility is crucial to survival and the trampling and grazing of cultivated fields before harvest consequently leads to the most common flash point initiating overt conflicts.

These conflicts have been particularly acute where competition for resources between managers of ethnically differentiated production systems is highest. Conflicts are found not only in conditions where complementary systems once co-existed, but they are also becoming more frequent among members of the same community. Inter- and intra-community conflicts increase as the natural resource base shrinks due to higher population pressure and diminishing annual rainfall. The disputed resources revolve around access to land, but are particularly focused on specific rights to cultivate or to graze, water and move livestock.

These conflicts are endemic to the transition between the extensive livestock and cropping systems characteristic of the precolonial past and the expanding, more intensive systems of the modern era. Traditional systems have used the natural resource base of the Sahel in multiple and complementary ways. With increasing pressure on the resource base, however, livestock and farming systems are coming into conflict over the schedule of land use by the different systems (Soumaré, 1996). For example, at one time near Nioro du Sahel in Mali, the farming Soninké and the herding Toucouleur benefited from an exchange of manure for grain and established a tradition of shared natural resource tenure. The Toucouleur eventually established a village in the region and continued their transhumance. Today, with decreased rainfall, land degradation and adoption of livestock raising by the Soninké, the exchange has lost its value. Although the Soninké rely on the Toucouleur for tending to their herds, the shared use of resources has become problematic. The herders have more difficulty accessing dry season crop residues for their herds, while the sedentary farmers feel their agricultural calendar is restricted by the incursion of the herds. Exchanges have become monetized, but there is no unified authority to regulate local land use. While farmers are governed by the village council, herders are governed by their clan, which has no fixed location.

Pairs of actors involved in natural resource conflicts are often used to develop a conflict typology. This usually involves pairing farmers and/or herders with like or other actors (fisherfolk, non-local private sector, migrants, the State or its agents). It may be more useful, however, to categorize conflicts in terms of whether they are within or between production systems. On the one hand, conflict within production systems usually involves attempts to maintain or increase the size of the actors' production units and is based on issues of inheritance, land boundaries and proprietorship. To the extent that unambiguous rules and procedures exist, these conflicts can normally be resolved within the village, but have become a source of socio-economic differentiation within rural communities.

On the other hand, conflicts between production systems are more serious, because they threaten not only the immediate livelihoods of the disputants, but also the way of life of distinct ethnic communities. When disputants are from different ethnic groups, a conflict may be further aggravated and larger constituencies drawn into the fray.

Increasing population, expansion of cultivated land (with the consequent disappearance of fallow), the introduction of cash cropping, and the accompanying degradation of the vegetative cover have conspired across the Sahel to create situations of confrontation between rural stakeholders. These types of conflicts, particularly between farmers and herders, are increasing (Souley, 1996; Maïga and Diallo, 1998; Blench, 1996).

On a larger scale, the government of Niger set a limit in the northern part of the country (called the Sorghum Line), north of which agricultural producers were not allowed. This region was reserved for rainy season grazing only. Today, migrant herders have begun to cultivate land around water points north of the sorghum line on lands once managed for livestock grazing (Sidikou, 1995). The official deadlines of 30 May (on which livestock should be moved out of the cultivation zone) and 30 November (the date for their return) are no longer respected. Furthermore, the changing rainfall regime has obliged herders to move north later, and consequently, they are tempted to take advantage of the new-growing grasses in the southern agricultural zone.

These conflicts can become extremely violent when disputants contesting the right to use a resource have a history of ethnic conflict. This was the case of the conflict between the Sosobe and the Salsalbe in the Niger Delta of Mali. Contested claims to a grazing territory dating back to 1936 exploded into a violent confrontation in December 1993, leaving 29 dead and 42 wounded (Maïga, 1996; Vedeld, 1994). Often, contested claims in the Niger Delta are linked to traditional ethnic relationships between the Peul and the Rimaibés (their ancient slaves). According to tradition, the Peul allowed the Rimaibés to farm Peul-dominated territory, while the Peul continued their transhumant practices (Maïga and Diallo, 1998). The advent of colonialism legally ended slavery. However, the Dina, a socio-religious land tenure organization established during Peul dominance, remained in effective operation. This allowed the Peul's former slaves continued access to the land they farmed, but also assured perceptions (at least in the medium term) of Peul proprietorship (Cissé, 1996).

Inter-state relations are also involved in natural resource conflicts. Nomadic or transhumant herders have little regard for national frontiers. Key informants indicate that this is often the case when herders arrive in northern Ivory Coast or Ghana from Burkina Faso or Mali, along transhumant routes between western Mali, Mauritania and Senegal, or between Niger, Nigeria and Benin. This type of conflict may provoke an international confrontation, as in the case of the Mauritanian military action killing two Senegalese (Soninké) cultivators following a confrontation between Mauritanian Peul herders and Soninké cultivators (Park *et al.*, 1993). This killing in turn led to a serious international incident involving the death of hundreds.

Lo *et al.* (1996) note the importance of distinguishing between conflicts which are horizontal and those which are vertical in orientation. Throughout the region conflicts are not only between cultivators and herders and among themselves and their communities (horizontal conflicts), but also between these communities and the state or specific development projects (vertical conflicts). Often the state, or a development project, is a background partner setting the stage for and/or reawakening latent conflicts. Privatization of the economy has led to access to forest reserves by the private sector aggravating this trend. In Senegal, despite encouraging the decentralized management of natural resources, the government has issued permits to harvest wood for charcoal production to urban-based charcoal merchants. For example, the Wolof and Mandinke communities of Maka Coulibanta attempted to assert their right to regulate the wood cutting trade only to have the Forestry

Service negotiate use rights to urban merchants (Kane and Winter, 1997). In Niger, the government denied access of local Toureg and Peul populations to the Gadabeji Reserve. In so doing, conflicts were created not only between the government and herders grazing their animals on Reserve lands, but also between herding communities as they concentrated around the only remaining watering holes and private wells (Sidikou, 1995).

While the most frequently recounted (and often more violent) conflicts are between herders and farmers, there are nearly as many less notorious affairs between farmers and farmers, or herders and herders. Cissé (1996) reports that 42% of the conflicts in the Niger Delta of Mali are between farmers and herders, and 40% are between farmers and farmers. Conflicts were equally divided between inter- and intra-village. An example of farmer–farmer conflict can be seen in the Burkina Faso village of Nébourou where conflict erupted between the original Nuni inhabitants and the in-migrant community of Mossi (Laurent and Mathieu, 1994). Nébourou is located in the land-rich, and once lowly populated, western region of Burkina Faso. With increasing soil degradation in the Mossi Plateau and worsening drought impacts, Mossi populations began immigrating to Nuni villages, particularly after the establishment of the Agrarian Reform of 1985, which opened all Burkina lands to those who farmed them. The Nuni chief distributed lands to the incoming Mossi through their Chief (the first Mossi settler). By 1986, the population of Nébourou was 74% Mossi. When an NGO presented a proposal to establish a village woodlot, the community council agreed. When it came time to plant the trees, however, the Mossi wanted to plant theirs separately. Because tree planting is considered sufficient to validate rightful 'ownership' of land, this was not acceptable to the Nuni, who did not want the Mossi to establish a claim to their land. Although the confrontation was diffused through informal mechanisms, tenure security was never established for either Nuni or Mossi and the latent conflict over land rights still exists.

At the heart of the perennial nature of most of these conflicts is the lack of a single accepted authority for the resolution of natural resource rights and associated tenure security in the management of natural resources (Touré, 1996; Maïga and Diallo, 1998; Lo *et al.*, 1996; Ngaido, 1996). For example, when local authorities and extension agents were asked to describe natural resource tenure in Senegal (Moore, 1996), they would begin by describing how the National Domain Law was supposed to operate. They explained that when the law was not explicitly followed it was due to lack of understanding of the texts and poor interpretation by illiterate local officials. However, when pressed, they admitted that traditional land allocation practices dominate: a household or compound head that requires additional land will turn to a neighbouring household head to borrow it. If, after some years or in the next generation, conflicting claims arise, village elders or the village head may arbitrate. Only when village authorities cannot resolve a dispute does litigation proceed to the Rural Council, and perhaps even to the *Sous-Prefet*. Even in these cases, resolution of the problem is dependent on the political power of the litigants as much as on the technical requirements of any law (see Blundo, 1996; or Laurent and Mathieu, 1994).

Conflicts over access to natural resources and their benefits between generations and between men and women in the same lineage have also been documented. Women's tenure security is often very weak. For example, in the village of Sikore (Mali) daughters receive only one-third of their father's inheritance while sons receive two-thirds. Even with a third share, it is unlikely that land will be passed on to women. One village Imam even attempted to seize the land inheritance of a deceased man who was survived only by a wife and daughter, which led to a conflict in which the whole village became involved (Maïga, 1996). In addition, the natural resource rights of pastoral women are rarely considered (Pointing, 1995). Young men, eager to earn money desire access to land for cash crops and resist attempts by outsiders taking what they feel is, by right, their land. Within the Nuni community of Nébourou (Laurent and Mathieu, 1994), the young men were the most concerned about the continued distribution of lands to incoming Mossi, because they felt that they were being dispossessed on behalf of people outside of the family.

The SANREM CRSP Response

The SANREM CRSP West Africa Project designed and implemented a programme to develop and test an approach to address issues surrounding decentralization, conflict, and NRM. The driving force behind the approach is the need to find long-term solutions to complex natural resource management problems. Although in dealing with conflict over natural resources, it is important to implement short term conflict resolution/management strategies, the approach also focuses on long-term consensus building and provision of social infrastructure as a platform for change and improved agricultural and natural resource decision-making. The same tools being used to resolve conflict and to develop a set of organizational skills (human capital investment) for consensus building and local governance are being used simultaneously to establish a new form of social infrastructure (social capital investment) to bridge relationships, build trust between local communities and link these communities with scientific research services to identify and introduce technologies that increase the productive capacity of the natural resource base. The working hypothesis of SANREM in Mali has been: When a local population is provided with: (i) methods for natural resource and conflict management, and (ii) an institutional vehicle for inter-village, inter-ethnic dialogue, the population can then become proactive in addressing major agricultural productivity and natural resource management issues.

Globally, the SANREM CRSP approach is based on a holistic view of NRM involving four pillars: participation; interdisciplinarity; multi-stakeholder involvement; and landscape/lifescape scales. This holistic, multidisciplinary approach integrates the fields of sociology, economics, agriculture, natural resources and animal science to provide decision-making tools for community-based management.

Local participation is crucial to the success of NRM projects of USAID, the World Bank, and numerous NGOs throughout Sahelian West Africa. Not simply participation, but the type and form of local participation are important for project success. This participation implies more than a simple presence in a situation or relationship. An active commitment by the 'participant' is necessary for the benefits of participation to be assured. This commitment is responsible for all successes using the participatory approach. Commitment can be seen when the local population specifies the needs to be addressed, defines the questions to ask, and sets the agenda to follow. These needs, questions, and actions must be locally meaningful; that is, they must correspond to circumstances as they are perceived within villagers' daily lives.

Interdisciplinarity draws the research process into a holistic dialogue across disciplinary boundaries. Without taking into account the sum of all that influences and shapes humankind's relationship with its environment, improved management is unlikely to occur without setting off a chain of unrecognized and unintended consequences. Working in disciplinary isolation, focusing on a single production system, emphasizing one resource over another, or assisting one ethnic group or village introduces a bias that, if left unchecked, leads to overall system imbalance.

However, the social and biophysical environment contains an infinite number of aspects, some perhaps more critical than others, depending on timeframe and situational interdependencies. Scientists or research managers are forced to choose, to isolate a single aspect in order to fully grasp it and manipulate it for the purpose of improving livelihoods. Disaggregating elements from the whole to isolate them for rigorous testing and investigation is the hallmark of the advances provided by modern science. Unfortunately, it also leads to monoculture production and the attendant weaknesses of an unbalanced landscape.

This tension between holism and particularity has pervaded implementation of the SANREM approach. It is characteristic of multi-institutional, multi-disciplinary teams; and it is also reflected in community relations.

Any process that involves multiple stakeholders must address how benefits can be optimized for the group as a whole. Stakeholder involvement implies not only participation, but also the recognition that not all participants have the same goals, or the same power to achieve them. Successful negotiation of a shared vision involves pooling of prior knowledge and reasoning processes and is predicated on a collective need or desire for results by and for the group. Confronting the reality of opposing stakeholder interests is required if the sense of ownership in a shared vision is to be created and the plan embedded in the social consciousness.

SANREM CRSP captures this holistic element by building on landscape concepts drawn from watershed management and completing the metaphor with a new term, lifescape. Landscapes are constructed realities, the mirror image of which is people constructing lifescapes for their livelihoods. In the global lexicon of SANREM, working at the landscape/lifescape scale means going beyond what has come to be known as *gestion de terroir* in the West African Sahel. As the landscape/lifescape scale increases, a wider range of stakeholders, often having no direct interest in specific local ecosystems become involved. SANREM research focuses on understanding both the complex biophysical and social processes within and across ecosystems.

For small groups, it is easy to gather in one location and quickly take these steps at one or two meetings. However, this is not possible in the context of decentralized administrative units, because the population of these communities can range from 2000 to 15,000 individuals. In this case, a standardized methodology is necessary to aggregate local (*terroir*/village) level participation into commune-wide participation. In addition to trial and error learning, substantial effort has been necessary to determine the most feasible mechanisms for aggregating village priorities and concerns. Local knowledge and practice, in fact, provide one of the most reliable ways to identify mechanisms for this aggregation.

The SANREM-West Africa Approach

The SANREM approach to improving natural resource and conflict management at the commune level is composed of five steps:

1. Ascertain local perceptions and priority needs;
2. Build commune consensus and establish local management capacity through NRMACs;
3. Build institutional capacity for impact:
- Training in literacy and governance;
- Training in natural resource management and conflict management;

4. Complement local knowledge with biophysical and socioeconomic research and development that lead to technologies and decision-making tools;
5. Monitor and evaluate.

These steps are summarized below.

Step 1: Ascertain local perceptions and priority needs

Implementing the SANREM process is predicated on the interests of local partners and local government. Further, local perceptions of the constraints and potentials of the community's landscape and lifescape must be understood. Generally, perceptions vary within communities and among stakeholders – farmers, pastoralists, village associations, NGOs and technical service providers. The best way to reveal and document these perceptions is with a rapid participatory survey at a multi-village level, such as the Participatory Landscape/Lifescape Appraisal (PLLA).

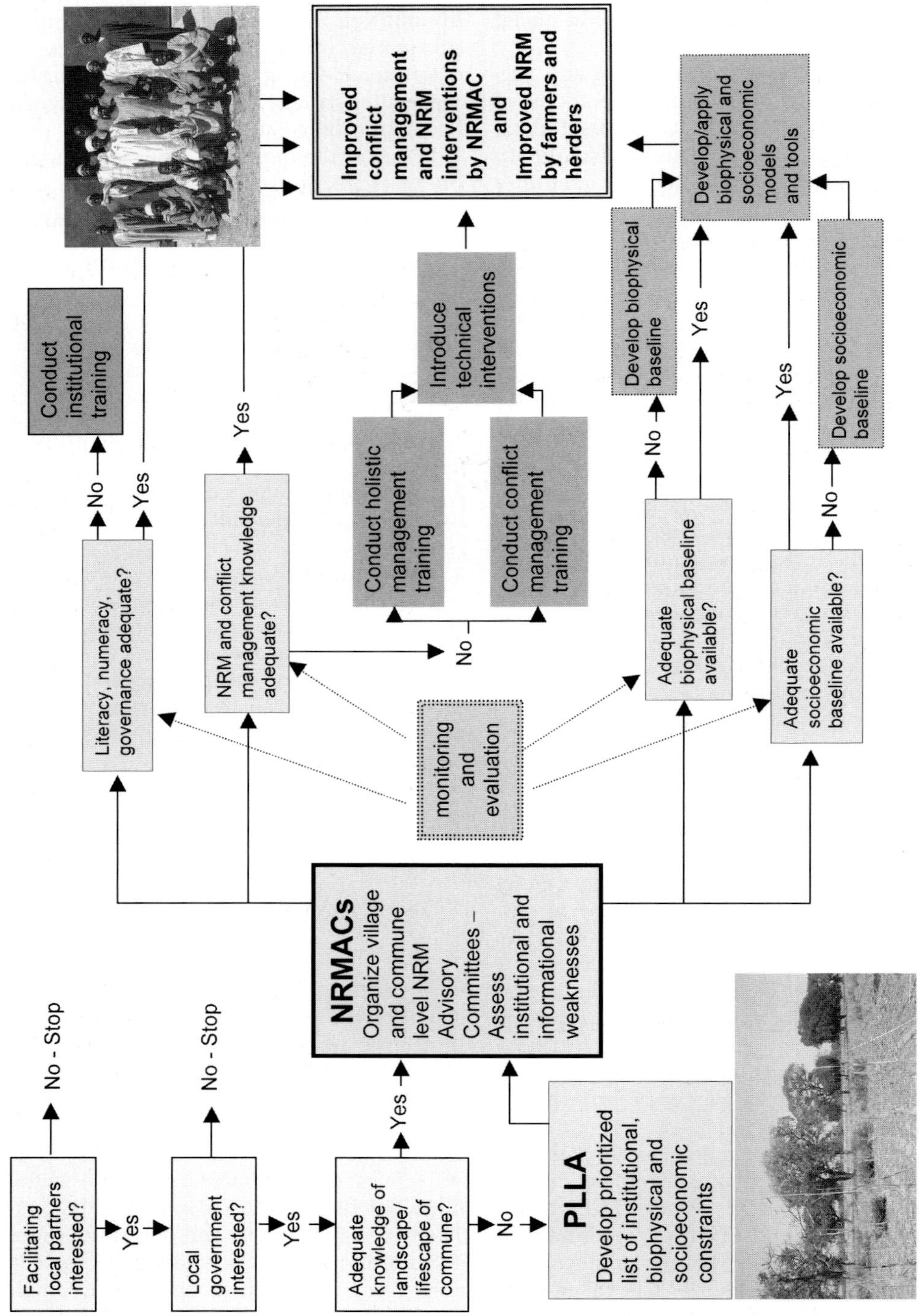

Fig. 1.1. SANREM West Africa process flowchart.

The PLLA is based on the proposition that any successful NRM project must be grounded in a balanced and thorough appreciation for the biophysical and socioeconomic milieu of the target community. The PLLA forms multi-disciplinary teams of researchers and local stakeholders to examine the natural resource base and the socioeconomic realities of the community. Rapid Rural Appraisal techniques identify key biophysical and socioeconomic constraints and opportunities. Groups and individuals from all strata of the local population become involved and a representative picture of the Commune becomes apparent during the course of the week-long exercise. Particular attention is paid to institutions and traditions that provide an interface between the landscape and the lifescape, because management decisions affecting both landscape and lifescape are made through them. The PLLA also provides the opportunity to inform the local population about the nature of the SANREM intervention and helps form realistic expectations within the community about what collaboration may mean in the future.

The PLLA in Madiama Commune took place in February of 1999 and identified three major NRM constraints: the poor and degrading condition of soil fertility, pasture, and water points for livestock. These constraints formed the basis for prioritizing the activities that became the work plans and research programmes prepared by the partners, including representatives from the local population.

Step 2: Build commune consensus and establish local management capacity through NRMACs

The roles of the NRMACs are:

1. To collaborate in:
- drawing up each year's work plan;
- choosing farmers and pastoralists to participate in field tests;
- monitoring and evaluating the results of field tests and progress towards fulfilling work plan objectives;
- disseminating awareness of these activities and any recommendations resulting from them;

2. To work with farmers, pastoralists, local chiefs, and other stakeholders to avoid and manage disputes over natural resources;

3. To act as a liaison between the local population and the commune's mayor and other government authorities regarding natural resource issues; and

4. To provide training and act as mentors for the development of holistic management and conflict management strategies in the commune's villages.

Great care was taken to ensure that all social strata in the commune were equitably represented on the advisory committee, particularly in terms of socio-professional groups, ethnicity and gender. Because perceived constraints and solutions vary considerably, any excluded group will not attach much legitimacy to decisions or advice coming from the committees. In Madiama, 18 representatives were elected to the commune level NRMAC from among representatives of each village-level user group. Each village committee has at least one representative. One-third of the commune level committee members are women. Although SANREM encourages diversity, the village committees are less diverse because the villages themselves are less heterogeneous in terms of ethnicity, socio-professional groups, and in their views of gender roles.

The NRMAC can only function well if village chiefs, the mayor and other authorities are fully informed, provide support, and understand that their own role in the local hierarchy will remain secure and undiminished by collaboration. In Madiama, the NRMAC has established good relations with the mayor who regularly participates in meetings, often officially opening and closing them. The *chef d'arrondissement* has also attended. Both have been outspoken in encouraging committee activities.

NRMAC sustainability is an important issue. Sustainability in this context is dependent on the participants acquiring the training and skills necessary to manage everyday problems and on the capacity to access and interact with deconcentrated technical service providers. The ever-changing landscape and lifescape of a commune means that new and difficult challenges will arise in the future. The NRMAC has been designed to build social capital within Madiama to meet these challenges, whether through the solidarity of horizontal bonds between villages and ethnic groups, or through bridges connecting the community to service providers.

Step 3: Build institutional capacity for impact

Training in literacy, governance, and financial management

To be effective, NRMAC members require significant training in organization, democratic organizational procedures, literacy and numeracy skills. The value of literacy and numeracy training is that it gives the NRMAC the capacity and confidence to act autonomously – gathering information, writing plans and contacting government authorities. Training is an essential element in the processes of decentralization and democratization.

Institutional strengthening was provided to the committee by the Djenné office of CARE/Mali through formal training workshops, informal exchanges, monitoring and technical assistance and study tours. The institutional reinforcement training programme was initiated with an institutional diagnosis that identified the strengths and weaknesses of the organization. Focusing on the identified weaknesses, a plan for institutional development was elaborated. Thus, the formal training programme applied functional literacy methodologies and focused on the principles of democratic governance and assured comprehension of national texts, laws and codes for NRM and decentralization. Subsequent training addressed financial management and accounting, strategic planning and lobbying. These workshops were designed to create a framework of exchange and dialogue between the NRMAC and the other actors implicated directly or indirectly in NRM. The NRMAC also benefited from the support of CARE in conducting inter-village negotiations for the establishment of agreements for wetlands management. Additional institutional coaching included assistance in developing statutes and rules of procedure, formal registration of the organization, financial management, the mobilization of the external resources, and the development of linkages with other local service providers and NGOs.

Training in natural resource management and conflict management

Holistic management (HM) concerns itself specifically with the degradation of natural resources, including how degradation occurs, and how to reverse it. As such, it is very appropriate to the West African Sahel.

HM as an approach to natural resource management was first conceived in southern Africa and has been applied throughout the world. It helps to foster skills necessary to maintain purposeful dialogue between stakeholders. HM introduces visioning techniques to help channel dialogue towards consensus, and conflict resolution skills to help negotiate and manage the inevitable conflicts. HM training and implementation in Madiama concentrated initially on the contentious issue of *bourgou*[2] management and is moving on with the help of the NRMAC to more common dryland pasture issues (see Chapter 13).

Conflict management training provides techniques for managing conflict within the contexts of scarcity, diversity and unequal power distributions. The training is well adapted to conflict avoidance and management where natural resources are concerned. NRMAC members have been very pleased with conflict management training and assert that it has had the most impact of all early SANREM interventions, because there are already fewer conflicts. Conflicts have been avoided or managed using the techniques learned during the training (Chapter 16). Conflict management tools used successfully help legitimize the NRMAC and the SANREM programme, particularly when combined with agricultural and NRM interventions planned in collaboration with the NRMAC and jointly implemented with local partners.

Step 4: Complementing local knowledge: biophysical and socioeconomic modelling

Science-based information and tools for decision makers at the commune level

SANREM's science-based tools include biophysical and socioeconomic models that can be used to generalize the experience of one commune to other, broader geographical areas. These models are described in Chapter 11. In Madiama, SANREM has applied CROPSYST, a cropping systems model that predicts the influence of climatic, soil and management practices on productivity and sustainability. CROPSYST serves three purposes:

- To simulate results with variables such as yield, soil fertility and soil erosion over a period of 20 years, allowing for long-term evaluation and comparison of current and alternative farming and NRM technologies without long-term field tests;
- to identify crops and technologies best suited for ecosystems in the region; and
- to provide a basis for dialogue with the NRMAC, local farmers, and herders.

The biophysical models require rainfall data, obtained from official meteorological statistics and supplemented with rain gauges placed by the project and monitored by local stakeholders. The models also use soil data collected by a cost-effective alternative to classic intensive soil surveys (see Chapter 3) that combines computer-based remote sensing analysis and Geographic Information System (GIS) technologies with limited field assessment and ground truthing to classify and map soil types, vegetative cover and agricultural land uses (Chapter 5).

The dialogue with farmers builds on the farmers and herders own considerable knowledge of the land and on prescriptions regarding crops and NRM techniques suggested by the modelling. Researchers correlate their scientific soil classes training with traditional soil concepts farmers have devised, and enter into discussion with farmers using the local terminology. Through this dialogue, researchers confront the local reality and farmers gain access to the researchers' knowledge of alternatives and the benefits they can bring. Chapter 10 provides a complete discussion.

Socioeconomic modelling

SANREM researchers have developed a social accounting matrix (SAM) for the commune of Madiama. The SAM is a simple model that traces the sources of income and demonstrates how that income is distributed among households. Consequently, it is a good tool for showing how economic growth affects different groups within a commune.

A SAM model can be developed at any level, from a village to a nation with the number of sectors it contains depending on the complexity of the economy being studied and the available data. Four different types of households were identified according to socio-professional status by community leaders as being most important for understanding the commune: farmers, whose income is mainly from crops; agropastoralists, who have a significant amount of livestock as well as crops, sedentary pastoralists; who raise crops but regard livestock raising as their primary occupation; and transhumants, who move their livestock seasonally to grasslands in or outside of the commune along traditional routes. Major findings illustrated how benefits from increased production in any activity are passed on, through repeated cycles of spending and income, to benefit other production activities and the incomes of the other household types. In Madiama, it was found that an increase in production in any sector, including livestock, benefits transhumants far less than the other three socio-professional groups. Thus interventions must take particular care to target transhumants or they will tend to fall further behind economically, a situation that could contribute to greater conflict. The Madiama SAM results are relevant to neighbouring communes with similar socioeconomic groups and natural resource bases. The SAMs are discussed in Chapter 12.

Step 5: Monitor and evaluate

The broad-based SANREM strategy requires an equally broad-based monitoring and evaluation (M&E) system to assess progress, and to identify and correct problems. Much of the M&E system is participatory, with indicators and measurements identified and followed by the local participants. To take advantage of local learning and apply it horizontally across communes and vertically to the regional level, the universal language of science must be employed. While local definitions of resources, indicators and results are indispensable for local applications, on their own they may not be widely applicable. The fundamental strategy SANREM uses to apply results to contexts beyond the commune is to ensure that biophysical and socioeconomic researchers work with those variables and categories that local populations understand and are familiar with. This choice of variables and categories is the responsibility of the researchers so as not to over-burden local participants with data gathering they do not understand or for which they have no practical use.

Overview of the Book

This book describes the process and results of the 5-year SANREM Phase II Program in West Africa, detailing its approach, components and outcomes. Working with our partners in Madiama, SANREM has succeeded in developing an approach that leads to sustainable social and economic development while minimizing NRM related conflicts. Evidence from the commune indicates that since SANREM began its activities, conflicts have been reduced as the community has begun to work together on priority problems. While still a work-in-progress, SANREM has developed a replicable process that will lead to successful locally controlled participatory decision-making and the decentralization of development services and community initiatives to improve resource management throughout the region. The contributors hope that this book will stimulate further work and development towards this important end.

The first part provides an overview of the landscape and lifescape in which the SANREM project intervened.

In Chapter 2, Cissé *et al.* introduce the social and historical context shaping the lifescape of the Inland Delta of the Niger, the core of Mali's 5th Region. Various ethnic groups have co-existed in this area for centuries, generating livelihoods with complementary systems of production. The chapter discusses how recent changes in agricultural and pastoral production systems have unbalanced this symbiosis and increased competition for scarce resources, thereby leading to land tenure confrontations that are not as easily resolved as they were in the past. In addition, decentralization and democratization have complicated the situation in which the modern state and civil society have been superimposed on the modified, but not displaced, customary governance systems. The discussion concludes, with some qualifications, that NGO-driven opportunities for local dialogue and problem solving hold considerable promise for rural Mali.

In Chapter 3, Badini and Dioni present a detailed description of the landscape types and soils of the Commune of Madiama. Combining knowledge gained from informal surveys, field observations, biophysical monitoring, transects, remote sensing and pit holes, the chapter characterizes soil types and distributions, climate and hydrology, cropping patterns, land use systems and potentials. The database on these landscape types, their location, potentials and constraints is at an appropriate scale for use by village, commune and regional level planners, as well as for providing input to biophysical models to evaluate technologies (as in Chapter 11).

In Chapter 4, Ballo and Ouattara describe the systems of animal husbandry within and around Madiama, thereby broadening the perspective of the landscape/ lifescape scale. The chapter characterizes each of three livestock management types (sedentary, semi-transhumant and transhumant) involving cattle, oxen, milk cows, sheep and goats. Pastoralists are either transhumant or resident, and often tend to the herds of local farmers much of the year. The pastoral resources available to the commune are limited and though traditional grazing of crop residues and fertilization of soil exists, there is increasing loss of organic matter. The chapter concludes that under Mali's new Pastoral Code, conditions could improve, but improvement will require the concerted efforts of agriculturalists and pastoralists.

In Chapter 5, Wynne *et al.* combine remote sensing and ground truthing in the analysis of land use change during a 50 year period (1952–2002) in the Commune of Madiama. The chapter documents the dramatic shift in land use from pastoral to crop-based production systems.

The second part of the book describes the elements of SANREM West Africa's multi-pronged approach to development intervention. The intention is not to provide a history per se, but to highlight the essential and potentially reproducible elements of the interaction between researcher and community. Each chapter in Part II presents components modelling the conduct of a successful community-mobilizing development intervention.

In Chapter 6, Earl and Kodio describe how the SANREM team introduced themselves to members of the Madiama community through the Participatory Landscape/ Lifescape Appraisal (PLLA), an informal set of participatory survey methodologies that bring out not only important information about the natural resources of Madiama and the production systems shaping the population's livelihood, but also its primary concerns and priorities. In particular, they note the desire of the communities for improved soil fertility and pasturelands in order to increase productivity and incomes, and the underlying concern about conflict.

In Chapter 7, Moore *et al.* describe the development of the commune-level NRMAC providing the social infrastructure that is both adapted to the exigencies of recent governmental decentralization and compatible with customary governance structures at the village level. They argue that it is not sufficient to simply assemble a group of men and women representing various ethnic groups and occupational categories, but it is also necessary to develop each member's individual capacities (functional literacy, leadership skills, association management, knowledge of codes and laws, etc.) in order to stimulate mutual trust and network building (social capital formation) between villages and clans and to help them to define their mission. Of particular importance is training in conflict management to build individual self-confidence and to provide a credible and valued service in the eyes of villagers.

In Chapter 8, Goebel *et al.* provide an overview of an alternative conflict management approach and discuss how it differs from common approaches to conflict resolution. The approach is based on building conflict management and consensus building skills rather than simple conflict management, per se. The training programme consists of a series of workshops focused on building skills and empowering local leaders by their learning a sustainable process for facilitation and management of diverse conflict situations. This process has been central to the evolution of the NRMAC from a group of village representatives to a committee with a Commune-level mission to improve NRM in the face of resource competition.

In Chapter 9, Bingham introduces the Holistic Management™ (HM) Model, an approach developed specifically for open range animal husbandry, but well adapted to provide intuitive analyses, insights and decision-making information for community-based management of natural resources. Consistent with the SANREM approach, this holistic methodology was introduced to scientists and community members in order to facilitate diagnostics of and behavioural change in the management of the natural resource base in the Commune of Madiama. He describes how HM has been a source of tension between scientists and HM promoters, but has also led to innovative attempts at changing resource management within the community.

Following on these capacity building interventions, the third part presents the results of various research activities involving the development and testing of technologies and decision-making tools appropriate for the community and environment of Madiama.

In Chapter 10, Crane and Traoré compare indigenous perceptions and models of soil fertility management with those of modern soil science. The differences between these perspectives are discussed along with the cultivators' perceptions of the scientific research conducted in Madiama. The authors argue for implicating end users in technology development from the beginning so that folk knowledge can guide scientific research and research findings can be better communicated to farmers to increase productivity.

In Chapter 11, Badini *et al.* demonstrate the use of a combination of results of data collected from 3 years of on-farm field trials and long term simulation with CropSyst modelling to evaluate crop rotations, and organic and inorganic fertilizer practices for yield efficiency, stability and soil sustainability over a simulated 30-year period. Organic fertilization produced the best results in the analysis and the chapter concludes by recommending increased efforts in the development of strategies to produce and apply more manure and other organic fertilizers.

In Chapter 12, Wyeth *et al.* pursue the issue of sustainability of technologies for enhancing soil fertility from economic and financial perspectives. This analysis takes into account the results of 3 years of on-farm trials and combines those findings with the output of the computer modelling analyses of Badini *et al.* (Chapter 11). Their results suggest that corralling livestock in the fields, spreading manure and micro-dosing with chemical fertilizer are adoptable within the range of farmers' risk preferences.

In Chapter 13, Brewster *et al.* model the economic linkages between different groups of natural resource users and analyse the effect of the potential growth strategies. Using data from the PLLA (Chapter 6) and an extensive household and enterprise level data set, a Social Accounting Matrix (SAM) model for the Commune of Madiama was developed. Application of this model demonstrated that the largest impacts from exogenous changes in demand are associated with rice and livestock sectors. Furthermore, impacts are shared differentially among socioeconomic groups with agropastoralists benefiting most and the transhumant group benefiting least.

In Chapter 14, El Hadj *et al.* describe research in response to a request by Peul village women to deal with *Cassia tora,* an invasive plant that is unpalatable for grazing animals. As pasture lands have degraded, *Cassia tora* has spread throughout their fields out-competing other plants. However, this noxious plant has potential as dry season forage when ensiled. The authors describe their analysis of the forage potential and their work with village women to develop ensilage techniques adapted to local conditions.

In Chapter 15, Abaye *et al.* investigated the regenerative potential of pastureland in two villages through a controlled experiment with tethered grazing of small ruminants. This work builds on the Holistic Management™ (Chapter 9) insight that animal impact is not simply a function of numbers of livestock or duration of grazing time in order to provide management indicators that optimize the potential of forage regeneration/biomass production rates, plant biodiversity, and animal performance. The chapter concludes that grazing vegetation down to a 3 cm height on any particular parcel is likely to limit forage regeneration.

In Chapter 16, Moore *et al.* address the issue of social capital as a factor of development within Malian civil society. Social capital is investigated with a focus on embedded and autonomous social relations at the commune and village levels. Using household survey data from the Commune of Madiama, they investigate the extent to which the NRMAC provides a social infrastructure on which to build community-level social capital. The analysis demonstrates the importance of building on traditionally valued social relationships and combining them with linkages across groups for the management of conflict situations. They conclude that the NRMAC does indeed provide a platform for building inter-ethnic and multi-village social capital.

In Chapter 17, Moore *et al.* review how the four pillars of SANREM (participation, landscape scale, multiple stakeholders and interdisciplinarity) were applied in the Commune of Madiama. In the list of lessons learned, they highlight the difficulties involved in establishing and maintaining full participation of and communication between all stakeholders in the context of power relations and traditionally excluded groups. Building social capital and co-management agreements is a long term and iterative process requiring that project and government development agents be well-trained and integrated within the community in order to empower the population to act on its own behalf.

Notes

1 The *terroir* is a biophysically and socio-culturally defined space for production and living generally associated with a village. *Gestion der terroir* is the development approach emphasizing community participation associated with this space.

2 *Echinoloa stagnina* or *bourgou* in French is a wetland grass that grows with the rising river waters during the rainy season. It is prized as a very nutritious animal feed.

References

Bary, H. (1996) *Etude sur la gestion des conflits liés aux aménagements pastoraux: cas de la zone pastorale de Sidéradougou.* Ministère de l'Agriculture et des Ressources Animales, Projet Régional d'Appui au Secteur de l'Elevage Transhumant (PRASET), Ouagadougou, Burkina Faso, pp. 1–42.

Benjaminsen, T. (1997) Natural resource management, paradigm shifts, and the decentralization reform in Mali. *Human Ecology* 25(1), 121–141.

Blench, R. (1996) Aspects of resource conflict in semi-arid Africa. *Natural Resource Perspectives* 16. Overseas Development Institute. http://www.odi.org.uk/nrp/16.html (accessed 1 May 2004)

Blundo, G. (1996) Gérer les conflits au Sénégal: le rôle de l'administration locale dans le sud-est du bassin arachidier. In: Mathieu, P., Laurent, P. and Willame, J. (eds) *Démocratie, enjeux fonciers et pratiques locales en Afrique. Cahiers Africain* 23–24, Institut africain/L'Harmattan, Paris, pp. 101–119.

Brown, M. (1993) *Non-governmental organizations and natural resources management in Africa's pastoral sector: Where to go from here?* United States Agency of International Development, The PVO-NGO/NRMS Project, Washington, DC, pp. 1–76.

Cissé, S. (1996) Problématique foncière et litiges fonciers en 5e région. In: Maïga, I. and Diallo, G. *Atelier de restitution et de reflexion sur les litiges fonciers dans la région de Mopti.* Groupe de Recherche-Actions pour la Développement/Institut International pour l'Environnement et le Développement, Bamako, Mali, pp. 24–26.

Dasgupta, P. and Serageldin, I. (2000) *Social Capital: a Multifaceted Perspective.* World Bank, Washington, DC.

Djalebord Diard, A. (1992) Le stratégies et le rôle du CILSS dans la sous-région. In: *La zone Agro-Sylvo-Pastoral au Sahel : Limites et Potentialités de Développement contre la Désertification. Actes de la Conference de Assissi.* Comité Inter-Etats de Lutte contre la Sechéresse au Sahel, pp. 25–30.

Grootaert, C. and van Bastelaer, T. (2002) *The Role of Social Capital in Development: an Empirical Assessment.* Cambridge University Press, Cambridge, UK.

Horowitz, M.M. and Little, P.D. (1987) *African Pastoralism and Poverty: Some Implications for Drought and Famine.* Cambridge University Press, Cambridge, UK.

Isham, J., Kelly, T. and Ramaswamy, S. (2002) *Social Capital and Economic Development: Well-Being in Developing Countries.* Edward Elgar Publishing, Limited, Cheltenham, UK.

Kane, O. and Winter, M. (1997) La gestion décentralisée des ressources naturelles au Sénégal. In: *La Gestion Décentralisée des Ressources Naturelles dans Trois Pays du Sahel: Sénégal, Mali et Burkina Faso.* Le Projet d'Appui au Développement Local au Sahel (PADLOS). Ouagadougou, Burkina Faso, pp. 13–131.

Laurent, P., and Mathieu, P. (1994) Authority and conflict in management of natural resources – a story about trees and immigrants in southern Burkina Faso. *Forests, Trees and People* 25, 37–44.

Lo, H., Sylla, C., Ndione, P. and Agne, A. (1996) Ressources forestières et conflits au Sénégal: quelles approches pour une typologie? *Bulletin Arbres, Forêts et Communautés Rurales* 8, 26–36.

Lowenberg-DeBoer, J., Kaboré, D. and Abdoulaye, T. (1994) *The opportunity cost of capital for agriculture in the Sahel: Case study evidence from Niger and Burkina Faso* (Staff Paper 94–2). Purdue University, Department of Agricultural Economics, West Lafayette, Indiana.

Lynam, J.K. and Blackie, M.J. (1994) Building effective agricultural research capacity: The African challenge. In: Anderson, J. (ed.) *Agricultural Technology: Policy Issues for the International Community.* CAB International, Wallingford, UK, pp. 106–134.

Maïga, I. (1996) *Etude sur la prévention et la gestion des conflits.* Centre Canadien d'Etude et de Coopération Internationale (CECI), Programme de Promotion de la Démocratie et des Droits de la Personne (PDDP), Bamako, Mali, pp. 1–55.

Maïga, I. and Diallo, G. (1998) *Les conflits fanciers et leurs gestion dans la 5ème Région du Mali* (Dossier No. 76). Groupe de Recherche Actions pour le Développement. International Institute for Environment and Development, Programme Zones Arides, London, UK.

McIntire, J., Bourzat, D. and Pingali, P. (1992) *Crop–Livestock Interaction in Sub-Saharan Africa.* World Bank Regional and Sectoral Studies, Washington, DC.

Messer, E., Cohen, M.J. and D'Costa, J.D. (1998) *Food from Peace: Breaking the Links between Conflict and Hunger* (Food, Agriculture and the Environment Discussion Paper 24). International Food Policy Research Institute, Washington, DC, pp. 1–44.

Moore, K.M. (1996) *Status report on CBNRM Project Monitoring System and NRM KAP study* (Summary Assistance Report No. 9). Ministry of Environment and Protection of Nature, Community-Based Natural Resources Management Project, Dakar, Sénégal, pp. 1–6.

Ngaido, T. (1996) Redefining the boundaries of control: post-colonial tenure policies and dynamics of social and tenure change in Western Niger. Ph.D. dissertation, University of Wisconsin, Madison, Wisconsin.

Niamir, M. (1991) *Traditional African Range Management Techniques: Implications for Rangeland Development.* (Pastoral Development Network Paper No. 31d). Overseas Development Institute. London, UK. http://www.odi.org.uk/pdn/papers/paper31d.html (accessed 1 May 2004)

Okai, M. (1987) Agricultural production, food security and poverty in West Africa. In: Asenso-Okyere, W.K., Benneh, G. and Tims, W. (eds) *Sustainable Food Security in West Africa.* Kluwer Academic Publishers, Boston, Massachusetts.

Painter, T., Sumberg, J. and Price, T. (1994) Your *terroir* and my 'action space': Implications of differentiation, mobility and diversification for the *approche terroir* in Sahelian West Africa. *Africa* 64(4), 447–463.

Park, T., Tabor, J., Frankenberger, T., Lynham, M., Baro, M., Ngaido, T. and Rogers, G. (1993) *Risk and Tenure in Arid Lands: the Political Ecology of Development in the Senegal River Basin.* University of Arizona Press, London, UK.

Pointing, J. (1995) Impact of social and economic change on pastoral women in East and West Africa. In: Stiles, D. (ed.) *Social Aspects of Sustainable Dryland Management* (United Nations Environment Program). John Wiley & Sons, Chichester, UK, pp. 253–265.

Pratt, D.J., Le Gall, F. and de Haan, C. (1997) *Investing in Pastoralism: Sustainable Natural Resource Use in Arid Africa and the Middle East* (World Bank Technical Paper No. 365). The World Bank, Washington, DC.

Preston, T.R. and Leng, R.A. (1994) Agricultural technology transfer: Perspectives and case studies involving livestock. In: Anderson, J. (ed.) *Agricultural Technology: Policy Issues for the International Community.* CAB International, Wallingford, UK, pp. 267–286.

Pretty, J. and Buck, L. (2002) Social capital and social learning in the process of natural resource management. In: Barrett, C.B., Place, F. and Aboud, A.A. (eds) *Natural Resources Management in African*

Agriculture. CAB International, Wallingford, UK, pp. 23–33.

Rochette, R.M. (1989) *Le Sahel en lutte contre la désertification: leçons d'expérience.* Gesellschaft für Technische Zusammenarbeit (GTZ), Weikersheim, Germany. pp. 1–59.

Sanders, J.H., Shapiro, B.I. and Ramaswamy, S. (1996) *The Economics of Agricultural Technology in Semiarid Sub-Saharan Africa.* Johns Hopkins University Press, Baltimore, Maryland.

Sanon, Y. (1996) *Etudes sur la gestion des conflits liés aux aménagements pastoraux: cas de la zone pastorale Samorogouan.* Ministère de l'Agriculture et des Ressources Animales, Projet Régional d'Appui au Secteur de l'Elevage Transhumant (PRASET), Ouagadougou, Burkina Faso, pp. 1– 36.

Shanmugaratnam, N., Vedeld, T., Mossige, A. and Bovin, M. (1992) *Resource Management and Pastoral Institution Building in the West African Sahel* (World Bank Discussion Papers Africa Technical Department Series No. 175). World Bank, Washington, DC.

Sidikou, H.A. (1995) *La gestion alternative des conflits liés à la gestion des ressources naturelles en zone pastorale: brève réflexion inspirée pas le cas de la reserve de Gadabeji (Nord Dakoro).* Paper prepared for Atelier national sur la gestion alternative des conflits liés à la gestion des ressources naturelles, Niamey, Niger, pp. 1–8.

Sommer, F. (1998) Pastoralism, drought, early warning and response. MA thesis. University of Sussex, Brighton, UK.

Souley, A. (1996) *Rapport d'Etude: Prévention et Gestion des Conflits au Niger.* Centre Canadien d'Etude et de Coopération Internationale, Niamey, Niger, pp. 1–52.

Soumaré, S. (1996) Pastoral land tenure: the exploitation of natural resources against a backdrop of actors from different ethnic groups, diverse production systems and frontier cohabitation. In: *Managing Land Tenure and Resource Access in West Africa.* Groupe de recherche et d'échange technologiques (GRET)/IIEE, Goré, Sénégal, pp. 173–178.

Toulmin, C. and Moorhead, R. (1993). *Local strength and global weakness: NGO experience with pastoralists in Africa.* Paper prepared for the PVO.NGO/NRMS-IIED-CCA. ONG Workshop on NGOs and Natural Resource Management in the Pastoral Sector of Africa – Strategies for Enhancing Performance and Impact. 22–28 February 1993, Mopti, Mali.

Touré, D. (1996) *Prévention et Gestion des Conflits.* Centre Canadien d'Etude et de Coopération Internationale, Ouagadougou, Burkina Faso, pp. 1–50.

Vedeld, T. (1992) Local institution building and resource management in West African Sahel. *Forum for Development Studies* 1, 23–50.

2 The Lifescape: Production Systems and Social Institutions

Salmana Cissé,[1] Keith M. Moore[2] and Charlene Brewster[2]

[1]Centre Régional de la Recherche Agronomique, Institut d'Economie Rurale, BP 205, Mopti, Mali and [2]Office of Intenational Research, Education and Development, Virginia Tech, Blacksburg, VA 24061–0378, USA

In this chapter, we outline the social and institutional context shaping the lifescape of the Inland Delta of the Niger, the core of Mali's 5th Region. This historical analysis of the region reveals the roots of many of today's conflicts across the Sahel. The lifescape of the Inland Delta can be viewed as an ethnic melting pot, whether in the inundated or rainfed zones, nested within complementary and conflictual production systems. The rules that regulate these production systems are organized into competing customary local and modern national governance systems. This arrangement complicates the livelihoods of the population. The social institutions governing ethnic and production relations have evolved over time with regard to the natural resource base (see Chapters 3, 4 and 5 for a discussion of the landscape), and have become ineffective for managing contemporary technological and population exigencies. Consequently there is an impasse in natural resource management (NRM) rule administration. This chapter describes the historical development of social and economic systems and relations between the State and civil society (an associational sphere intermediary between the family and the state).

Historical Development of Productive and Ethnic Relations

Three general production systems have characterized the Sahel: agro-sylvo, agro-sylvo-pastoral, and sylvo-pastoral (Sunzini, 1992). The Inland Delta of the Niger has a fourth, aquaculture. Historically, the numerous ethnic groups in the region (Bambara, Peul, Bozo, Bobo, Malinké, Marka, Dogon, etc.) are principally associated with different principle production systems, based on millet, rice, livestock, fish, etc. These systems, however, are in transition as herders, farmers and fisherfolk seek out new means of survival in a changing ecology. Herders and fisherfolk are becoming sedentary; agro-pastoralists are increasingly using trees for animal feed and soil fertilization; and farmers are raising more livestock. Today, ethnic flexibility with respect to production practices can be found in all production systems of the two zones of the 5th Region. An individual can be a farmer, herder or other regardless of ethnicity. In addition, significant and increasing numbers of each of these groups seek urban employment, thereby reducing the available labour pool and shifting management responsibilities to already overburdened and under-prepared women.

Traditional production systems in the Sahel have existed for centuries in various proportions dependent on available resources and ethnic groups. Although Blench (1996) disagrees, Watts (1987), Park *et al.* (1993), and Pratt *et al.* (1997) all agree that sustainable NRM was a conscious, if not always explicit, practice. The production systems were primarily extensive with a low ratio of output to input. New fields were cultivated nearly every year, while large areas were left fallow or used for animal grazing. Herds were moved across vast expanses in a seasonal pattern of migration opportunistically seeking fresh pastures. Fishing and gathering were also extensive, each utilizing a relatively large natural resource base (water or forest). Describing similar conditions in the Senegal River Basin, Park (1993) has pointed out that the key adaptive strategy across production systems was to maximize returns to labour on a selective temporal and spatial basis.

This intersection of production systems in time and space led to a wide variety of social groups in the 5th Region and shaped intra-community and inter-ethnic relations over the centuries (Turner, 1992). A complex and hierarchical social structure, composed of successive settlements of Sorongo fisherfolk (Bozo), Nono (Marka) rice cultivators, and Fulbe (Peul) pastoralists, was established in the delta region by the time of the Mali Empire (14th century). Central authorities (from the Mali and Songhay Empires through those of the Moroccans and Bamanan (Bambara)) were paid tribute but had limited authority over the mostly self-governing populations of the Inland Delta. Local Peul families and clans tended to dominate the region through their slave cultivators (often Bozo and Marka). The successive historical waves introducing each group (Bozo, Marka, Bobo, Peul, Bambara, Maures, Dogon) culminated in the Hamdallahi regime (1818–1862) that formalized the political, economic and social structure and created a land tenure and pasture management code called the 'Dina'.

Under the Dina, competition for space between different productive activities was minimized due to the abundance of land, as well as the spatial and temporal distribution of different production activities. The Dina divided the delta region into 37 pastoral territories (*leydi*) each under the administration of a local pasture and herd master (*jowro*), in order to maximize mobility and coordinate opportunistic grazing in the pastoral management system. The *leydi* included not only the pasture lands, but also farmer and fisherfolk villages and their production spaces. Although village chiefs (*jom*) and land chiefs had control over local uses of the village *terroir* (the physical and social territory attributed to a village or group of villages), *jowros* administered pasture use by regulating access rights for different groups, scheduling use and negotiating annual exchanges of pasture with other *jowro* (Crowley, 1991).

The Dina provided the minority Peul (20% of Inland Delta population) with a mechanism to protect their transhumant herds, pastures and routes, maintain local dominion and regulate relations with complementary production systems. This form of social organization was able to persist, because all the local inhabitants, whether Peul herders, Marka farmers, or Bozo fisherfolk, were considered part of the same patrilineal extended family or clan (*suudu baaba*) in relation to other groups and in the face of various social or other adversities (Crowley, 1991). This 'family' nature of relationships between different groups of the same *leydi* was founded on political accords, marriage alliances or simply on particular social relations where cohabitation could produce feelings of sharing in the same social entity.

A unified Peul State dominated social and economic relationships until the Toucouleur invasion of the mid-19th century. The Peul loss of political power reduced the state's formal administrative structure supporting the Dina. Successive new states were designed with a territorially fixed administrative structure favouring the sedentary (farming) majority and their local leaders. Nevertheless, the Dina's economic and non-politicized social organization of production relations has persisted despite the abolition of slavery by the French at the turn of the 19th century and the attempted institutionalization of a sharecropping system.

The scale of management for each system was related to its need for natural resources. Organizing their production on a concentric basis within their *terroirs*, sedentary farmers first cultivated fields around their dwellings, then in the village, and finally in the bush. Fisherfolk migrated along the river with the fish after the flood wave each year (Ambach, 1996). Operating on a larger scale, pastoralists went from zone to zone seeking pastures for grazing and water for their animals according to seasonal availability (Pratt *et al.*, 1997). Following the rainfall and searching for fresh pasture, this transhumance shifted between dry and humid regions. During the rainy season they moved out of the Delta into the rainfed grasslands and returned during the dry season as the pastures and water holes dried out. It was only in the 1980s that the Peul became a predominantly sedentary population, expanding their pastoral production system into agricultural production. Transhumance still occurs but now all extended families provide members with a permanent residence.

The current land management and pastoral movement is embedded in the Dina-imposed social structure despite the form of the governing state shifting from Toucouleur to French colonial rule and later to national independence. Some ethnic groups, including the Marka, Bambara and Dogon, occupy and control a clearly defined territory for cropping and pursuing other livelihoods. Other groups, including the Peul and Bozo, occupy space in a more transitory fashion, with shifting campsites scattered through space and time according to the seasonal availability of natural resources. This space is open and, according to the season, may engulf or demarcate a village space where relations between social groups are either complementary or conflictual, depending on the relations of power: demographic relations (age and gender), and social relations (classes), and political relations (dominant clan groupings).

Crop and Livestock Production Relationships

Although farming and herding have diverged as different specializations, there is a long history of complementary interaction between the two groups through livestock sharing and bartering milk for grain (Sturm, 1996). Under this system, the products and by-products of one system constituted an important input for the other, and there was coordinated action between the actors. For example, cultivators have obtained animal manure and labour in return for providing cash, crop residues or water rights (wells) to herders (Heasley and Delahanty, 1996).

In the case of manure, a formally negotiated contract has traditionally allowed herders to exchange manure from their herds as compensation in kind for cereals, animal drinking water, clothing, etc. Although the initiative for such a contract often came from the farmer who simply provided a well in his field for the animals, herders in need of millet stocks or rice straw left in the fields after harvest could also initiate contracts. In this case, milk, or more rarely an animal, would be offered in exchange. In all cases, however, the size of the herd and/or the value of the pasture was an important element in the contract negotiations. These relationships have decreased as crop farmers have increasingly come to manage their own cattle herds and herders become sedentary (Toulmin, 1983).

In recent decades, the advent of harnessed traction for cultivation set in motion a dramatic transformation in the relationships between rural production systems in the 5th Region. The passage from hand cultivation with the *daba* (a short handled hoe) to the animal drawn plough has resolved a number of problems (while often creating others) related to bottlenecks in the labour supply. It set in motion significant modifications in the organization and management of field space, including encroachment cultivation on agricultural trails, designation of pastures reserved for working cattle and recruitment of permanent herders to manage this newly created herd category (see Chapter 4). Harnessed traction has also had an impact on the transport of agricultural inputs to the fields and field products to the family granary or the market.

Although herders and farmers have traditionally used what appeared to be two separate landscapes, livestock now occupy a significant place in sedentary farming systems (McIntire *et al.*, 1992). In addition to the need for traction animals, farmers are diversifying into livestock production to reduce the risk of crop failure. While diversification in sedentary agriculture is taking place, livestock mobility – the comparable risk-averting behaviour in pastoral systems – has been restricted as potential grazing lands are converted to crop lands and herding families have established permanent villages for crop production and other livelihood needs. Such crop/livestock integration and accompanying intensification of production systems is generally viewed as a positive development for both productivity and environmental conservation (Mortimer and Turner, 1994; Williams *et al.*, 1995; Sturm, 1996). However, Wolmer (1997) points out that such integration may not be environmentally sustainable and rejects the notion of a single diversification development trajectory for farmers and herders.

In Mali, there is strong evidence that users are concentrating in areas of higher quality resources (e.g. the Inland Delta of the Niger), and it is apparent that competition is replacing the more complementary and cooperative resource use of the past. Also increasing is competition between farmers and sedentary or transhumant herders for crop residues and grazing areas. The traditional complementary exchange of stubble grazing for improved soil fertility has disappeared, while competition for woody vegetation used as fuel or fodder has increased.

Competitive Relationships

Schematically, three principal land-based production systems can be identified in the zone – an agricultural system, a pastoral system and an agro-pastoral system which combines elements of the other two. However, it is very difficult to encounter a pure production system that has evolved through space and time on its own. All the systems have evolved in the same space, using the same human tools, with the same general objectives to satisfy fundamental human needs. Consequently, there is a history of multiple confrontations over access to space and labour.

Competition for space

Both between and within production units, pastoral spaces are being lost to crop production. Even within the same production system, competition can be found between two coincidental practices. One, the confrontation between production practices in a given space, can be partially explained by the willingness of one social group to exclude another. For example, successive use of pastures by herds of different origins can create conflicting uses when the first herd is late to leave the space and impedes the entrance of the next. Competition, however, involves not only spatial occupation of the same nature (agriculture/agriculture or livestock/livestock), but also of different natures (agriculture/livestock). This latter competition is more commonly perceived, because it is easily discernible in the destruction of crop fields by animals and the destruction of animal corridors by non-herders. The immediate consequences of this competition have typically been manifested in violent confrontation (Maïga, 1996).

Competition for labour

A division of labour exists within households, because members are typically assigned specific tasks by age and gender. As a general rule, planning and decision making are left to the senior members of the household – most frequently the male household head (Cissé, 1985). For the most part, women are charged with household and reproductive tasks (food preparation, collecting water and firewood, childbearing, childcare and so on). They also tend the livestock kept around the homestead and, in particular, are charged with milking and processing dairy products. In nomadic households, women are also often responsible for setting up and taking down their temporary household structures. Men's tasks include clearing and preparing land for cultivation, herd management (herding, castration, vaccination and slaughter), and digging wells (Pointing, 1995). However, as population increase puts pressure on the limited resource base – intensifying livestock production, establishing agro-pastoral production systems, and increasing migration in search of employment – women appear to be taking on more of these traditionally male tasks (Pointing, 1995).

Family labour allocation decisions are made between rainfed and irrigated crop production activities, or between herding and cropping activities, or between fishing and cropping activities. For the organization of production, a double consequence can be noted: on one hand, the periodic appearance of bottlenecks assures that optimal production strategies are not followed, and, on the other, the use of unqualified labour and inappropriate production implements results in inefficient use of resources. Competition for labour is not limited to intra-family allocation, but can also involve an entire village or group of villages.

Social differentiation

The level of household wealth determines vulnerability to such disasters as drought, disease epidemics and subsequent famine. Wealth may be indicated by housing materials, size of holding or size and composition of the herd. Poor farmers may be forced to sell off productive assets, while richer farmers may even gain because of such a disaster (Frankenberger and Lynham, 1990). The poorer the pastoralist, the more likely he is to disappear as an independent herdsman after such a disaster, thereby widening intra-ethnic disparities (see Brewster *et al.* in Chapter 13). Some marginalized and impoverished populations have established permanent settlements and raised crops as a survival mechanism. Traditional support mechanisms such as cow sharing no longer function (Bovin, 1990). Permanent settlement reduces herd size and restricts the economic returns from raising livestock (Horowitz and Jowkar, 1995). As men migrate in search of seasonal employment, women's workload increases.

Specialization, such as in the intensification of milk, meat or horticultural production, has led to further individualization and inequities in wealth. However, another type of entrepreneurialism is growing in West African countries: the absentee herder or farmer, who owns livestock or a farm but lives in town. This category of entrepreneur has been created by ongoing changes brought on by drought and consequent economic problems that have impoverished rural people relative to urban-based government employees, merchants, and so forth. These new urban entrepreneurs hire poor herders, who lost their livestock during previous droughts and have been unable to restock their herds, to tend their livestock (often the same animals bought from the herders during the crisis) in return for a salary or other in-kind arrangements.

Village solidarities have traditionally mitigated against lost access to resources. Increased competition for resources, however, is expanding the gap between rich and poor. This differentiation is growing both between, and within, villages. Resource tenure for village livelihoods has become highly insecure and conflicts over resources have increased in recent years. In the next section, we will see how the social institutions have or have not evolved to manage resource access and conflict.

Customary Authority, State and Civil Society

Contested terrain: modern vs. customary law

In recent years, Sahelian governments have tried to reconcile the realities of customary law at the village level with the need for standardized national laws (Bohrer and

Hobbs, 1996; Delville, 1998). Those involved in conflict resolution relating to these laws cite either inappropriate laws or failure to apply them correctly (Maïga and Diallo, 1996; Ngaido, 1996; Touré, 1996; Moore and Thiongane, 1998). In fact, what is typically considered the region's land tenure framework is no more than a spatial mosaic of productive activities, a physical manifestation of the social organization, or '*logiques sociales du territoire*' (Mathieu, 1996). This mosaic is rarely linked to modern land tenure status. In order to secure their livelihoods, the majority of actors completely ignore all legal documentation, thereby causing frequent renewed legal challenges, despite the administration's efforts to settle these disputes. This explains why, until recently, administration officials have had little respect for peasant traditions and practices.

There once might have been coherence between production practices and politico-administrative systems. Resource tenure in West Africa has traditionally balanced the land needs of various households within low population production systems (Cissé, 1985; Sanders *et al.*, 1996). Use rights to a resource varied seasonally, providing the flexibility needed for multiple production systems. Resource conflicts were mostly resolved amicably. At the core of this customary law[1] is a process of reconciliation between disputants, rather than a winner-take-all consequence, as in modern western law. Procedures in traditional courts were simple and informal, although not easily accessible to women or younger men. In modern courts, while access is universal in principle, limits still exist due to transportation costs and the need for a lawyer to advocate one's case. Currently, the simultaneous existence of customary and modern legal systems offers multiple opportunities for contestation if resource users are dissatisfied with the existing determination of their rights. This legal pluralism often contributes to, rather than resolves land tenure conflicts.

In the Inland Delta region, customary rights to the use of natural resources tend to vary from season to season and from resource to resource. Access to land for crop production is determined at the village or clan level, first through the lineage, then through the household. Access to pastureland has depended on the *jowro*. Leaders of the original or conquering lineage or ethnic group are often the only people deemed capable of 'owning' property (Cissé, 1985; Ngaido, 1993). For the most part, male household heads inherit the use right to the majority of land they farm (Cissé). Some may borrow from others in the same lineage or from outside the lineage (Matlon, 1993). Temporary use rights may be arranged through sharecropping or rental. Slaves may access land through their masters (Cissé, 1985; Ngaido, 1993). Women and other subordinate family members have traditionally had access to lands through the male household head. Inheritance of land by women, although occasionally practiced, is rare as it tends to fragment household and lineage land holdings (Cissé; Ngaido). Within pastoral systems, women are not authorized to own cattle; however, childbearing may establish rights in, and responsibility for, milk production (Pointing, 1995).

Initially, the modern state did little to change customary land tenure arrangements in rural areas. Docking (1999) notes that despite French and Malian government attempts to manage local government in Mali, village-based decision making processes changed little. In 1934, the contemporary context was set by the Supreme Court of French West Africa which decreed that the African village was 'a legal entity with customary rights and the village chief was the defender of those rights' (Mamdani, 1996, p. 114). An intermediary infrastructure of cantons was adapted as the mechanism for organizing relations between the central state and the villages. Moore *et al.* (2003) contend that this preservation of customary rights defined a subordinate, but partially autonomous state apparatus. The colonial administration in Mali sought to govern indirectly by relying on local authorities to implement colonial policies while limiting the organizational integrity of Malian civil society through the singular administrative binding of villages to the state.

As the demand for scarce resources increased, conflict could only be avoided if access was structured according to mutually agreed on criteria of legitimacy. Asserting its authority, the state saw this as a matter of defining a single system of legal ownership and usufruct. Such a universal system does not account for the multiple uses and sustained sharing of the resources that have customarily governed resources at the local level. Customary law, on the other hand, is based on local practice and the local legitimacy of traditional leaders (herd masters, land chiefs, village chiefs, etc.). These leaders, who are perceived to represent the best interests of the parties involved in any conflict over natural resources, have been able to resolve many conflicts through adaptation and interpretation of customary law. In the eyes of the state, however, customary law and traditional leaders have often been perceived as an impediment to progress, even when customary laws and authorities are officially recognized.

In the Niger Delta, overlapping spatial claims are seen among livestock production systems at the intersection of pastoral territories (*leydi*) managed exclusively by herders and the village pasturelands (*harima*) managed by villagers. This competition between the *leydi* and the *harima* has occurred as the size of village *terroirs* has increased at the expense of pastoral *terroirs*. It is a manifestation of the power of village chiefs, legitimized by the support of village elders and legalized by the laws of the republic. This process has reduced the power and customary legitimacy of the herd masters (*jowro*) since their legality has not been recognized by the Malian administration. With this de-legitimization of their roles, local herd masters and other resource masters (land and water) are finding it more justifiable to privatize the resources for which they have historically held a public responsibility (Keïta, 2003).

National policy and institutional context

It is worth considering the macro-institutional context in which these local legal struggles have been unfolding, because how the state ultimately resolves the contradictions inherent between legal texts and customary practices has important implications for the livelihood of the rural population. This is a profoundly political issue, involving relations between the modern state, the customary local state it has nurtured and a nascent rural civil society (Ribot, 1999).

During the socialist era after independence, the party and government were essentially fused in the state. The totalizing, and thus centralizing, approach of the Keïta regime (1960–1968) tried to bypass the local customary state, because it severely limited the associational sphere of both urban and rural society. Canton chiefs were eliminated and villages came into direct relationship with state power. However, as Rawson (2000) notes, 'the party misjudged the power of village elders' (p. 271), who helped to move conditions at the national level towards increasing anarchy. Consequently, the modernizing mission of the socialist regime was thwarted at the rural level. In response, the military regime of Traoré (1968–1991), which toppled the Keïta regime, instituted an extensive reform of territorial administration. This administration sought to co-opt the customary system of local governance by incorporating it into the state. 'The *chef* [*de village*] was now 'the representative of the *chef d'arrondissement'* and hence of the territorial administration' (Rawson, 2000, p. 273). Thus, local autonomy was severely restricted, much as it was under French colonial rule. Customary authorities were treated as an appendage of the state. Civil society under the Traoré regime began to open up, but participation without the benefit of democratization was restricted primarily to urban populations.

Both national regimes attempted to end the phenomenon of a state within a state by changing national law, either by denying customary authorities effective recognition (Keïta) or by incorporation (Traoré). Within this modern legal framework, the village chief was to function as an appendage of the central state. Village elders would propose a candidate for village chief to the *chef d'arrondissement* who would either accept or reject him. The traditional counter-power of secret societies and elder councils played a moderating role in customary decision-making. But, in all matters in which the State was concerned, the village chief acted as the local representative of the *chef d'arrondissement*, executing decisions handed down by him. Autonomous voices of rural civil society were not to be heard beyond the village *terroir*.

Decentralization/deconcentration

It is argued that Mali's rich history of complexity and differentiation has provided a long and healthy tradition of civil society (Fox *et al.*, 2001). Rawson (2000) claims that 'Associational activity has existed in varying degrees of autonomy from political authorities for centuries; civil society is not a new idea in Mali' (p. 269). Docking (1999) suggests that 'rural associational groups are key actors within the village level political structure and form the backbone of rural communal life' (p. 24). However, he notes that 'the concept of civil society as articulated in the West fails to account for the dynamics unfolding at the grassroots level in Mali' (p. 395). Social relations in rural civil society, such as they are, have been largely coincidental with the nearly autonomous local state at the village level and external relations have been structured through the network of the national state apparatus.

Opportunities for the growth of a vibrant civil society were given new life with the downfall of the Communist regimes of Eastern Europe and the Soviet Union. Indeed, the fall of the Traoré regime can be dated from this historic world event. The new regime in Mali took past lessons to heart and instituted, during the 1990s, a policy of administrative decentralization and democratization throughout the country (Rawson, 2000). Substantial powers, including land tenure decision-making, were devolved to newly created Rural Communes. However, regulations specifying how this power transfer will be implemented have yet to be drafted. The delineation (*découpage*) of these territorial collectives involved frequent consultations at the local (or regional) level and was often controversial. Contestation between villages and within the national assembly was common. Despite serious attempts to organize 'communes based on criteria of social solidarity, economic viability, and logic of geography and space' (Rawson, 2000, p. 278), regional government officials imposed many commune boundaries, often re-establishing the territorial limits of the cantons during the colonial period (Cissé, 1997).

Nevertheless, the current Malian government has accepted the principle that the central government is poorly placed to make appropriate decisions for the local level and has moved to implement decentralization policies (i.e. relinquishing decision-making authority over local matters, particularly land use and natural resource allocation; *Mission de Décentralisation*). Ostensibly, this increases the local populations' role in deciding appropriate land uses and ownership/usufruct issues. Such local control is expected to optimize the use of natural resources and encourage investment. However, the reality is that local decision-makers have very little legal discretion. The local institutions established to govern natural resources rarely have budgetary power, local legitimacy or a mandate for more than awareness building. Furthermore, the central government, while delegating the responsibility for decision making to local authorities, has required that they enforce national laws and provided few resources for implementing their decisions (Johnson, 2001; Ribot, 2002). In effect, the current policy of decentralization has become largely a matter of deconcentration of state power (i.e. strengthening the national government at the local level).

Local Institutions: *Gestion de Terroir* and Beyond

Recognizing the Malian State's positive step towards democratization, the donor community (United States Agency for International Development, the World Bank, etc.) has promoted the role of non-governmental organizations (NGOs) in order to encourage the growth of civil society. This effort builds on NGO experiences working within urban civil society under the Traoré regime. However, since 1991, the customary local state, presided over by village chiefs, has not been affected by these transformations (Davis, 2000), including the election of rural commune councils in 1999.[2]

Many NGOs, donors and national governments have implemented the *gestion de terroir* approach using socio-professional groups to represent civil society at the local level. Local here is defined, by default, most often as a single village. Indeed, what passes for a local community has remained largely unexamined (for example, see Fox *et al.*, 2001). Common practice when a 'stranger' (state or NGO development agent) desires to work with villagers is for them to address the village chief requesting entry into the village and explaining their mission. The chief then calls together the village association or a collection of representative elders to discuss the matter. After this consultation, the chief either indicates the village's acceptance of the mission by designating villagers to participate with the 'strangers' or declines the offer.

The local management committees developed in the context of these projects do afford a level of participation and local control over decision making, but are still too few to have a major impact on NRM or rural civil society. Although these forms of local management show promise for the future of NRM, the approach has been criticized for its limited scope, avoidance of conflict, and bias against pastoral management (Benjaminsen, 1997; Painter, 1994). As a mechanism for improving local NRM governance, local management committees still need considerable support in the form of literacy and budgetary training. Diversity of committee membership is encouraged, but the tendency is for local elites to dominate and for women, youth and pastoralists to be excluded (Bohrer and Hobbs, 1996).

This recent promotion of rural civil society has been based on donors' normative expectations in evaluating the democratic (and oppositional) attributes of Malian political life. However, democratic voting behaviours are essentially unknown in Malian village traditions (Kassibo, 1997). Failure to recognize the significance of this fact has led to a blurring of the distinction between the organization and practices of the customary local state, on the one hand, and rural civil society, on the other. Despite concern about fostering an intermediary associational sphere between individuals and the national state, 'rural civil society' in donor lexicon has come to be equated with NGO-supported extensions of the customary state (Ribot, 2002).

Nevertheless, decentralization has created opportunities for local dialogue and social adaptation. Recently, the development of local agreements for collective, multi-village management of natural resources has begun (Bocoum *et al.*, 2003; Hilhorst and Aarnink, 1999). Confronted by real and shared natural resource-based conflicts affecting the reproduction of their livelihoods, local populations are learning to take advantage of the organizing space created by NGO-led efforts to develop and apply innovative, locally controlled, problem-solving approaches. Local open-forum dialogues may concern one or several villages. Dialogue at the village level tends to reduce divisions within the community which draw on rights of first settlement, etc. by an explicit recognition of rights afforded to all engaged in productive activities. When this dialogue concerns a group of villages, the moment of communal management of space is introduced. These discussions advance towards accords which are the fruit of vigorous discussions and negotiations.

Although civil society in rural Mali is weak, decentralization has created conditions ripe for growth. Civil society, however, is confronted with a subordinate customary state's clan-based and ethnic solidarities that can serve as both a constraint and a building block for the social capital needed to strengthen rural civil society (see Moore *et al.*, Chapter 16). Functional horizontal linkages between socio-professional groups across villages are beginning to be established, and inter-group interaction will be required in order to strengthen those ties. Given the objective of trying to build civil society in rural Mali, effectiveness requires that social relationships be built that extend beyond the local customary state, because by monopolizing the trust and reciprocal expectations of rural social capital, it inhibits the development of viable alternative social networks and relationships that can promote a viable consensus concerning access rights and equitable management of natural resources.

Notes

1 While one cannot speak of African customary law as a single body of legal customs applicable to all of the different societies found in Africa, it can be used 'as a generic term for more or less similar traditional systems of law, or standards of behavior' (Rugege, 1995, p. 1). This is how we use it here. However, we recognize that customary law in West Africa includes Islamic influences and is a living law, not dictated by rigid tradition.

2 In contrast, urban unions and syndicates have experienced considerable separation from the organs of the state and their networks of social relations are expanding.

References

Ambach, B. (1996) Nomads on the river: Mali's fishermen threatened by change. *Development and Cooperation* 3, 20–23.

Benjaminsen, T. (1997) Natural resource management, paradigm shifts, and the decentralization reform in Mali. *Human Ecology* 25(1), 121–141.

Blench, R. (1996) Aspects of resource conflict in semi-arid Africa. *Natural Resource Perspectives* 16. Overseas Development Institute. http://www.odi.org.uk/nrp/16.html (accessed 1 May 2004)

Bocoum, A., Cochrane, K., Diakité, M. and Kane, O. (2003) *L'inclusion sociale pour une gestion équitable et durable des ressources naturelles: deux expériences au Mali* (Securing the commons No. 7). SOS-Sahel International Institute for Environment and Development, London, UK.

Bohrer, K. and Hobbs, M. (1996) *Post-Praia Progress Towards Tenure Security and Decentralization: Review of CILSS-member country legislative reforms.* University of Wisconsin, Land Tenure Center, Madison, Wisconsin, pp. 1–41.

Bovin, M. (1990) Nomads of the drought: Fulbe and Wodabee nomads between power and marginalization in the Sahel of Burkina Faso and Niger Republic. In: Bovin, M. and Manger, L. (eds) *Adaptive Strategies in African Arid Lands.* Uppsala, Sweden, pp. 29–58.

Cissé, S. (1985) Land tenure and development problems in Mali: the case of the Niger Delta. In: Hill, A.G. (ed.) *Population, Health and Nutrition in the Sahel.* Kegan Paul, London, UK, pp. 140–151.

Crowley, E.L. (1991) *Resource Tenure in Mali: an Anthropological Analysis of Local Institutions.* Club du Sahel/OECD, The Hague, Netherlands, pp. 1–65.

Davis, J.U. (2000) Classrooms of democracy? The educational prospects of Malian civil society. In: Bingen, R.J., Robinson, D. and Staatz, J.M. (eds) *Democracy and Development in Mali.* Michigan State University Press, East Lansing, Michigan, pp. 289–320.

Delville, P.L. (1998) Harmonizing formal law and customary land rights in French-speaking West Africa. In: *Politique des Structures et Action Foncière au Service du Développement Agricole et Rurale.* Groupe de Recherche et d'Echanges Technologiques (GRET), Paris, France, pp. 1–29.

Docking, T.W. (1999) International influence on civil society in Mali: the case of the cotton farmers' union, SYCOV. Ph.D. dissertation. Boston University, Boston, Massachusetts.

Fox, L., Ouattara, R., Boly, A., Ba, M.B., Konaté, Y. and Charlick, R. (2001) *Francophone Civil Society Organization Assessment: Mali and Côte d'Ivoire Synthesis Report* (West Africa: Civil Society Strengthening for Conflict Preventions Study). Associates in Rural Development (ARD). United States Agency for International Development Broadening Access and Strengthening Input Market Systems (BASIS), Burlington, Vermont.

Frankenberger, T.R. and Lynham, M.B. (1990) Household food security and coping strategies along the Sénégal River Valley. In: Park, T.K. (ed.) *Risk and Tenure in Arid Lands: the Political Ecology of Development in the Senegal River Basin.* University of Arizona Press, Tucson, Arizona, pp. 51–86.

Heasley, L. and Delahanty, J. (1996) Politics of manure: resource tenure and the agropastoral economy in Southwestern Niger. *Society and Natural Resources* 9, 31–46.

Hilhorst, T. and Aarnink, N. (1999) *Co-managing the Commons: Setting the Stage in Mali and Zambia* (Bulletin 346). Royal Tropical Institute, Amsterdam, Netherlands.

Horowitz, M.M. and Jowkar, F. (1995) Gender and participation in environment and development projects in the drylands. In: Stiles, D. (ed.) *Social Aspects of Sustainable Dryland Management* (United Nations Environment Program). John Wiley & Sons, Chichester, UK.

Johnson, C. (2001) Local democracy, democratic decentralization and rural development: theories, challenges and options for policy. *Development Policy Review* 19(4), 521–532.

Kassibo, B. (ed.) (1997) *La decentralization au Mali: état des lieux* (Bulletin 14). LitVerlag, Hamburg, Germany.

Keïta, N. (2003) *Décentralisation et responsabilité dans les modes de gestion des ressources naturelles et des redevances y afférentes: les bourgoutières Yallarbé de Youwarou* (Working Paper). Centre Universitaire Mande Bukari/World Resources Institute, Bamako, Mali.

Maïga, I. (1996) *Etude sur la prévention et la gestion des conflits.* Centre Canadien d'Etude et de Coopération Internationale (CECI) Programme de Promotion de la Démocratie et des Droits de la Personne (PDDP), Bamako, Mali.

Maïga, I. and Diallo, G. (1998) Les conflits fonciers et leurs gestion dans la 5ème Région du Mali (Dossier No. 76). Groupe de Recherche Actions pour le Développment (GRAD), International Institute for Environment and Development, Programme Zones Arides, London, UK.

Mamdani, M. (1996) *Citizen and Subject: Contemporary Africa and the Legacy of the Late Colonialism.* Princeton University Press, Princeton, New Jersey.

Mathieu, P. (1996) La sécurisation foncière entre compromise et conflits: un processus politique? In: Mathieu, P., Laurent, P. and Willame, J. (eds) *Démocratie, Enjeux Fonciers et Pratiques Locales en Afrique. Cahiers Africains,* No. 23–24. Institut africain/L'Harmattan, Paris, pp. 26–44.

McIntire, J., Bourzat, D. and Pingali, P. (1992) *Crop-Livestock Interaction in Sub-Saharan Africa.* World Bank, Washington, DC.

Moore, K.M. and Thiongane, S. (1998) *Review and Analysis of CBNRM Activity Implementation* (Summary Assistance Report No. 22). Ministry of Environment and Nature Protection, Community-Based Natural Resource Management Project, Dakar, Sénégal.

Moore, K.M., Cissé, S. and Touré, A. (2003) Building social capital in the Sahel. *Rural Sociological Society Annual Meeting.* Montreal, Quebec, Canada, pp. 1–36.

Mortimer, M. and Turner, B. (1994) *Crop–Livestock Farming Systems in Semi-arid Zone of Sub-Saharan Africa: Ordering Diversity and Understanding Change* (Agricultural Research and Extension Network Paper No. 46). Overseas Development Administration, London, UK.

Ngaido, T. (1993) *Land Use conflicts in Western Rural Niger: Kollo and Tillabery Arrondissements* (Land Tenure Center Discussion Paper No. 1). Madison, Wisconsin, pp. 1–30.

Ngaido, T. (1996) Redefining the boundaries of control: post-colonial tenure policies and dynamics of social and tenure change in Western Niger. Ph.D. dissertation, University of Wisconsin, Madison, Wisconsin.

Painter, T., Sumberg, J. and Price, T. (1994) Your *terroir* and my 'action space': implications of differentiation, mobility and diversification for the *approche terroir* in Sahelian West Africa. *Africa* 64(4), 447–463.

Park, T., Tabor, J., Frankenberger, T., Lynham, M., Baro, M., Ngaido, T. and Rogers, G. (1993) *Risk and Tenure in Arid Lands: the Political Ecology of Development in the Senegal River Basin.* University of Arizona Press, London, UK.

Pointing, J. (1995) Impact of social and economic change on pastoral women in East and West Africa. In: Stiles, D. (ed.) *Social Aspects of Sustainable Dryland Management* (United Nations Environment Program). John Wiley & Sons, Chichester, UK.

Pratt, D.J., Le Gall, F. and de Haan, C. (1997) *Investing in Pastoralism: Sustainable Natural Resource Use in Arid Africa and the Middle East* (World Bank Technical Paper No. 365). World Bank, Washington, DC.

Rawson, D. (2000) Dimensions of decentralization. In: Bingen, J.R., Robinson, D. and Statz, J.M. (eds) *Democracy and Development in Mali.* Michigan State University Press, East Lansing, Michigan.

Ribot, J.C. (1999) Decentralization, participation and accountability in Sahelian forestry: Legal instruments of political-administrative control. *Africa* 69(1), 23–65.

Ribot, J.C. (2002) *African Decentralization: Local Actors, Powers and Accountability* (Governance and Human Rights, Paper 8). United Nations Research Institute for Social Development, Programme on Democracy, Geneva, Switzerland.

Rugege, S. (1995) Conflict resolution in African customary law. *Africa Notes*, 1–3.

Sanders, J.H., Shapiro, B.I. and Ramaswamy, S. (1996) *Economics of Agricultural Technology in Semiarid Sub-Saharan Africa*. Johns Hopkins University Press, Baltimore, Maryland.

Sturm, H. (1996) Traditional crop-livestock-interaction systems as a basis for sustainable farming systems in the subhumid savanna. In: West, N.E. (ed.) *Rangelands in a Sustainable Biosphere: Proceedings of the Fifth International Rangeland Congress*, Vol. 1. Society for Rangeland Management, Denver, Colorado, pp. 536–537.

Sunzini, P. (1992) Limites et potentialités pour un développement soutenable dans la zone agro-sylvo-pastoral au Sahel. In: *La zone Agro-Sylvo-Pastoral au Sahel: Limites et Potentialités de développement contre la Désertification. Actes de la Conference de Assissi.* Comité Inter-Etats de Lutte contre la Sechéresse au Sahel, pp. 51–62.

Toulmin, C. (1983) *Herders and Farmers or Farmer–Herders and Herder–Farmers* (Agricultural Adminstration Unit, Pastoral Network Paper 15d). Overseas Development Institute, London, UK.

Touré, D. (1996) *Prévention et gestion des conflits.* Centre Canadien d'Etude et de Coopération Internationale, Ouagadougou, Burkina Faso.

Turner, M.D. (1992) Living on the edge: Fulbe herding practices and the relationship between economy and ecology in the Inland Niger Delta of Mali. Ph.D. dissertation. University of California, Berkeley, California.

Watts, M.J. (1987) Drought, environment, and food security: some reflections on peasants, pastoralists and commoditization in dryland West Africa. In: Glantz, M.H. (ed.) *Drought and Hunger in Africa.* Cambridge University Press, Cambridge, UK.

Williams, T.O., Powell, J.M. and Fernández-Rivera, S. (1995) Manure utilization, drought cycles, and herd dynamics in the Sahel: implications for cropland productivity. In: Powell, J.M., Fernández-Rivera, S., Williams, T.O. and Renard, C. (eds) *Livestock and Sustainable Nutrient Cycling in Mixed Farming Systems of Sub-Saharan Africa*, Vol. II, Technical Papers. Proceedings of an International Conference, Addis Ababa, Ethiopia.

Wolmer, W. (1997) *Crop–Livestock Integration: the Dynamics of Integration in Contrasting Agroecological Zones, a Review* (Sustainable Livelihood Program Working Paper 63). Institute of Development Studies, University of Sussex, Brighton, UK.

3 Landscape and Soil Characterization of Madiama Commune

Oumarou Badini[1] and Lassana Dioni[2]

[1]*International Programs/Development Cooperation, Washington State University, Pullman, WA 99164–5121, USA; and* [2]*Centre Régional de la Recherche Agronomique-Sotuba, Institut d'Economie Rurale, BP 258, Bamako, Mali*

Knowledge of the relationships between landscape, soil types and land use systems is essential to improving sustainable natural resource management. This chapter uses computer-based remote sensing and geographic information systems (GIS), combined with field observations, to describe, classify and map the landscape of the Commune of Madiama. This supervised classification is conducted at a scale of 1/40,000 and is suitable for commune and village, but not farm level decision making. Most existing work of this nature is only applicable for national level planning, but this verified characterization of spatial variability and distribution of landscapes and soil types provides a basis for optimizing land uses down to the village level. It becomes possible to recommend the best way to plan, manage and use different landscape and soil types, and to identify appropriate agricultural practices and technologies. In addition, the output and information generated can serve as input for biophysical models for extrapolation to other similar sites or through time.

The chapter begins with a general description of the commune, followed by an account of the historical formation of the physical and biological landscape (geology, agroecology, climate and hydrology). The chapter ends with a characterization of each landscape type and associated soils according to their importance in the commune, as well as their locations, potentials and constraints.

General Setting and Description of the Commune

Site localization

Madiama commune is located in the extreme southern portion of the south-Sahelian Zone (less than 650 mm isohyets). The commune is about 25 kilometres from Djenné (capital of the administrative *Cercle*) and 120 kilometres south of Mopti (capital of the 5th Region of Mali) (Fig. 3.1). It lies between latitude 13° 45 N to 13° 52 N, and longitude 4° 22 W to 4° 30 W in the north-central part of Mali. Madiama commune covers an area of some 16,970 ha (169.7 square kilometres) gradually rising from the Bani River flood plain in the west to plateaus and occasional rock outcrops in the east. In 2001, there were about 7973 inhabitants (with a density average of about 47 inhabitants per km^2) comprising people of numerous ethnic groups. The dominant ethnic groups are Bambara, Marka and Peul.

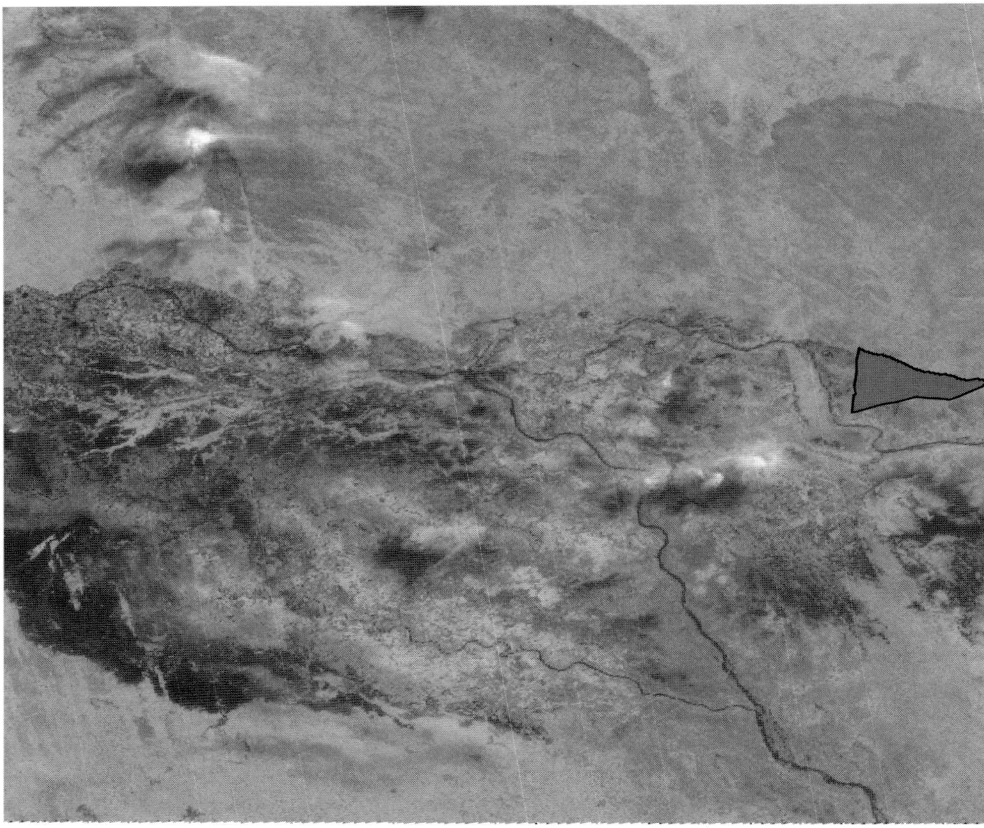

Fig. 3.1. Physical location of Madiama Commune in the 5th Region of Mali (Landsat 7 Image from November 1999).

Agricultural systems

Reduced vegetative cover, shorter fallows, declining soil fertility, and increased erosion are consequences of the climate changes and demographic pressures affecting the Sahel (Taylor-Powell, 1991). Expansion of farming on marginal lands and conflicts over limited resources are common in the study area. Like other areas of the West African semi-arid Sahel region, the study zone has suffered diminished food production on a per capita basis since the early 1970s. Long-term rainfall throughout the region declined dramatically in the early 1970s and has not returned to earlier levels. Although exacerbated by population growth, the fundamental problem is physical. Unfavourable rainfall patterns, soil degradation, and declining soil fertility are considered the primary constraints to crop production across the Sahel (Van Keulen and Breman, 1990).

Agriculture is the primary economic activity. Millet (*Pennisetum* spp.) and sorghum (*Sorghum vulgare*), generally intercropped with cowpea (*Vigna unguiculata*), dominate the cropping pattern. The lowland areas near the Bani River are used for rice cultivation. More favourable microenvironments of the sandy and loamy plains with favourable moisture and nutrients conditions such as the lower areas (micro basins) are cultivated in sorghum, watermelon (*Citrullus lanatus*), okra (*Hibiscus esculentus*) and dah (*Hibiscus sabdarifa*). Based on availability, organic fertilizer (manure) is used in dryland farming to a great extent. Chemical fertilizers are used mostly in irrigated rice plots and watermelon cultivation. In general, all crops respond well to fertilizers, but fertilizer use is insignificant (applied by less than 2% of farmers) and the rates applied are very low (e.g. 20 kg/ha nitrogen diammonium phosphate (DAP) for watermelon and 50 kg/ha of urea on rice). Farmers in the commune dry seed or seed with the first sufficient rains after ploughing in mid- to late June. Reseeding is often necessary due to early season drought. Planting can continue through July in most years. Two weedings (*sarclage*) are customary, although these are dependent on labour availability and the prospect of potential crop yield. Harvesting is usually done in late October to mid-November for millet and sorghum and late December for rice. Overall, the crop calendar varies depending on the rainfall pattern. Final crop yields are strongly dependent on time of planting, with higher yields obtained in years of early planting dates.

Animal husbandry is the second major economic activity in the commune. Nearly all households own livestock, principally mixes of cattle, sheep and goats. Animals are an important source of income and provide energy for fieldwork (e.g. tillage), transport, manure and food. Animal traction is used in about 80% of dryland farming, and the remaining 20% is cultivated using hand-held hoes. Pastures are not under specific management regimes, except that herders move their animals from place to place depending on the seasons in order to maximize the grazing of the animals. Most available land within village limits is used for agricultural production regardless of soil fertility, and the size of grazing lands is diminishing due to increased field sizes.

Principal production problems are related to the geoecological changes affecting the Sahel region due to inadequate and irregular rainfall, low soil fertility and soil degradation, lack or insufficient animal manure, and poverty. Land degradation and low soil fertility are common features of the landscape. The traditional practice of long bush fallow as a means to restore soil productivity is breaking down. Fallow is either non-existent (the same parcel is cultivated year after year without crop rotation) in some villages or substantially reduced to 2 or 3 years in others.

Factors that influence agricultural production decisions and natural resource management strategies include: anticipated rainfall (based on duration and intensity of the cold season), soil types, availability and access to land, family size, farming equipment, market prices for cash crops, and social factors (e.g. bride price, funerals, etc.). The major natural resource conflicts in the commune are between farmers and herders and occur because increased areas under cultivation are causing reduced pasture areas, reduced animal passage routes, and competition for seasonal wetland and bourgou.

Physical and Biological Environment (Landscape)

The study zone, which is part of the Inland Niger Delta region, is characterized by a diverse potential dependent on the site specific combination of water, soils and plants. The following section describes the landscape of the commune of Madiama in regard to its geology, geomorphology, physiography, agroecology, climate and hydrology.

Geomorphology and geology

Like much of Mali, the commune of Madiama has a monotonous relief, consisting of mainly flat with low plateaux and basins, and a mean altitude of 279 m. It is part of the Inland Niger Delta flood plain and influenced by the Bani River (an affluent of the Niger), which is on its western side towards the city of Djenné.

The Commune is located in the sedimentary basin of the Bani River that was formed by alluvial deposits during the Quaternary period (1.8 million years ago to present – the most modern geological period). Infra-Cambrian sandstone called the Bandiagara sandstone formed the area, which is relatively elevated with rock outcrops on its eastern border. The inner valley of the Bani and Niger rivers started its formation during the beginning of the Quaternary with the Bandiagara sandstone covered with ferruginous hardpan (cuirass). Remnants of this cuirass are still found at the borders of the commune. Sea level oscillations as well as climatic variations during the Quaternary produced diverse alluvial deposits that still determine the soil types and landscape of the study area today.

The soils of the Inland Delta region were developed over a time sequence of alternating subhumid and hyper-arid climatic periods during the Quaternary. The time sequences of the alluvial sedimentation and the impact of sea oscillations and climatic variations have shaped the final makeup of the landscape. These variations explain the general characteristics of the pedogenesis and the potentials of soils found in the region. Details concerning this geomorphology can be found in many previous works (Blanck, 1967; Gallais, 1967; Bertrand, 1973).

Agroecological characteristics

The agroecological zone is defined here as a homogeneous geographic entity from the standpoint of its climate, geology, geomorphology, soil and vegetation. In Mali, the Project d'Inventaire des Ressources Terrestres (PIRT) project (1986) identified 49 agroecological zones and 14 natural regions. As part of the south Sahelian Bio-climatic zone, most of the Madiama commune is located in the Inland Niger Delta region, but its extreme east stretches to the Bandiagara-Hombori Plateau (cf. Carte du Zonage Agro-écologique du Mali, PIRT, 1986). Therefore, from east to west of the commune, the agroecological zones are: the *Bas-Plateau Bobo*, the *Moyen Bani-Niger*, and the *Delta Vif*.

The Bas-Plateau Bobo (the Low Bobo Plateau)

Higher elevation with indurated laterite and soils on hardpan (cuirass) and rock outcrops on non-arable lands characterizes the relief in this zone, in which intermittent streams and rivers are found. The arable lands have loamy sand, sandy clay loam textures or laterite clay. The vegetative cover is diverse and its ligneous component is dominated by small shrubs such as *Combretum glutinosum*, large shrubs such as *Pterocarpus erinaceus* and *Terminalia* spp. and other woody species such as *Parkia biglobosa* (African locust bean or *néré* in French), a tree with pods whose seed and inner gelatin are eaten. The herbaceous stratum is mainly composed of annual grasses such as *Loudetia togoensis*. The forage potential is average. The main crop is millet. However, sorghum can be planted in low-lying areas.

The Moyen Bani-Niger (the Bani-Niger Midlands)

This transition zone to the floodplain is characterized by soils with slightly sandy texture and average natural fertility. The main ligneous plants are small shrubs such as *Piliostigma reticulatum* and other woody species such as the nitrogen fixing *Acacia albida* (apple-ring acacia, ana tree, winter thorn), a widely used tree well documented for increasing the yields of crops grown under it. The herbaceous strata are diverse and dominated by *Loudetia togoensis* in non-inundated areas and vetiver grass (*Vetiveria nigritana*) in the often flooded zone. The forage potential is very high. Rice cultivation is possible on low flooding lands and sorghum is cultivated on the higher elevations.

The Delta Vif (the Active Delta)

In the commune, the *Delta Vif* is composed of the floodplains of the Bani River (an affluent of the Niger). These lands are part of the lacustrian zone seasonally flooded by the rising waters from the Bani. The type of vegetation found here is a function of the length and patterns of flooding. On the soils where floodwaters remain longer are found some perennial grasses such as *bourgou*[1] (*Echinocloa stagnina*). On the soils with shorter duration of floodwaters are found vetiver grass (*Vetivera nigritana)* and woody species such as *Mitragyna inermis*. The forage (pastures) potential is very high. The receding floodwaters permit the cultivation of rice.

Climate and Hydrology

Brief climatic history

Like all the earth's regions, the Sahel has been subjected to repeated climatic fluctuations throughout its history. Cold, warm, dry and moist periods have altered the face of the region many times. These fluctuations have taken place over the last 18,000 years. Approximately 20,000 years ago the Sahara expanded southward into the current Sahel zone (including the study area) during a pronounced dry period. Rivers dried up, vegetation died out, and sand dunes formed. About 15,000 years ago, climatic conditions began to improve. Moist tropical air masses advanced far to the north during the summer months, bringing a considerable increase in rainfall (Leisinger and Schmidt, 1995).

This favourable climate lasted for about 5000 years (from about 8000–3000 BC). The Sahara desert receded to small, isolated areas in northern Africa. An enormous drainage network came into existence in the regions south of the Sahara. Rivers deposited sediments over broad areas such as in Inland Delta region of the Niger, where land use today is based on extensive irrigation. Previous arid, desolate areas came to life again, and the tropical rain forest advanced far to the north. Sand dunes were gradually covered with vegetation and sand masses no longer migrated southward. Red, fertile tropical soils were formed in areas where it is virtually impossible for soils to be formed today due to aridity. Humankind is still profiting today from the favourable climatic conditions of that time, because soils that were formed then continue to benefit present-day agriculture.

After this relatively 'wet' period, the climate once again worsened. Precipitation became highly variable and the decreased amount of rainfall caused lakes, river systems and rain forests to recede, which led to a renewed advance of the desert. Today climatic history still determines to a considerable extent the potential for agriculture in the Sahel. In many areas, people live in an environment that took many thousands of years to develop. If these areas continue to be degraded, they will probably not recover under today's arid conditions.

Climatic zone and rainfall patterns

The study area is part of the South Sahelian zone (below 650 mm isohyets) with a short rainy season, considerable variability in the rainfall amount and distribution, severe drought due to high evapotranspiration, and high daily and annual temperature. There are no noticeable seasons that can be defined by temperature as in temperate zones. Differences in temperature are greater between day and night than between summer and winter.

The 3–4 month rainy season occurs during the summer from June to September, with rainfall ranging from 400 to 600 mm. However, because of high temperatures and intense solar radiation, much of this precipitation evaporates before it can be used for agriculture. There is sufficient humidity for agriculture for only 2–4 months during the year, with dry months predominating the rest of the time. The annual rainfalls, the long-term mean and the general rainfall trend in the area of Madiama in the period from 1950 to 2000 are plotted in Fig. 3.2.

The annual rainfall in the commune and surrounding area varies considerably from year to year, a very common characteristic in the semi-arid tropics. From 1950 to 2000, as shown in Fig. 3.2, the mean annual rainfall is 544 mm (Badini, 2001). The lowest annual rainfall of 274 mm was recorded in 1987, while the highest annual rainfall in the past 51 years was 914 mm, received in 1957.

The average rainfall for the period 1950–1969 was 636 mm compared to an average of 482 mm for the period 1970–2000. The average rainfall loss of around 154 mm between the two periods highlights the declining rainfall over the last three decades (Fig. 3.2). The consequences for this drop in rainfall amount are increased aridity and more risks for crop failure due to limited water availability for rainfed crops

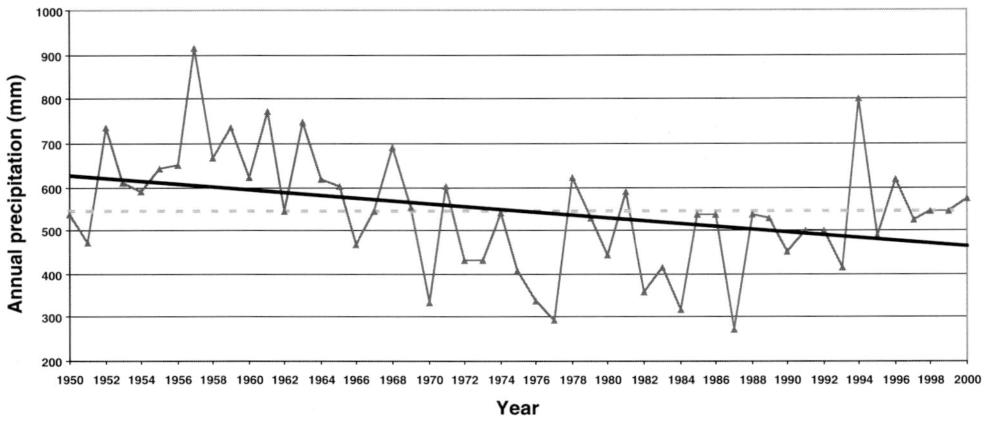

Fig. 3.2. Precipitation distribution, mean and trend from 1950 to 2000 in Djenné-Madiama (Service National de Meteorologie, Bamako, Mali).

and diminishing pastures. These same constraints caused by the drop in rainfall are common for the many parts of the Sahel region since the droughts of the 1970s.

Due to the seasonal movement of the Inter Tropical Convergence Zone (ITCZ) that regulates the rainfall pattern, precipitation in Madiama and in the Sahel region in general is monomodal (Sivakumar *et al.*, 1984). Compared to the wetter zones in West Africa, rainfall in July (125 ± 53 mm) and August (155 ± 68 mm) shows a pronounced peak in the Madiama area (Fig. 3.3). Short but heavy rainstorms and strong gusts of wind distinguish the beginning and the end of the rainy season, whereas more gentle rainfalls that tend to endure for longer periods of time characterize the peak period in August. The wide annual variability in rainfall is characteristic of the whole Sahel region and represents a main agroclimatic constraint to crop production.

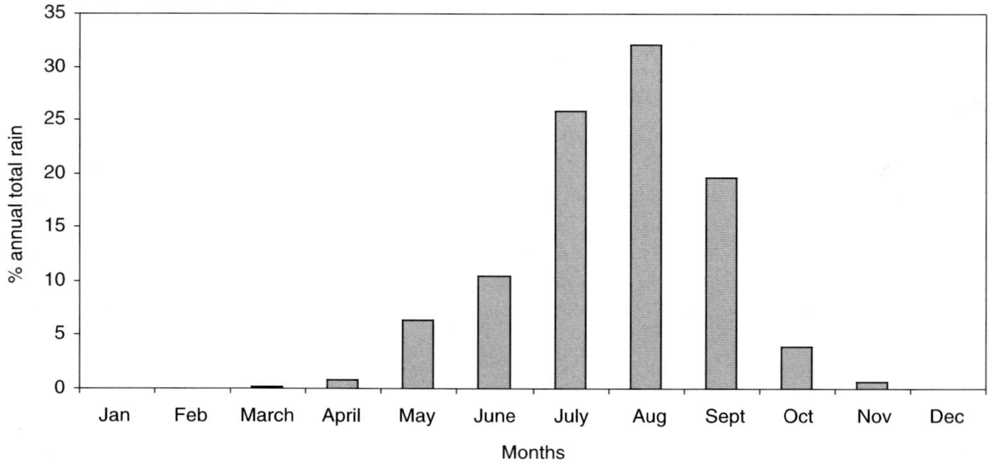

Fig. 3.3. Monthly rainfall distribution in Madiama.

Hydrology

In the study zone, flooding follows the start of the rainy season, but the onset date for the rise in the water levels varies from year to year (Gallais, 1967). The water rise and flooding pattern in the study area are normally slow and gradual. The waves of the rising water from the Niger at Ké-Macina (upstream from Madiama) take up to 2 or 3 weeks before reaching Belentieny and then Sofara (downstream from Madiama).

Due to its location in relation to the Bani River, the Commune of Madiama is mainly an intermediary zone lying between the flood plain in the west and the ironstone plateau on the east. Although many rivers and streams (called *yamé* in *Fulfulde* and local *Bambara*) run through the commune, the most important are the seasonal streams flowing through Promani-Tatia to the Bani River at Baratou village. Another less important stream runs through Kessédougou and Toumadiama. At the extreme south of the commune, a third, more or less steep-sided stream is found.

Landscape and Soil Types Characterization

Methodology

A morpho-pedological approach combining computer-based remote sensing and GIS analysis with field observations of landscape and soil types made it possible to describe, classify, and map land and soil resources of the Commune of Madiama (Badini and Dioni, 2001). This approach takes into account the interactions between the geomorphological characteristics (i.e., landform, relief, nature of soil parental material), the hydrology, the soil profile and the pedogenesis. The analysis of these interactions determined the specific landscape and soil types of the area.

The objective of the soil survey investigations (Badini and Dioni, 2001) was to provide data for the rational planning and management of land use. Also, the results from this work met the data requirements for biophysical modelling (see Chapter 11) to further the assessment of land qualities and the impact of agricultural practices on productivity and the environment. The data consisted of geo-referenced physical and chemical characteristics of soils that were recorded in the field, determined in the laboratory and/or extracted from remotely sensed imagery. They were used in interactions with the geomorphology (e.g. landforms) and the hydrology of the area to resolve landscapes into mappable areas in which the soil is less variable than the overall landscape (Beckett and Webster, 1971). These areas, are presented as landscape types, and comprise one or two dominant soil types.

The area covered by each examination point (mapping unit or pixel) is 16 ha. In the commune, a farmer's parcel size is far less than the 4 ha found in the Sahel region in general. An examination point of 4 ha is only achievable with a detailed soil survey and large-scale map of 1/10,000 or less. Although these large-scale maps could better assist in farm level planning and even detailed fertilizer recommendations dependent on field variability, their implementation is labour intensive and very expensive. To compensate for these limitations, the approach has been to combine semi-detailed soil survey (1/40,000) and biophysical modelling with field studies to evaluate the impact of agricultural practices on productivity and the environment (see Chapter 11). The existing wealth of knowledge derived from the landscape and soil characterization study of the commune that allows for a description of the landscape and a determination of the soil types, distribution, potentials, constraints and suitability should contribute to sustainable natural resource development and better land use planning and decision-making.

The present chapter describes how soils with different potentials and constraints are distributed within the commune boundaries. It also provides a semi-detailed map (Fig. 3.4) and analyses of agricultural suitabilities and ecological vulnerabilities of the classified landscape types (see Tables 3.1 and 3.2). Because this work identifies the precise nature and distribution of the land and soils in the commune, it can serve as a basis for the planning and management of agricultural projects. Furthermore, it also provides a framework for specifying appropriate land management techniques related to each land type.

Description of Landscape and Soil Types

For the total land area of 16,970 ha of the commune, eight landscape types are identified and characterized according to specific landform, pedogenesis, hydrology, topography, land use and soil cover. They range from hydromorphic flood plains on the west side near the Bani River to plateaux and some rock outcrops on the east side outside the inundated area of the commune. Based on the age of each formation, and starting from the youngest to the oldest (in geologic terms) landforms, the following landscape types are shown in Fig. 3.4 and Table 3.1. In each landscape type, one or two main soil types are identified and described. The soil texture, fertility level, production constraints, and overall agricultural or pastoral suitability are described for each landscape type unit (Tables 3.1 and 3.2).

Recent (young) formations

Hydromorphic floodplains (Unit Ci)

This landscape, called 'lateral basin' or 'inundation basin' (or in French *Cuvette d'inondation – Ci*), is the most recent alluvial formation. These hydromorphic floodplains located in the upper west side of the commune are the youngest geological landform. Their formation is still taking place today. They are part of the lacustrian zone seasonally flooded by the rising waters from the Bani River. They cover a total land area of 746.8 ha or 4.4% of the commune and are called 'goubé' by the people in the commune.

The type of vegetation found in this landscape is a function of the length and patterns of flooding. On the soils where floodwaters remain longer are found some perennial grasses such as *bourgou* (*Echinocloa stagnina*). On the soils with shorter duration of floodwaters are found vetiver grass (*Vetivera nigritana*) and woody species such as *Mitragyna inermis* and *Acacia* (Fig. 3.5).

Two types of soils are found in this landscape: the pseudo gley hydromorphic soils (or in French *sols hydromorphes à gley réduit*) in the low-lying areas, and the amphigley hydromorphic soils (*sols hydromorphes à amphigley*) in higher areas. They can be classified as United States Department of Agriculture (USDA) Aqualfs (Fig. 3.4 and Table 3.1) (Survey Soil Staff, 1999). These are deep (95 cm) silty clay loam hydromorphic soils with gleying at the bottom of the profile. Because of their loamy texture, they have a high water holding capacity. Organic matter (OM) levels vary from 0.4% (very low) to 1.4% (very high) with an average of 0.9% (high). The carbon/nitrogen (C/N) ratio of around 10 shows that high nitrogen mineralization is taking place. The soils are acidic in surface (pH 5.0 on average), with low assimilable phosphorus content of 4.3 ppm, and low nitrogen content between 0.02 and 0.04%.

The major constraints to production (Table 3.2) are due to the hydrology of these lands. The underground water is at shallow depth, with very high risks of inundation and waterlogging excluding any possibility of rainfed agriculture. The receding floodwaters permit the cultivation of rice. The soils can also be used for gardening and horticulture. The forage (pasture) potential is very high.

Hydromorphic alluvial levees or 'sand banks' (unit Tr)

This landscape called 'recent alluvial terrace' (or in French *terrasse récente – Tr*) is a relatively recent alluvial formation. It is characterized by alluvial levees or 'sand banks' forming a convex strip around the edges of the inundation basin (*unit ci*) above. The landform is slightly elevated and made of coarse sandy soils often covering some pebbles of quartz. It covers a total land area of 1391.5 ha or 8.2% of the commune.

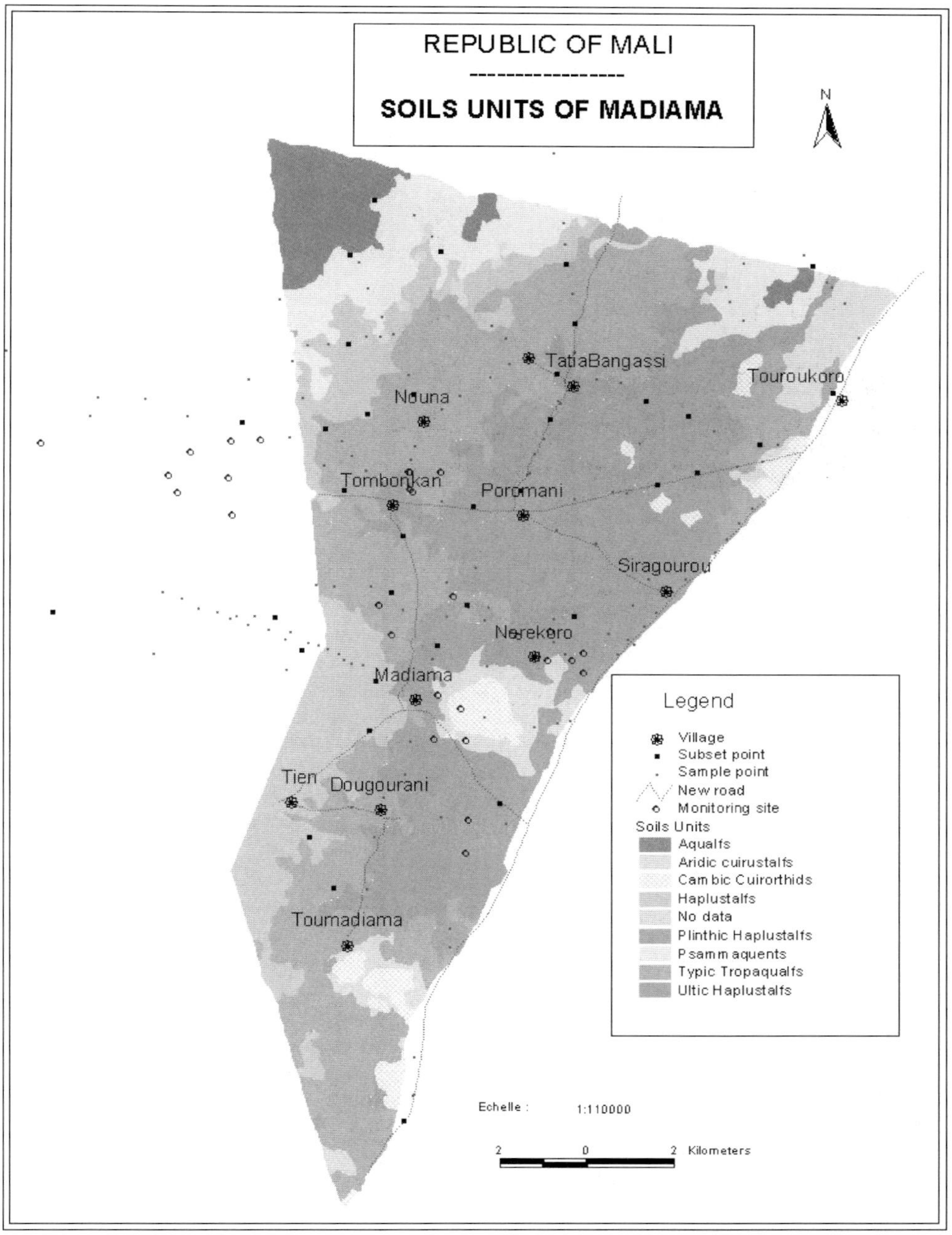

Fig. 3.4. Land and soil units of the Commune of Madiama.

Fig. 3.5. Hydromorphic flood plains.

The main ligneous plants are small shrubs such as *Piliostigma reticulatum*. The herbaceous layer (Fig. 3.6) is mainly constituted of *Ipomea repens* and vetivers (*Vetiveria nigritana*).

The dominant soils found in this landscape are the soils of limited development from alluvial deposits (or in French *sols peu évolués d'apport alluvial*). They can be classified as USDA Psammaquents (Fig. 3.4 and Table 3.1). The soils of this landscape are very deep (120 cm) with sandy clay loam texture covered by alluvial sand deposits. These soils are acidic and have a very low chemical fertility.

Besides their low fertility, the major constraints of the soils of this sand bank landscape are their low water holding capacity and risks for temporary inundation (Table 3.2). Their sandy texture permits a

Fig. 3.6. Sand banks.

Table 3.1. Legend and landscape/soil types classes.

	Landscape types						
	Names						
Code	Français	English	Bambãna	Soil classification (French CPCS)	Soil classification US (tentative)	Area (ha)	Total Area (%)
ci	Cuvette d'inondation latérale	Hydromorphic flood plains	Farah-lèh dougoukolo	Sols hydromorphes à gley réduit et à amphigley	Aqualfs	746.8	4.4
tr	Terrasse récente (banc de sable)	Hydromorphic alluvial levees (sand banks)	Tientien ou Bahtientien	Sols peu évolués d'apport alluvial sableux à texture grossière	Psammaquents	1391.5	8.2
t1	Terrasse subactuelle	Old levees and alluvial terraces	Tientien-fing	Sols peu evolués d'apport alluvial faiblement hydromorphes sableux et sols argileux hydromorphes à gley de profondeur	Typic Tropaqualfs	5243.7	30.9
cl	Cuvette limoneuse/ dépression ou ancien bras mort	Hydromorphic loamy plains	Bona dougoukolo	Sols hydromorphe à gley réoxydés et sols hydromorphes à engorgement de surface	Haplustalfs	2104.3	12.4
t2	Terrasse ancienne	Plains of sandy materials	Nanganblen ou Tientienblen	Sols ferrugineux tropicaux non à peu lessivés et Sols ferrugineux tropicaux peu lessivés à tâches et concrétions	Ultic Haplustalfs	4327.3	25.5
ca	Cuvette ancienne de la terrasse t2	Plains of silty and loamy materials	Boua dougoukolo ou Djè	Sols hydromorphes à tendance vertique et nodules calcaires	Plinthic Haplustalfs	1612.1	9.5
vi	Glacis- versant induré (Regosols)	Laterized indurated pediment	Fara-dèbè dougoukolo	Sols peu évolués d'erosion sur cuirasse ou régosols	Aridic Cuirustalfs	882.4	5.2
bk	Bas plateau ou croupe cuirassée (Lithosols)	Ironstone plateau	Fuga dougoukolo	Pratiquement pas de sol sinon sol squelettique très peu profond ou Lithosol	Cambic Cuirorthids	526.1	3.1

high permeability and does not allow good retention of water and nutrients. These soils have a very low agricultural value and their improvement would necessitate costly investments. The use of fertilizers on these soils must be accompanied by soil amendment in order to improve nutrient retention and decrease soil acidity. However, these soils exhibit some potentialities for the development of fire wood trees and pastures.

Hydromorphic loamy plains (unit Cl)

This landscape, called 'loamy basins' (or in French *Cuvette Limoneuse – unit Cl*), is located below the old levees and alluvial terraces (unit t1). The relief is uniform and regular. Mounts of sandy soil trapped underneath small shrubs called 'nebkass' characterize the present morpho-dynamic. This landscape type, the third largest in the commune, has a total land area of 2104.3 ha or 12.4% of the commune (Table 3.1). The natural vegetation is dominated by *Acacia albida* and grasses (Fig. 3.7).

Two types of soils are found in this landscape. They are called hydromorphic soils with reoxydated gley and hydromorphic soils with surface saturation (or in French *sols hydromorphes à gley réoxydés* and sols *hydromorphes à engorgement de surface*), which corresponds to USDA Haplustalfs. The soils in this landscape are deep (100 cm). Their texture is sandy loam to clay loam in the surface horizon with pebbles of quartz at a depth around 60 cm where the clay content is between 35% to 40%. Due to their texture, they have a high water holding capacity (15–20%), and OM content averaging 0.5–1%. The C/N ratio is around 10 due to a very active mineralization process.

Major constraints to production (Table 3.2) in these soils include possibilities of flooding, soil compaction, difficulties for tillage due to high soil adhesivity and weed (adventices) invasion. However, these soils are deep with a silty clay loam texture and a good capacity to hold water. Besides grazing, rainfed rice cultivation is possible in years of average rainfall.

Levees and alluvial terraces (unit t1)

This landscape, called 'young alluvial formation' (or in French *formation alluviale subactuelle – t1*) was formed during the moist period that succeeded the dry period 15,000 years ago (see climatic history in previous section). This land type, which covers 5243.7 ha or 30.9% of the land area in the commune dominates the commune (Fig. 3.4

Fig. 3.7. Loamy plains.

Fig. 3.8. Recent alluvial terraces.

and Table 3.1). The landscape is made of low-lying peneplains (terraces) with depressions (micro-basins) of loamy clayey texture alternating with oval-shaped sandy levees. It has an undulating relief with beige-coloured materials located between the sand banks (unit tr) and the old alluvial terraces (unit t2 below). The vegetation cover is composed of *Acacia albida* and *Piliostigma reticulata* on the higher elevation (sandy levees) and *Mitragyna inermis* and *Diospyros mespiliformis* in low lands (micro-basins) (Fig. 3.8).

Two dominant soil types are found in this landscape: alluvial sandy soils with limited development and hydromorphy (or French *sols peu evolués d'apport alluvial faiblement hydromorphes sableux*) on the higher elevations (sandy levees), and the clayey hydromorphic soils with gley (or French *sols argileux hydromorphes à gley de profondeur*) in micro-basins areas. These soils, which can be classified as USDA Typic Tropaqualfs, are deep (105 cm) sandy clay loam soils with gleying at lower depth. OM content is between 0.5% and 1%, and a C/N ratio of 10. The soils are only slighly acid with an average pH of 6.2. Phosphorus content and nitrogen content in the surface are 9 ppm and 0.02% (very low), respectively.

Major constraints to production in these soils include soil crusting, limited soil fertility due to acidity and low soil nutrients reserve. The landscape type covers much of the commune and the soils are suitable to agriculture with a good capacity to hold water. These soils allow the rainfed cropping of an array of crops, including millet, sorghum, groundnuts and cowpea and rainfed rice cultivation.

Old formations

Old alluvial plains and terraces (unit Ca)

This landscape, called 'old basin' (or in French *cuvette ancienne – Ca*) is characterized by slightly obliterated depressions with a regular relief. The present morphodynamic is shaped by mounds of sand and silt trapped underneath the small shrubs and forming what is called 'nebkass'. This land type covers a total land area of 1612 ha or 9.5% of the commune (Fig. 3.4 and Table 3.1). The natural vegetation is dominated by thorny shrubs and trees such as *Acacia seyal* and *Balanites aegyptiaca* and parkland trees such as *Tamarindus indica*.

The dominant soils in this landscape are slightly 'vertique' hydromorphic soils (or French *sols hydromorphes à tendance vertique avec nodule calcaire*), which are classified as USDA Plinthic Haplustalfs. They are deep (105 cm) with a texture varying from sandy loam to clay with depth. As a result, the water storage capacity is high, with soil water availability of about 20% on average. The soil OM content is low (0.5%) and decreases with depth. Temporary flooding and waterlogging due to runoff water, as well as high soil compaction and glaciation are the main constraints. These soils are, however, suited to sorghum and rice, and have high pastoral value (Table 3.2).

Table 3.2. Agropastoral constraints and potentialities.

Code	Constraints	Agricultural suitability	Grazing suitability	Forest suitability
ci	Very high risk of inundation and water logging	Restricted only to rice cropping	Good	Marginal
tr	High grounds, soils with coarse sandy texture, high permeability, low fertility	Marginal, sandy soils with low fertility	Good	Good
t1	Seasonal water logging and risk of inundation	Good to sorghum and rainfed rice	Good	High
cl	Risk of inundation and water logging, costly conservation needed for improvement	Moderate; suitable for sorghum; more crops can be cultivated if soil water conservation and management techniques are implemented	Good	Good
t2	Very fragile soils very highly prone to erosion, deep water table and low fertility	Moderate overall; suitable for millet; more crops can be cultivated if soil water conservation and management techniques are implemented	Good	Good
ca	Possible slight inundation due to runoff water, weak permeability due to high soil compaction	Suitable for rice and sorghum	Good	Good
vi	Shallow soil on hardpan (cuirass), very low water holding capacity, problems of roots penetration, high risk of erosion	Marginal for millet and sorghum	Marginal	Good
bk	Very shallow soil with rock outcrops	Unsuitable	Unsuitable	Unsuitable

Plains of sandy to loamy materials (unit t2)

This landscape called 'old alluvial terrace' (or in French *terrasse alluviale ancienne – t2*) is made of sandy to sandy loam peneplains (terraces) found on the eastern and western edges of the Delta. It covers a land area of 4327.6 ha or 25.5% of the commune (Fig. 3.4 and Table 3.1). It is almost never flooded (inundated) due to its higher profile. The parent material is made from wind blown sandy and silty materials from the Quaternary period. The land can be distinguished by its reddish (due to rubefication) overall colour. Natural vegetation is dominated by protected parkland trees such as *Acacia albida* (apple-ring acacia) and *Vitellaria paradoxa* (shea-nut tree) (Fig. 3.9).

Two types of soils are found in this landscape: the tropical ferruginous soils with limited development (or French *sols ferrugineux tropicaux non à peu évolués*) associated with tropical ferruginous soils with concretions and 'tints' (or French *sols ferrugineux tropicaux peu lessivés a taches et concrétions*). The soils are very deep (115 cm) with sandy loam texture. The water storage capacity is low in surface to average in deeper layers. Soil OM content is low (0.5%) on average, and the pH varies from 4.6 (acid) to 6.0. Average phosphorus content is 3.3 ppm. Nitrogen content is low with an average of 0.02%.

These soils have no major physical constraints (Table 3.2) except their low water

Fig. 3.9. Plains of sandy and loamy materials.

holding capacity. Constraints are mainly chemical with low pH (high acidity) and deficiency in phosphorus and nitrogen. Soil amendments as well as organic and mineral fertilization are needed in order to improve productivity. These soils constitute the second largest rainfed cultivated unit in the commune. Cultivated crops include millet, groundnuts and cowpeas. Sorghum is only planted in the low-lying areas of the field. To ensure a sustainable management of this unit, trees and productive pastures could be planted.

Land underlain by laterite (unit Vi)

This landscape type, called 'laterized indurated pediment' (or in French *versant glacis induré – Vi*), is made of peneplains (terraces) with skeletal soil of loamy clayey texture on a substratum of rocks or cuirass found at a shallow depth of 60–70 cm. It covers an area of 882.4 ha or 5.2% of the commune. On the map in Fig. 3.4, it can be seen as ring-like shape surrounding the ironstone plateau (*unit Bk*) with a slight slope towards the landscape type t2. Natural vegetation is dominated by diverse small shrubs (Fig. 3.10).

The main soils are soils of limited development on cuirass (hardpan) or Regosols (or French *sols peu évolués d'érosion sur cuirasse*), classified as USDA Aridic Cuirustalfs. These are shallow (60 cm) gravelly and laterized soils. Their hydrology is characterized by a diffuse and intense runoff that causes sheet erosion. These soils are prone to erosion with a slope of about 2–3%. Soil crusting and surface glaciation in addition to the shallowness of the soils are the main constraints. Cultivation of sorghum and millet is possible but yields are generally low. These fragile soils must be protected and soil conservation measures as well as appropriate cultural practices should be implemented.

Ironstone plateau (unit Bk)

This landscape called 'low hardpan plateau' (or in French *bas plateaux cuirassés – Bk*) covers a total surface area of 526.07 ha or 3.1% of the commune. The relief in this zone is characterized by higher elevation, with indurated laterite and soils on hardpan (cuirass) and rock outcrops on non-arable lands. The vegetation is stunted and dominated by diverse small shrubs.

There is almost no soil in this landscape except skeletal and thin gravels covering a cuirass found at a very shallow depth of about 10–20 cm. These very shallow soils are called Lithosols or USDA Cambic Cuirorthids soils. It has the highest elevation (300 m) in the commune. The hardpan

Fig. 3.10. Laterized indurated pediment and ironstone plateau.

(cuirass) is covered by gravelly materials from broken indurated laterized rocks.

Major constraints for production are the shallowness of the soil, the presence of cuirass, the very low water holding capacity, and the high susceptibility to water erosion. These soils are unsuitable to agriculture. If agriculture has to be attempted on these soils there is a need for improved and labour intensive cultural practices, including the removal of gravel, tillage techniques allowing an increase of soil thickness and organic and mineral fertilization.

Conclusion

This chapter has shown how land and soils with different potentials and constraints are distributed within the boundaries of Madiama Commune. Soil properties for making agricultural and pastoral land use decisions are identified. Information and soil parameters generated from this study have also served as input for the biophysical models (see Chapter 11) and helped in the assessment of agricultural practices and technologies best suited to the various soils and landscapes of the commune (see Chapter 12). Community members and their officials as well as researchers and development agents can use the information to plan land use and identify special practices needed to ensure proper performance.

From a total of eight landscape types, four types, representing a total 13,197 ha (67.0% of the commune), are suited to rainfed agricultural production with varying degrees of constraints as described in this chapter. These landscapes types are the old levees and alluvial terraces (*unit t1*), the old alluvial plains and terraces (*unit Ca*), the hydromorphic loamy plains (*unit Cl*), and the plains of sandy to loamy materials (*unit t2*). The hydromorphic floodplains (*unit Ci*) and alluvial levees, representing a total of 2141 ha (12.6% of the commune), are partly used as grazing lands, and in some cases for irrigated rice cropping due to their proximity to the Bani river. The two last landscape types, found mostly on the east side of the commune, are lands underlain by laterite and hardpan (*unit Vi*), and Ironstone Plateau (*unit Bk*). They both represent a total of 1408 ha or 8.3% of the commune. They are degraded (about 40% bare soil) with stunted vegetation, constituted mainly of annual pasture grasses such as *Loudetia* and *Schoenefeldia*, and used as wet season pastures.

Note

1 The term *bourgou*, derived from the Peul word bourgou meaning forage, refers to a grass growing in deep basins of the Active Delta that produce a lush (rich) pasture as the flood waters recede. These flood retreat pastures (pâturages de décrue) become the focal point for transhumant herds during the dry season.

References

Badini, O. (2001) *Agro-climatic Assessment (Description and Analysis) of Madiama Commune in Mopti Region, Mali (West Africa).* Washington State University, International Programs, Pullman, Washington.

Badini, O. and Dioni, L. (2001) *Etude Morphopédologique de la Commune de Madiama, Cercle de Djenné* – Technical report SANREM West Africa. Washington State University, International Programs, Pullman, Washington.

Beckett, P.H.T. and Webster, R. (1971) Soil variability: a review. *Soils and Fertilizers* 4, 1–15.

Bertrand, R. (1973) *Systeme de Paysage des Plaines du Delta Vif et Moyen du Niger: Etude Morphopédologique de la Quelques Plaines du Delta* (Tome I et II). Institut de l'Economie Rurale and Office de la Vallée du Niger, Bamako, Mali.

Blanck, J.P. (1967) *Etude Morphologique de la vallée du Niger.* Université de Strasbourg (under the direction of Pr Tricart), Strasbourg, France.

Gallais, J. (1967) *Le Delta Interieur du Niger*, No. 70 Tome I. Institut Fondamental d'Afrique Noire (IFAN) Dakar, Sénégal.

Leisinger, K.M. and Schmidt, K. (1995) *Survival in the Sahel: an Ecological and Developmental Challenge.* International Service for National Agricultural Research (ISNAR), The Hague, Netherlands.

PIRT Project (1986) *Projet d'Inventaire des Ressources Terrestres. Zonage Agroecologique du Mali.* Ministere de l'Elevage et de l'Environnement, Bamako, Mali.

Sivakumar, M.V.K., Konaté, M. and Virmani, S.M. (1984) *Agroclimatology of West Africa: Mali* (Information Bulletin, 19). International Crops Research Institute for the Semi-Arid Tropics, Bamako, Mali.

Taylor-Powell, E. (1991) Integrated management of agricultural watersheds: Land tenure and indigenous knowledge of soil and crop management. *TropSoils Bulletin* No. 91-04.

United States Department of Agriculture Soil Survey Staff (1999) *Soil Taxonomy. A Basic System of Soil Classification for Making and Interpreting Soil Surveys*, 4th edn. (USDA-NRCS Agricultural Handbook 436). Washington, DC.

Van Keulen, H. and Breman, H.B. (1990) Agricultural development in the West African Sahelian Region: a cure against land hunger? *Agriculture, Ecosystems and Environment* 32, 177–197.

4 Investigation into the Movements of Cattle, Sheep and Goat Herds through the Commune of Madiama*

Adama Ballo and Abdoul Karim Ouattara

Centre Régional de la Recherche Agronomique, Institut d'Economie Rurale, BP 205, Mopti, Mali

The Commune of Madiama suffers from a lack of pasture land, particularly rainy season pastures (ESPGRN/SANREM-CRSP, 1999). This lack of rainy season pastures is so extreme in certain villages that some locally owned sedentary and semi-transhumant herds must leave the commune for pastures in neighbouring communes or go on transhumance. However, other *terroirs* of the commune are sites of livestock concentration during the return from rainy season pastures before crossing into the *bourgoutières*.[1] These livestock concentrations vary in duration from year to year and often cause multiple conflicts between farmers and herders.

This chapter focuses on the system of animal husbandry within and around Madiama. Within the commune reside approximately 5100 cattle, 2550 sheep and 2100 goats. Although nearly all of the commune's families own livestock, the type of livestock ownership, and thus husbandry, can be broken into three broad categories. Those farmers mainly invested in crop production own a predominance of small ruminants and oxen that hired pastoralists herd during the growing season, and that are left to graze on crop residues in the owners' fields during the dry season. Those farmers mainly invested in livestock production own both cattle and small ruminants, and labour is distributed within the family between cultivating crops and herding livestock. Finally, there are a small number of residents that rely only on livestock, and pastoralists that visit the commune with their herds during the crossing to and from the rich grazing available in the *bourgoutières* of the Niger delta during the dry season. Within these broad generalizations are a multitude of specialized husbandry techniques, relationships between resident and non-resident herders and farmers, traditional pastures within and around the commune for both resident livestock and visiting pastoralists, and rights of access to resources such as water and grazing.

To document the circumstances surrounding these pastoral issues in the Commune of Madiama, a survey was made of all livestock production systems and accompanying strategies. This survey helps to identify and propose sustainable techniques for managing pasture resources and raising livestock. The survey was carried out in two stages. The first stage involved focus groups, assembled in each of the ten villages

* Translated and adapted by Keith M. Moore.

of the commune. In the second stage, a village level inventory of livestock resources and management practices was conducted. This chapter reports the results of these surveys in two parts. The first part provides a general description of the area's livestock and pasture management practices. The second part of the chapter focuses on transhumant herd management during their movement through the various *terroirs* of the commune.

There are three systems of livestock management in the commune of Madiama: the sedentary, semi-transhumance and transhumance.

Sedentary Herd Management

Cattle herds of Madiama commune

Table 4.1 lists the cattle herds by villages of the commune.

The village of Nérékoro has approximately 33% of the cattle population in the commune, followed by Madiama (15%) and Bangassi (12%). Torokoro and Toumadiama have the fewest cattle (2%). Nérékoro has the highest average number of cattle per household (31.48), followed by Nouna[2] (with 14.29), Promani (3.47) and Bangassi (3.25). The villages of Nérékoro, Nouna and Promani are primarily of the Peul ethnic group with raising livestock as their principal activity. A few villagers are employed in ancillary activities related to agriculture or small trade, and others work in small craft industries in these villages. In contrast, in Bangassi a significant share of the cattle population consists of oxen owned by farmers for ploughing and cultivation of crops during the rainy season and for transport in the dry season. In the villages of Téguégné, Siragourou and Toumadiama, the livestock are nearly all oxen. Milking cows are found primarily in Nérékoro, Promani and Nouna, where milk production is a major activity.

Cattle and Pasture Management

Dry season

Feed

From the end of the millet and sorghum harvests in November to the first rains in June, local herds of sedentary cattle graze on crop residues (millet, sorghum and rice straw) remaining after the passage of the transhumant cattle. Bush and shrub fodder growing on rare pockets of uncultivated land and on fallow land are also grazed. The grazing of crop residues is done collectively within each village *terroir*. All animals have access to these resources once harvests are completed. Towards the end of the dry

Table 4.1. Number of cattle by village and average per household.

Villages	Principal activity	Number of cattle*	Number of households	Average number of cattle per household
Torokoro	Agriculture	100	74	1.35
Nérékoro	Livestock	1700	54	31.48
Bangassi	Agriculture	630	194	3.25
Promani	Livestock	520	150	3.47
Nouna	Livestock	500	35	14.29
Téguégné	Agriculture	165	58	2.84
Siragourou	Agriculture	410	82	5.00
Tombonkan	Agriculture	200	76	2.63
Toumadiama	Agriculture	100	150	0.67
Madiama	Agriculture	750	365	2.05
Estimated total		5075	1238	

* The number of heads indicated in the table is an estimate made in village assembly.

season (May–April), the availability of these resources becomes limited and the animals suffer from malnutrition. To minimize the consequences of under-feeding during this period, some livestock owners harvest and store stocks of millet and sorghum straw for feed before the animals graze the harvested fields. In addition to storing these crop residues, straw from shrubs and bushes are stored, as are *bourgou*, cowpea and groundnut hay. The agroindustrial feed *Bétail Huicoma* (ABH) is also used as a feed supplement. In general, the feed supplementation strategy observed in all the village is to distribute ABH to production animals (dairy, pasture-fattened animals, oxen, asses and horses), and to weakened and sick animals. However, in the villages of Téguégné and Siragourou, ABH is not used (Table 4.5). This could be explained by the weak purchasing power of these agropastoralists. It should be noted that these types of dry season feeding practices apply to sedentary, semi-transhumant and even certain transhumant herds that do not enter the *bourgoutières*.

Watering

The animals are watered at the river and at wells. The number of wells per village varies from two (Nérékoro) to 20 (Siragourou). In the villages of Bangassi and Promani, in addition to the wells, the animals water themselves at the river (Table 4.2). In the village of Nouna, livestock are watered only at the river. The village of Nouna is located closest to the river.

Herding of the livestock

In the dry season, herdsmen under contract or young boys of the village lead some of the livestock. Other animals are left to themselves. In many cases, herdsmen are contracted for the rainy period until the animals enter the *bourgoutières*. Generally Peuls, the herdsmen, are remunerated in kind and/or in cash, and according to the season. The methods differ from village to village (Table 4.3). For instance, in the case of milk cows, milk production is used to compensate the herdsman.

Stabling

In the dry season, the animals are kept in enclosures at night in order to collect their manure. In Madiama, a villager will often take on the cost of the herdsman in order to benefit from the manure produced while the animals are stabled. Some livestock owners

Fig. 4.1. Cattle grazing on the Bani flood plain.

Table 4.2. Number of watering points per village.

Villages	Traditional wells	Well with large diameter	Bore hole wells	Observations
Torokoro	11	2	1	All functional
Nérékoro	2	–	1	All functional
Bangassi	16	–	1	All functional
Promani	11	2	1	All functional
Nouna	–	–	–	The animals water themselves at the river
Téguégné	10	–	–	All dry out during the dry season
Siragourou	20	–	–	All functional
Tombonka	2	–	–	Functional
Toumadiama	7	–	2	Functional wells; 1 bore hole in good condition
Madiama	11	–	6	Well with low flow and 3 drillings in good state
Total	90	4	12	

Table 4.3. Methods of remuneration for herdsmen by village.

	Dry season		Rainy season		
Villages	Cash (fcfa)	In kind (kg millet)	Cash (fcfa)	In kind (kg millet)	Other remuneration
Torokoro	125/head/mo	0.5/head	1000/head		Development of 0.5 ha cereal
Nérékoro	9000/herd/mo	–	15,000/herd		
Bangassi	100/head/mo	1.0/head	500/head		Purchase of boots, torch and raincoat for the shepherd
Promani	125/head/mo	0.5/head	5000/herd		
Nouna	7500/herd/mo	–	15,000/herd		
Téguégné	5000/herd/mo	45/herd	7500/herd		Purchase of boots, torch and raincoat for the shepherd;
Siragourou	125/head/mo	0.5/head	1250/head		Production of 1 ha of cereal
Tombonka	125/head/mo	1.0/head	1250/head		
Toumadiama	600/head/mo	–	12,500/herd		
Madiama	150/head/mo	–	1000/head		Two working days in the field of the shepherd

stake their oxen in their compounds and provide them with feed supplements.

Health

Both cattle and small ruminants are vaccinated annually against principal diseases. Other treatments such as the de-parasiting are done sporadically.

The rainy season

Feed

During the rainy season, cattle graze fodder resources on the occasional patches of grass, fallow land and the borders of fields. These animals are primarily oxen that are kept in these areas until the end of field cultivation (August). At this time, they are led out of the *terroirs* and join the other animals on neighbouring or distant rainy season pastures. Only the animals of Téguégné, Toumadiama and Bangassi remain in their respective *terroirs* during this period. In Promani, Nouna and Nérékoro, some milking cows are kept close to the village to insure family milk consumption throughout the rainy season. The milking cows graze on the field borders and in a few pockets of undeveloped land. All categories of livestock in Bangassi and part of the animals of Promani are taken to a 257 ha wooded pasture, called *wala-wala,* in the *terroir* of Torokoro and remain there for the rest of the rainy season. However, to avoid a concentration of animals in the *wala-wala*, Torokoro livestock do not graze these pastures. The use by others of Torokoro's *terroir* constitutes a regular source of conflict.

Watering

The animals primarily water themselves in temporary ponds existing on the grazed pastures.

Shepherding

During the rainy season the animals are under the supervision of herdsmen in order to avoid damage to the growing crops. The cost of employing a herder varies from 5000 to 7500 FCFA/herd/month in the dry season and from 7500 to 12,500 FCFA in the rainy season (Table 4.3).[3]

The Number of Small Ruminants in the Commune

The number of small ruminants (sheep and goats) in the commune is provided in Table 4.4.

Small Ruminant Herd and Pasture Management

Dry season

Feed

Natural pasture resources and crop residues found in the fields provide dry season feed for small ruminants. Feed supplements are rarely given to small ruminants, except for the animals, particularly sheep, being fattened. These sheep receive a wide range of feed resources, including groundnut, cowpea and millet hay, millet grain, ABH, *bourgou* straw, and kitchen wastes.

Table 4.4. Number of sheep and goats by village and average per household.

Villages	Principal activity	Number of small ruminants*		Number of households	Average number per household	
		Sheep	Goats		Sheep	Goats
Torokoro	Agriculture	250	200	74	3.38	2.70
Nérékoro	Livestock	300	400	54	5.56	7.41
Bangassi	Agriculture	200	250	194	1.03	1.29
Promani	Livestock	250	150	150	1.67	1.00
Nouna	Livestock	200	300	35	5.71	8.57
Téguégné	Agriculture	100	160	58	1.72	2.76
Siragourou	Agriculture	250	100	82	3.05	1.22
Tombonka	Agriculture	400	200	76	5.26	2.63
Toumadiama	Agriculture	200	100	150	1.33	0.67
Madiama	Agriculture	400	200	365	1.09	0.55
Estimated total		2550	2060	1238		

* Estimates made in village assembly.

Watering

Like cattle, small ruminants are watered at wells in all the villages except at Nouna where watering is done at the river. Small ruminants are watered twice a day.

Herding

The small ruminants are left on their own to browse and graze where they will. However, in the evening, they are brought back to the household compound, typically by the small children (boys and girls) of the village.

Stabling

In both the dry and rainy seasons, small ruminants are stabled in the compounds of the owners in enclosures made of thorn-bushes or they are tied to stakes.

The rainy season

Feed

Small ruminants in the villages of Bangassi, Tombonkan, Promani, Siragourou, Toumadiama, Nouna and Téguégné spend the rainy season in their respective terroirs grazing on field borders, fallow lands and undeveloped spaces (Table 4.6). In the evening they are tied to the stakes or put in the enclosures. In the villages of Nérékoro and Torokoro, some are maintained by the women and graze at field borders tied to stakes. Others are semi-transhumant in the neighbouring *terroir* of Timissa for most of the season. In Madiama, two-thirds of the herds graze during the day in the *terroirs* of Timissa and return to the village in the evening.

Watering

During the rainy season, animals are watered in temporary ponds along their pathways to and from grazing areas.

Shepherding

Unremunerated villagers, generally young boys and/or girls, often herd small ruminants. Herding duties are rotated between various family members in Toumadiama. The herder might also be a Peul (local or

Fig. 4.2. Small ruminants staked in a village.

Table 4.5. Cattle feed supplements, rainy season pastures, watering points and markets frequented by village.

Villages	Feed: Supplements distributed in dry season	Feed: Rainy season pastures	Watering points (frequency/day dry season)	Trade (markets attended)
Torokoro	Crop residues, ABH * (April–June)	Timissa	Well 2 times	Madiama-Sofara
Nérékoro	Crop residues, ABH (March–June)	Tominian, Timissa, Fangasso, Ouan	Well 2 times	Madiama
Bangassi	Crop residues, ABH (April–June)	Wala-Wala **	Well and river 1 to 2 times	Madiama
Promani	Crop residues, ABH (April–June)	Wala-Wala, Timissa, Fangasso	Well and river 1 to 2 times	Madiama-Djenné
Nouna	Crop residues, ABH (April–June)	Timissa, Fangasso, Ouan, Toumidian	River 2 times	Madiama-Djenné-Sofara
Teguégné	Crop residues (April–June)	Fallow land	Well 2 times	Madiama-Konio
Siragourou	Crop residues (April–June)	Laramba Timissa	Well 2 times	Madiama
Madiama	Crop residues, ABH (March–June)	Timissa	Well 2 times	Madiama-Sofara-Djenné
Tombonkan	Crop residues, ABH (April–June)	Timissa, Fangasso	Well 2 times	Madiama-Sofara
Toumadiama	Crop residues, ABH (April–June)	Fallow land	Well 2 times	Madiama

* ABH = food Huicoma cattle; ** Wala-Wala = forest installation pertaining to the *terroir* of Torokoro (257 ha).

non-local), recruited according to a contract, with varied remuneration. In Promani, in addition to villagers planting a millet field for him, the local Peul herder also receives 4 kg of millet per animal. Non-local Peul herders receive 8 kg of millet per animal, without profiting from labour and the millet field. Generally, payment for herding is only calculated for the ewes and does having had their first labour. The herding of males (rams and bucks) is not remunerated. The payment is made in kind and in cash. For example, in Madiama, the herder receives 350 FCFA/head of ewe or doe plus 500 g millet and 25 FCFA (for the cost of condiments). In Téguégné, the females that have had their first labour cost 200 FCFA per animal per month plus seven bowls of millet (equivalent to 3.5 kg) per month during the rainy season.

Semi-transhumant Herd Management

Semi-transhumance is defined in this context as a low frequency of movement by cattle and/or small ruminants into and out of the *terroir*. The animals leave their *terroir* for neighbouring *terroirs* only when in search of rainy season pasture.

Cattle

Semi-transhumance is practised in the villages of Madiama, Nérékoro, Torokoro, Promani, Nouna, Tombonkan and Siragourou. With the first rains (June, July), most of the cattle except oxen are moved to the neighbouring Commune of Timissa in the search of better pastures (Table 4.5). The cattle are the first to leave their respective *terroirs* for these rainy season pastures. Once the field cultivation is completed, the oxen and the few dried-up milk cows join the other animals. The duration of the stay varies from 3 to 4 months for oxen and from 4 to 5 months for the rest of the herd. During this period, the cattle of seven villages of Madiama are found on the pastures of Timissa where they spend the whole rainy season exploiting the fodder and waterways. This zone experiences a strong concentration of the animals during this period.

Timissa is not the only space exploited by the animals. Animals of the villages of Nérékoro, Promani, Nouna and Tombankan exploit pasture resources of other neighbours in the Communes of Fangasso, Ouan and Tominian. They remain there throughout the rainy season until the harvest or the water of the temporary ponds dries up. The return to their respective *terroirs* starts in November with the beginning of harvesting.

In addition to oxen kept for crop cultivation until August, some milking cows are maintained in the villages to ensure family milk supply as well as provide milk for marketing. Maintaining a core of milk cows is a common practice among the Peuls in the villages of Nouna, Promani and Nérékoro.

Sheep and Goats

Herds of small ruminants are also found on the same rainy season pastures of neighbouring communes as the cattle. They are primarily animals from Madiama and Torokoro (Table 4.6). In these villages, the livestock are stabled for the night in enclosures of thorn-bush to avoid crop damage. Nevertheless, sometimes crop damage does occur and friendly solutions are always found. The return of small ruminants to their respective *terroirs* occurs at the same time as the cattle.

Transhumant Management

Local herds

Cattle

Some cattle herds of the commune of Madiama are continually under transhumant management. During the rainy season, these animals are led up on to the plateau in search of fresh grazing and watering points in the pastures of Seno (in the *cercles* of Bankass, Koro and Douentza) and of Konari (*cercle* of Mopti). Some travel as far as the frontier with Burkina-Faso. These transhumant herds are generally mixed and made up of dry cows, heifers, bulls and calves from the villages of Nérékoro and Nouna. This management practice, characteristic of these two villages, is partly related to the pastoral vocation of the dominant ethnic group.

The animals leave for these rainy season pastures in June–July (according to the early or late onset of the rainy season) and return in November–December (period of harvests). These herds return at the same time as the herds of non-local transhumants. During this passage, all herds graze the same spaces (fallow fields and the borders of fields and roads) and use the same watering points. This period lasts for approximately a month on average, until the rice fields are opened for crossing into the *bourgoutières*. Among the locally based transhumant herds, 75% continue in the *bourgou*, another 5% are milking cows and calves that remain in the village *terroir* to ensure the family milk supply, and 20% remain all year in the dry zone. Ten per cent of the animals that go into the *bourgou* are dairy cattle, oxen, or weak or sick animals that graze the young growth of *bourgou* for a month. Once the harvests are finished, these animals go up to the millet fields to graze on the crop residues until the next season of rains. Those remaining in the *bourgoutières* leave with the non-local transhumant herds at the beginning of the rainy season.

Sheep and goats

The local small ruminants do not practice transhumance.

Non-local herds

During October and November, the *terroirs* of certain villages of the commune become zones of livestock concentration for local

Table 4.6. Sedentary small ruminant management practices by villages of Madiama.

Villages	Control			Vaccination	Sheep fattening for cash income	Observations
	Dry season	Rainy season				
		Day	Night			
Torokoro	Free roaming	Staked out around village	Staked out near the house	once/year	Women	A part goes to Timissa
Nérékoro	Free roaming	Plain	Staked out near the house	once/year	Men and women	
Bangassi	Free roaming	Wala-wala	Staked out or stabled	once/year	Women	
Promani	Free roaming	Space Promani Siragourou	Staked out or stabled	once/year	Women	
Nouna	Free roaming	Plain	Staked out or stabled	once/year	Men and women	The evening boys and girls stake them out
Téguégné	Free roaming	Fallow land	Staked out or stabled	0	Women	
Dougoukourani	Free roaming					
Siragourou	Free roaming	Laramba	Staked out	0	Women	
Tombonkan	Free roaming	Fallows	Staked out	once/year	Women	
Toumadiama	Free roaming	Fallows	Staked out	once/year	Women	
Madiama	Free roaming	Timissa	Staked out	once/year	Men and women	During rainy season a semi-transhumant herd remains in Timissa

and non-local herds returning from rainy season pastures in the direction of the *bourgoutières*.

Cattle

The feed and water points of the *terroirs* of all the villages are used by the non-local herds.

ORIGIN OF THE ANIMALS The herds come from neighbouring communes (Konio, Sofara Timissa), certain *Cercles* of the area (Bankass, Bandiagara, Koro) and even from Burkina-Faso. The majority of herdsmen are young Peul men. Dogon, Sore and Mossi ethnic groups are also involved (Table 4.7).

PERIOD OF ARRIVAL OF THE ANIMALS IN THE COMMUNE OF MADIAMA The herds generally arrive at the moment of harvests during November. This date is not fixed since the descent of the animals is a function of the unavailability of fodder and water resources on the plateau. During the years of bad rainfall, the animals go down towards the delta earlier than normal because water becomes the limiting factor on the rainy season pasture.

FLUX OF THE ANIMALS IN THE COMMUNE The pressure of the animals on the fodder resources of the various *terroirs* is related to the proximity of these *terroirs* to the livestock corridors and river crossing points. The principal passage of the animals through the commune follows this path: Siragourou-Nérékoro-Promani-Nouna. There is also a passage through Konio-Kansara-Kessédougou-Toumadiama-Téguégné. The number of herds that pass daily through these *terroirs* varies. In the villages of Nérékoro, Promani and Nouna, an average of up to 300 herds per day may pass with the number of heads per herd varying between 40 and 200. In the villages of Bangassi, Siragourou, Toumadiama and Téguégné, an average of 20 herds pass by daily in the direction of the *bourgoutières*. The duration of passage in the *terroirs* varies from 1 day to 15 days, with an average of 7 days for the passage.

RESIDENCE TIME OF THE ANIMALS IN THE COMMUNE Each year the date of crossing is fixed by the administration and technicians, in collaboration with the local authorities of the various zones concerned. This date is communicated and diffused by all forms of media. Prior to the date of river crossing to the *bourgoutières*, the duration of stay varies according to village (Table 4.7), with the animals spending an average of 1 month while waiting for the crossing. Before the end of the harvests, the animals graze on the borders of the fields of millet and rice, undeveloped lands, fallow lands and residues from harvested fields. The watering takes place in the plains. During this period, there are numerous conflicts, causing the villages of Torokoro, Tombonkan, Bangassi and Promani, to organize their populations into monitoring groups to prevent livestock damage to the crops.

To reach the *bourgoutières*, the only point of crossing in the commune is Nouna. Three groups of herds can be distinguished in this moment in particular:

Group 1 (95%) herds cross to reach the *bourgoutières* and remain until the first rains, a period lasting from 7 to 8 months (November to June–July).
Group 2 (3%) herds, after having crossed, return at the end of 1 or 2 weeks to graze the fields of millet of the terroirs of Nouna, Torokoro and Bangassi, and remain in the villages during the rest of the dry season (November to June–July).
Group 3 (2%) herds do not cross and remain in the area only to graze crop residues in the millet and rice fields of the commune.

Once harvesting is completed, local animals as well as transhumants graze the crop residues (millet and rice) of all *terroirs*. Cattle are watered initially at the river and ponds, and later at the wells.

The livestock return from the *bourgoutières* during June and May with the coming of the rains and the rising of the river. However, the stay in the *bourgoutières* can be prolonged until August if the herbaceous biomass and watering points are

Table 4.7. Transhumant cattle herds movement by village in the Commune of Madiama.

				Duration of residence		
Villages	Origin	Number of herds able to pass per day at the time of the crossing	Average size of herd	In the Commune before the date of the crossing (days)	In the *bourgoutières* before return to graze on crop residues (months)	Period of return from *bourgoutières* to rainy season pastures
Torokoro	Burkina-Faso, Séno, Bankass, Bandiagara, Timissa, Sofara	5	100	1–7	3 December–February	June–July
Nérékoro	Burkina-Faso, Ouenkoro, Bankass, Koro	500	100–150	30–60	–	May–June Stay (10–15 days)
Bangassi	Bankass, Bandiagara, Koro, Sofara, Konguéna	20	50–200	7–15	6 January–June	July–August
Promani	Burkina-Faso, Bankass, Ouenkoro, Sofara, Togo, Tiocki	400	50–300	10–15	2 May–June	June–August
Nouna	Burkina-Faso, Séno, Sofara, Fangasso, Konio	100–200	50–200	2–7	7 December–June	June–July
Téguégné	Burkina-Faso, Séno, Wonikoro	15	40–60	30	–	May–June Stay (15 days)
Siragourou	Burkina-Faso, Bankass	25	400	60	3 February–April	June–August Stay (2–3 days)
Tombonkan	Burkina-Faso, Bankass	30	30–200	15–20	–	June Stay (1–2 days)
Toumadiama	Bankass	1–5	20–100	30	–	June Stay (30 days)
Madiama	Séno, Bankass, Gonkoro, Burkina-Faso, Kessédougou	1000	50–100	30	–	June

Fig. 4.3. Cattle grazing stubble in a farmer's field.

late in occurring in the rainy season pastures. Under these conditions, damage is noted on the crops traversed by the few late herds (1–2%). In contrast, when the rainy season pastures are established early, the animals leave the *bourgoutières* before planting.

The livestock return from the *bourgoutières* is often disorderly with each herder taking his own path. The duration of this passage of the animals varies according to *terroir*. In the majority of the villages, the passage is short, not exceeding two to three days. However, the duration of the passage through the villages of Téguégné, Nérékoro and Toumadiama can reach 15–30 days if the rainfall comes early to these villages. The village of Siragourou constitutes a special case. Indeed, some transhumant herds can remain in the *terroir* on the pastoral zone, Darilalan, some 2–3 months. Some herds can remain during the entire rainy season in the *terroir* on the pastoral zone, Laramba. The longer stay of these animals in the village *terroirs* generally leads to conflicts related to crop damage.

Small ruminants

Like cattle, small ruminants also exploit the resources of the commune as they traverse the *terroirs* in the direction of the *bourgoutières*.

These small ruminants arrive from the *Cercles* of Koro, Bankass, Bandiagara and Korientzé, the *Cercle* of Niafunké in the area of Tombouctou, and from Burkina Faso. The herders are either Peul or Dogon. Normally, the small ruminants precede the cattle and generally arrive during September, before the harvest begins. These herds of small ruminants are not held to await the date of crossing.

The livestock corridor is the same as the one followed by the cattle leading to the *bourgoutières*. All the categories of herds are found: mixed herds (sheep and goats), herds of sheep and herds of goats. The passages of the goat herds are announced in Madiama, Siragourou and Téguégné. Three to 20 herds of sheep can pass each day in some villages with the number of head per herd varying from 50 to 400. For goats, the number of herds per day varies from 5 to 30, with 30–400 head per herd.

While the duration of passage for a single herd is quite brief, normally lasting only as long as the time it takes to walk through, which rarely exceeds 3 days, the complete passage of all of these herds can last from 1 to 3 months. During this time, the borders of fields are grazed.

Table 4.8. Small ruminant transhumant movements by village in the Commune of Madiama.

Villages	Month of arrival	Origin	Number of herds passing per day at the moment of crossing		Average herd size		Days in Commune before crossing	Date of return from *bourgoutières*
			Goats	Sheep	Goats	Sheep		
Toroko	September	Douentza, Koro, Bandiagara, Sofara	20	–	50–100	–	–	–
Nérékoro	October	Burkina-Faso, Saraféré, Séno-Bankass, Me Binga, Sarè-Yamou	40	30	60–150	100–300	–	June (staying 3 days)
Bangassi	September or October	Zone of Kounari and of Guimbala	30	20	15–80	50–200	15	–
Promani	September	Niafunké, Gouma, Douentza, Borated, Korientzé, Konna, Fatoma	6	8	100–150	20–100	–	–
Nouna	Mid-October	Gouma, Douentza, Borated	2–5	5–10	30–100	100–300	15	–
Téguégné	September October	Zone of Kounari, of Guimbala and Bandiagara	10	3	60–100	60	1–30	May
Siragourou	October	Bandiagara, Koro, Borated, Douentza	3–10	–	40–200	–	–	May–June
Tombonkan	September	Burkina-Faso, Borated Bankass, Koro	40	20	800	800	–	May–June
Toumadiama	November	Séno-Bankass, Koro	5	3	200	100	–	June
Madiama	September	Séno-Bankass	1–10	–	50–100	–	–	–

Table 4.9. Frequently used rainy season pastures, average stay, number and destination of the transhumant herds for eight villages in the Commune of Madiama.

Village	Most common rainy season pastures (%)	Average stay in						Number of livestock						Existence of social bond between herdsmen and the village (%)		
		Rainy season pastures (months)			*Terroir* (days)			Cattle		Sheep		Goats				
		Cattle	Sheep	Goats	Cattle	Sheep	Goats	Herds	Head	Herds	Head	Herds	Head	Cattle	Sheep	Goats
Madiama	Timissa (28) Seno (24) Koula (14) Who (14) Fangasso (53)	5	5	5	15	21	81	24	2680	15	2800	11	1025	4	33	55
Toumadiama	Timissa (27) Ouan (20)	5	5	5.5	3	3	3	14	1140	8	640	6	465	29	25	50
Torokoro	Timissa (38) Seno (36) Koula (9) Soufroulaye(9)	4	4	4.5	34	14	21	44	4326	34	3943	36	3585	56	38	44
Tombonkan	Timissa (62) Seno (38)	4.5	4.5	5	39	45	50	7	532	6	329	7	422	71	50	71
Bankassi	Timissa (26) Konari (23) Saréyamu (13) Seno (13)	3	3	3	18	23	19	23	2175	16	1130	12	810	39	42	67
Nérékoro	Timissa (78) Burkina (5)	5.5	5.6	5.5	3	3	4	83	9770	69	3490	64	2170	100	100	100
Téguégné	Timissa (47) Burkina (33)	5	4.5	5.5	10	1	10	56	5550	2	700	41	3600	0	14	23
Nouna	Timissa (53) Seno (16) Kougué (9) Burkina (6)	5	–	–	25	–	–	126	12,519	–	–	–	–	48	–	–
	Commune (average/total)				18	16	27	377	38,692	150	13,032	177	12,077	43	43	59

Because the passage of the small ruminants causes less crop damage than cattle and the pressure of the small ruminants on the natural resources of the commune is much less compared to that of the cattle, the conflicts related to their passage are limited. Indeed, the small ruminants pass through all the village *terroirs* in the commune. After the crossing, the herds remain in the *bourgoutières* until the beginning of the rainy season. As with the descent into the *bourgoutière*, the small ruminants return quickly to their rainy season pastures, although each shepherd follows the path of his choice.

A detailed study on the movement of transhumant livestock was conducted in eight of the ten villages of the Commune of Madiama. The study involved collaborating members of the Natural Resource Management Advisory Committee (NRMAC). A collaborator villager was chosen in each village and provided with a book for data collection. The data collected included the number of herds (cattle, sheep or goats), the approximate number of animals in each herd, duration in the *terroir*, location during the rainy season, destination during the descent and finally the existence of a social bond between the herdsmen and villagers. This investigation was designed to complement the previous data collection discussed in the first part of this chapter. The survey data cover the period from harvest until the return of the livestock from the *bourgoutières*. The results are summarized below by village.

Madiama

Among the rainy season pastures most frequented by the livestock encountered in the *terroir* of Madiama are Timissa (28%), Douentza (24%), Koula (14%) and Woh (14%). All species stay on these pastures for an average duration of 5 months. The duration of the stay in the *terroir* of Madiama varies according to the animal species, with sheep remaining nearly 3 months, cattle only 52 days, and goats just 2 weeks. The minimum duration in the *terroir* was 2 days for all species and the maximum stay 180 days for a mixed herd of sheep and goats.

During the days of the survey, 24 herds of cattle, 15 of sheep and 11 of goats were listed. The total number of livestock was approximately 2680 head of cattle, 2800 of sheep and 1025 of goats, amounting to 2527 UBT (tropical cattle unit: 1 cattle = 0.8 UBT; 10 sheep/goats = 1 UBT). All transhumant cattle herds listed moved to the *bourgoutières* after their stay in the *terroir*. As for the small ruminants, 32% of the sheep and 59% of goats remained to use the resources of the *terroir* for the following 6 months. The transhumant livestock pressure of cattle on the resources of the *terroir* is less than that of the small ruminants. In addition to the transhumant herds, it is necessary to add the local herds that amount to 577 head of cattle and 60 of goats. These cattle and goats remain in the *terroir* for a full 8 months of the year. The majority of the cattle (96%) and sheep (67%) herdsmen do not have a local landlord in Madiama, while less than half of goat herders (45%) do not have a local landlord. This could partly explain the prolonged stay of the goat herds.

Toumadiama

Transhumant livestock listed as passing through this village use the rainy season pastures of Fangasso (53%), Timissa (27%) and Ouan (20%). The average time of stay on these pastures is 5 months for cattle and sheep and 5.5 months for goats. Contrary to the *terroir* of Madiama, Toumadiama constitutes a crossing point of the transhumant livestock towards the *bourgoutières*. Indeed, the time spent in the *terroir* does not exceed 3 days. During the investigation, 14 herds with a total of 1140 head of cattle, eight herds totalling 640 sheep and six herds totalling 465 goats were listed. This corresponds to 1023 UBT. The locally owned animals included a cattle herd with 70 head, one sheep herd with 45 head and one goat herd with 30 head. All the herds listed as transhumant pass through in the direction of the *bourgoutières*. Therefore, the fodder resources of the village *terroir* of

Toumadiama do not experience high animal grazing pressure.

Torokoro

The livestock passing through the terroir of Torokoro are most likely to have come from rainy season pastures in Timissa (29%), Bankass (24%), Douentza (12%), Koula (9%), Tominian (9%), Soufroulaye (9%), Bandiagara (5%) and Dialassagou (3%). The cattle and sheep herds remain in Torokoro for an average of 4 months and the goat herds for 4.5 months. During 43 days of investigation, there were 44 cattle herds with a total of 4326 head, 34 sheep herds totalling 3943 head, and 36 goat herds with 3585 head. This totals some 4214 UBT. The fodder resources of this *terroir* are grazed on average 34 days for cattle, 14 for sheep and 21 for goats. The minimum time of stay is 1 day, with a maximum stay of 270 days for cattle and 60 days for small ruminants. After their stay on the *terroir*, all of the cattle, and two-thirds of the sheep and goats head into the *bourgoutière*. The remainder of sheep and goats migrate towards the prefecture of San. However, some animals remain in the *terroir* for up to 8 or 9 months. These include 2% of transhumant goats in addition to local livestock consisting of 156 cattle, 57 sheep and 96 goats. Many herdsmen have social bonds with the village. Indeed, 56% of cattle herdsmen, 38% of sheep herdsmen and 44% of goat herders have a landlord in Torokoro.

Tombonka

The results of the investigation show that the pastures most frequented by transhumant animals on the *terroir* of Tombonka are those of Timissa (31%), Tominian (31%), Bandiagara (23%) and Koro (15%). These rainy season pastures are used on average for 4.5 months by the cattle and sheep and 5 months by goats. There are seven cattle herds estimated at 532 head, six sheep herds of 329 head, and seven herds of 422 goats. These 20 herds are estimated at 500 UBT. The time of stay of the animals on the *terroir* of Tombonkan is an average of 39 days for cattle, 45 days for sheep and 50 days for goats. The minimum time of stay is 5 days for cattle and sheep and less than a day for goats. The maximum duration of stay is 180 days. After this stay on the *terroir* of Tombonkan, the majority (82%) of the cattle and sheep moves down to the *bourgoutière*. The remainder of the cattle and sheep stay in the *terroir*. The goats are dispersed between the *bourgoutière* (47%), the *terroir* of Madiama (47%) and Tombonkan (6%). In addition to those transhumants that make a prolonged stay on the *terroir* of Tombonkan, there are also local herds of 125 cattle, 55 sheep and 55 goats. Nearly three-quarters of goat and cattle herdsmen and half of the sheep herders maintain social bonds with the villagers.

Bankassi

The herds passing through Bankassi come from the rainy season pastures of Timissa (17%), Fatoma (15%), Saréyamou (13%), Diallassagou (13%), Tominian (9%), Jamakan (9%), Konari (8%), Bankass (6%), Niakoungo (4%), Farimaké (4%) and Burkina Faso (2%). The residence time on these pastures is on average 3 months for all livestock. During the 34 days of the survey, transhumant herds identified included 23 cattle herds of 2175 head, 16 sheep herds of 1130 head and 12 goat herds of 810 head. These animals, which correspond to 1934 UBT, graze the forage resources of the *terroir* for an average of 18 days for cattle, 23 days for sheep and 19 days for goats. The minimum time of stay is 7 days for cattle and sheep and less than 1 day for goats. The maximum stay is 28 days for the cattle and sheep and 30 days for goats. All the animals listed in the *terroir* during the period of investigation enter the *bourgoutières*. The cattle (39%), sheep (42%), and goat (67%) herdsmen maintain relations with villagers.

Nérékoro

Cattle, sheep and goat herds counted on the *terroir* of Nérékoro spend the rainy season on pastures primarily in Timissa (78%) and Burkina (5%). These pastures are grazed for 5.5 months by all livestock species. During the period of the survey (37 days), 83 cattle herds totalling 9770 head, 69 sheep herds totalling 3490 head, and 64 goat herds totalling 2170 head were listed. These numbers yield a total of 8382 UBT. The herds remained on the *terroir* for a brief time, 3–4 days, with a minimum stay of 1 day and a maximum stay of 7 days for sheep, 12 days for goats and 15 days for cattle. All the transhumant animals descend into the *bourgoutière*. All the herdsmen (100%) have a social bond with the village. In addition to the transhumant herds, the local livestock includes seven herds of cattle totalling 138 head, three herds of sheep totalling 28 head and two herds of goats totalling 90 head, accounting for a total of only 141 UBT. These dairy animals remain on the *terroir* during the entire year. This is for both autoconsumption and sale of milk and dairy products.

Téguégné

The rainy season pastures of Timissa (47%) and Seno (33%) are the source of transhumant livestock identified in the *terroir* of Téguégné. The sheep stay here 4.5 months, the cattle 5 months and goats 5.5 months. On the *terroir*, the average stay of transhumant animal is 10 days for cattle and goats and less than 1 day for sheep. The maximum duration of stay is 30 days. There are 56 cattle herds (5550 head), seven sheep herds (700 head), and 41 goat herds (3600 head). These correspond to a total of 4870 UBT. All the cattle descend into the *bourgoutière,* while the small ruminants stay on the non-flooded pastures. The vast majority of herdsmen do not have any social ties with the villagers.

Nouna

The data collected in the hamlet of Nouna (part of the officially twinned village of Tatia Nouna) refer only to cattle. The transhumant cattle counted on the *terroir* of Nouna arrive from the rainy season pastures of Timissa (53%), of Seno (16%), Kougué (9%) and Burkina Faso (6%) where they have been during the preceding 5 months. On the whole, 126 herds amounting to 12,519 head, or about 10,015 UBT, were listed. The duration of the stay on the *terroir* averages 25 days with a minimum duration of 1 day and maximum of 120 days. All of these cattle descend into the *bourgoutière*. Half of the herdsmen do not have any social ties with villagers of this *terroir*.

Conclusion

The rainy season pastures of the Commune of Madiama are insufficient. Only the livestock in three villages out of the ten remain in their *terroir* (sedentary livestock management) for the rainy season. In the other villages, the few ruminants and oxen that remain survive on small pockets of undeveloped land, the rare fallow, and the borders of the fields. The vast majority of commune-owned livestock are outside of their own *terroirs*, either in semi-transhumance or in full transhumance.

Many non-local transhumant herds of cattle, sheep and goat herds collectively graze the crop residues being used as feed supplements for livestock during the dry season. Their herders generally camp in proximity to the croplands, often before the end of harvesting, thereby creating many conflicts between their herders and local farmers. During the dry season, some transhumant herds continue their movement through to the *bourgoutières*, while others remain in the *terroirs* of the commune to exploit the crop residues. The majority of the transhumant animals inventoried in the Commune of Madiama from November to December 2001 return from the rainy season pastures of Timissa and Seno where they had been for an average of 5 months.

The length of time that these herds remain in the commune varies from 3 to 50 days, depending on the *terroir*. The average stay of the cattle and sheep herds in the commune is 17 days, which is lower than that of goats (27 days). The number of UBT and the time of stay allow for a crude estimate of livestock pressure on the feed resources of the commune. The *terroir* of Nouna experiences the greatest pressure, followed by the *terroirs* of Torokoro, Téguégné and Nérékoro.

The Commune of Madiama is a livestock corridor for passage from rainy season pastures to the zones of *bourgou*. Indeed, nearly all of the inventoried livestock pass through the commune to the *bourgoutières*. This position as a zone of transit exposes the commune's feed resources to intense exploitation. Indeed, more than 30,000 UBT remain for an average of 20 days within the *terroirs* of the commune. A rate of consumption of 2 kg dry matter/UBT corresponds to the removal of 1200 tonnes of feed resources (dry matter weight). Only a small part of these resources is restored to the soils through manure deposited mainly on the croplands. During their stay in the commune, the transhumant herdsmen may work under contracts that allow them to stable their own livestock in the fields. However, this is most often dependent on the existing relationship between herdsmen and the landlords of the various *terroirs*. In this regard, goat herders have more and better relationships with the villagers in these *terroirs* than cattle or sheep herdsmen.

The results showed that during the rainy season the natural pastures of the rural district of Timissa are of cardinal importance to the management of the transhumant herds in the region as well as the vast majority of the locally owned livestock from the Commune of Madiama. This region deserves further investigation in order to better understand the management of these natural pastures and the characteristics of the transhumant herds that impact the commune.

The natural resources available for livestock production within the Madiama Commune are limited. Water and grazing resources are scarce, and there is significant conflict over crop damage by herds as they cross the commune for the delta. Although the traditional relationship still exists between some agriculturalists and pastoralists in which the herder spends time with his herd on the farmer's fields to graze residues and fertilize the soil, the amount of time spent by herds on Madiama's fields is decreasing, resulting in a loss of organic matter. With the adoption of Mali's Pastoral Code in 2001, rules that resemble the historic consultative process of the Dina have been set in place to determine the dates for entrance and exit from the Niger Delta. These dates are based on harvest and planting. This pastoral code begins to address this conflict between agriculturalists and pastoralists in the crossing of Madiama into the delta. However, to improve soil fertility, there remains a need to encourage the traditional relationships between the two groups for fertilizing agricultural field.

Notes

1 The *bourgoutière* is where *bourgou, Echinochloa stagnina,* is grown. *Bourgou* is a highly productive plant that grows underwater with the seasonal rise of the Niger River and the associated system of lakes and marshes. As the water falls the *bourgou* becomes the major source of fodder for a wide region.

2 Nouna is the Peul dominated hamlet of the administratively twinned village of Tatia Nouna.

3 FCFA refers to the currency in Mali, franc de la Communauté Financière d'Afrique (franc of the African Financial Community). The exchange rate varies, but usually about 500 FCFA = $1.00.

Reference

SANREM CRSP – West Africa (1999) *Participatory Landscape/Lifescape Appraisal: Synthese du diagnostic participatif réalisé à Madiama, Nérékoro, et Tombonkan du 1er au 14 février 1999* (Sustainable Agriculture and Natural Resource Management Collaborative Research Support Program – West Africa Site, Working Paper No. 99–02). Office of International Research, Education, and Development, Virginia Tech, Blacksburg, Virginia, pp. 1–94.

5 Land Use Changes in Madiama Commune

Randolph H. Wynne,[1] Mohamed Sidi Mohamed Touré,[2] Nina Sengupta[1] and Adama Ballo[2]

[1]*Center for Environmental Applications of Remote Sensing, Virginia Tech, Blacksburg, VA 24061–0324, USA;* [2]*Centre Régional de la Recherche Agronomique, Institut d'Ecomonie Rurale, BP 205, Mopti, Mali*

A thorough understanding of the transformation of opportunistic grazing management by transhumant herders is critical to understanding and avoiding natural resource-based conflicts as well as to ensuring economic and food security. Unfortunately, however, there is very little quantitative information on the magnitude and type of changes in land use that have occurred in recent decades. This is particularly problematic given that recent transformations in governance in the region have created a cadre of newly empowered local-level decision makers who are tasked with natural resource management, yet are often only partially aware of important changes in the natural resource base. High quality information on land use trends is, of course, equally important to scientists and managers focused on a range of scales.

Further exacerbating the potential problems facing managers of a natural resource base that is changing in ways not fully understood is the often untested assumption of unidirectional expansion of fields on to marginal land once reserved for other uses. As noted by Reenberg *et al.* (1998) the dynamics of cultivation patterns are quite complex, and are affected by both biophysical and socio-economic factors at a range of scales. In addition, there are often: (i) substantial lags between historical policies and current land use (see Wardell *et al.*, 2003) and (ii) unpredictable random elements resulting in sudden changes (see Reenberg, 2001).

Having a sound picture of land use pathways in this region is, therefore, important for a variety of reasons. The remainder of this chapter elucidates the methods used to determine land use change over an approximately 50 year period, reports the results, and discusses the implications of the findings.

Methods

A field campaign was carried out in August 2002. In situ data collected at that time focused on documenting the location and type of land use at a minimum of 100 locations in and around each of the project study areas. Two QuickBird images were acquired for Madiama, one in 2002 (Fig. 5.1) and a second in 2003. Fortunately, there was also an historical high quality panchromatic stereo photograph of the Madiama area acquired on 30 January 1952 (Fig. 5.2). Such imagery provided a truly unique opportunity to infer land use and carbon change over 50 years.

A team from the United States Department of Agriculture–Agricultural Research

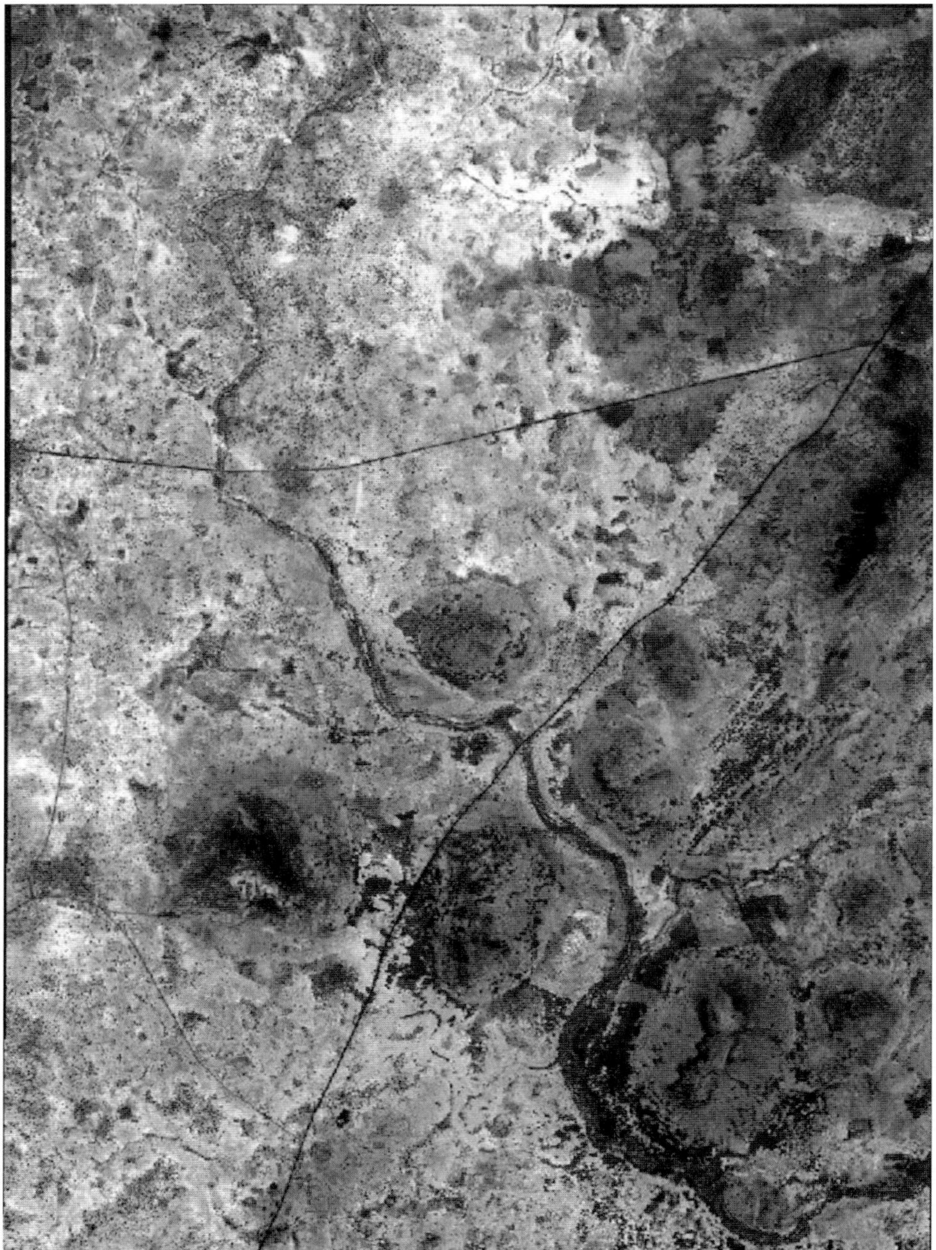

Fig. 5.1. QuickBird image of Madiama and vicinity acquired 11 November 2002.

Service (USDA–ARS) performed a supervised classification of Landsat 7 ETM+ images acquired over Madiama (Wynne *et al.*, 2003). Unfortunately, due to the small field sizes, the availability of only a single date of QuickBird imagery, and 2002 rainfall that led to poor yields or outright crop failure, crop separation was very difficult. The team

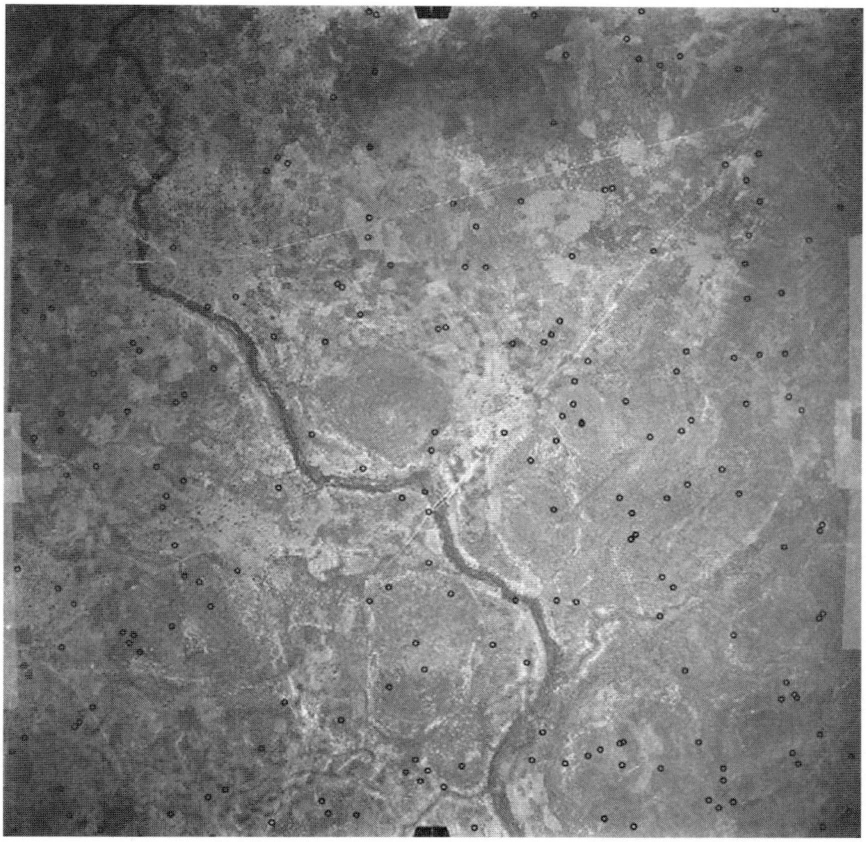

Fig. 5.2. Panchromatic aerial photograph of Madiama and vicinity acquired 30 January 1952. (Note: The small black circles are the random points at which land use was interpreted on this photograph, obtained from the Institut Geographique National of France.)

used a radiative transfer modelling approach to develop leaf area index maps for all crops (aggregated), which are useful for gaining a rough understanding of the spatial distribution of biomass. None the less, it was clear from the 2002 results that the land cover in the region was not very amenable to even quite advanced semi-automated classification techniques. Possible issues contributing to this include the following: (i) average field area, whether farmed or fallow, is at least an order of magnitude smaller than comparable agricultural features in most of the developed world, (ii) frequent interspersion of farmed land, trees and cultural features is the rule rather than the exception, thereby leading to a very poor fit of classification schemes assuming the separation of these categories in space.

The panchromatic photography was not suitable for any sort of spectrally based discriminant analysis of land use or cover since it only contained one spectral band (black and white). Furthermore, because there are so few fixed cultural features (building footprints, roads, etc.) suitable as horizontal ground control points in either the photos or the 2002 and 2003 QuickBird images, accurate image co-registration was not feasible. The aerial photos could be viewed in stereo, which (for the trained analyst familiar with local conditions) slightly offset the disadvantage of not having information in colour.

The classification scheme was based on the work of Badini *et al.* (2001; see Chapter 3), which provided detailed information concerning the soils and land use practices in the commune. The United States Geological Survey (USGS) Level I classification scheme (Anderson *et al.*, 1976) was used as a point of departure, but the USGS numbering scheme and class descriptions were amended to better reflect local conditions. The classes used are as follows: bare soil/soil with litter, brush/brush land, cultivation, park agroforestry, riparian vegetation, road/settlement, pasture, denuded area, river/water and stone quarry.

The protocol developed required careful implementation by the Malian analysts who visually interpreted the land use in these broad categories at 300 random points on the 1952 aerial photos (Fig. 5.2) and the 2002 QuickBird image, with different random samples used for each date. The number of points in the first data set was later circumscribed to the 205 points that fell on the photograph that covered nearly all of the study area due to some difficulties in utilizing the point grid on adjacent photographs. While this reduction to 205 points had a random element, the fact that the data sets were slightly different in size, even given the same study area, is a clear limitation. Field visits to the random points used for the 2002 imagery using global positioning system (GPS) receivers for navigation improved the quality of the interpretations by allowing the analysts to learn how particular land uses *in situ* appeared on the imagery.

Results

The results are shown in Table 5.1.

Rounded to the nearest per cent, the principal trends are as follows: bare (uncultivated) soil has decreased by 23%, woody vegetation has decreased by 27%, cultivation has increased by 40%, park agroforestry has increased by 10%, the area of degraded/denuded soil has tripled (2% to 6%) and pasture has decreased by 29% (from 32% to 3%). In this case the general observation that increasing population has increased the cultivation of marginal lands once reserved for other uses has been verified and quantified.

Discussion

With the reduction of pasture and brushland from 63% of the landscape to less than 8% of the landscape, opportunistic grazing management by transhumant or resident herders is no longer sustainable in the commune. Furthermore, careful use of the

Table 5.1. Land use change in Madiama: 1952 and 2002–2003.

	1952		2002–2003		
Land use	Frequency	Per cent	Frequency	Per cent	Per cent difference
Bare soil	21	10.3	37	12.3	2.0
Brush/brush land	65	31.7	15	5.0	−26.7
Cultivation	35	17.1	170	56.7	39.6
Park agroforestry	2	1.0	34	11.3	10.3
Riparian vegetation	2	1.0	4	1.3	0.3
Road/settlement	5	2.4	5	1.7	−0.07
Pasture	65	31.7	8	2.7	−29.0
Denuded area	4	2.0	18	6.0	4.0
River/water	6	2.9	6	2.0	−0.9
Stone quarry	0	0.0	3	1.0	1.0
Total	205	100.0	300	100.0	

remaining grazing resources will be required to avoid increasing the denuded/ degraded area even further. The qualitative observations indicate that many of the remaining wooded areas are so marginal in soil quality that they are unlikely to be cultivated even with additional population pressure. However, they are likely to face increasing grazing pressure that may impact their ability to provide a wide variety of forest products, including firewood, construction materials, and traditional foods and medicines. Reducing grazing pressure through community-managed rotational grazing is therefore indicated from the perspective of both long term food security and natural resource management. In terms of cultivation, one bright spot is that approximately 8% of the current cultivated area is currently fallow. However, this is minimal given past practices of 15 year fallows. It was not possible to determine whether this percentage of land left out of cultivation has changed in the last 50 years, but the fact that traditional practices ensuring long term soil viability for agriculture have not been abandoned even in the face of population pressure is encouraging.

In summary, with excellent cooperation between US and Malian scientists, we have quantified trends in land use in the Madiama Commune, Mopti Region, Mali. The information provided by this study has implications for sustainability, food security, and (potentially) income from evolving carbon/climate exchanges.

References

Anderson, J.R., Hardy, E.E., Roach, J.T. and Witmer, R.E. (1976) *A Land Use and Land Cover Classification System for Use with Remote Sensor Data* (United States Geological Survey Professional Paper 964). US Government Printing Office, Washington, DC.

Badini, O. and Dioni, L. (2001) *Etude morphopédologique de la Commune de Madiama, Cercle de Djenné – Technical report SANREM West Africa.* Washington State University, International Programs, Pullman, Washington.

Reenberg, A. (2001) Agricultural land use pattern dynamics in the Sudan-Sahel: towards an event driven framework. *Land Use Policy* 18, 309–319.

Reenberg, A., Nielsen, T.L. and Rasmussen, K. (1998) Field expansion and reallocation in the Sahel: land use pattern dynamics in a fluctuating biophysical and socio-economic environment. *Global Environmental Change* 8(4), 307–327.

Wardell, D.A., Reenberg, A. and Tøttrup, C. (2003) Historical footprints in contemporary land use systems: Forest cover changes in savannah woodlands in the Sudano-Sahelian zone. *Global Environmental Change* 13, 235–254.

Wynne, R.H., Moore, K.M., Doraiswamy, P.C., Badini, O., Touré, M.S.M. and Ballo, A. (2003) *Remote Sensing of Land Use and Soil Carbon Changes in a Semi-arid Agropastoral System: Case Study in the Madiama Commune, Mopti Region, Mali* (Technical Papers, Africa GIS, 3–7 November 2003). Dakar, Sénégal.

6 Participatory Landscape/Lifescape Appraisal

Julia Earl[1] and Amadou Kodio[2]

[1]Preventing Harm Minnesota, Minneapolis, Minnesota 55407, USA; [2]Centre Régional de la Recherche Agronomique, Institut d'Economie Rurale, BP 205, Mopti, Mali

In keeping with SANREM CRSP's participatory research methodology developed during its initial years of field work and investigation, the SANREM West Africa project employed the Participatory Landscape/Lifescape Appraisal (PLLA) in Mali. The PLLA is a method for participatory rural appraisal that goes beyond the village level to the more inclusive landscape scale. It was used to gain an initial understanding of the issues facing the farmers and herders in Madiama Commune, Mopti Region, Republic of Mali, an area of increasing resource scarcity and growing conflict. The chapter begins with an overview of the PLLA methodology, followed by a description of how it was applied in Mali. A synthesis of the findings then leads into a brief introduction to the activities that have been simulated and provide the basis for the rest of the book.

The Methodology

SANREM CRSP Global Programme developed the PLLA methodology during its initial years of operation in the Philippines, Burkina Faso and Ecuador in the early 1990s, and SANREM investigators subsequently used it in a variety of research sites. The approach was based on four philosophical and methodological cornerstones: participatory research, interdisciplinary teams, inter-institutional collaboration and a landscape scale of analysis. This platform was designed to bring scientific and local knowledge systems together with ecological, agricultural and social scientists and development and research practitioners to address natural resource and agricultural problems. SANREM selected the landscape focus, because it encompasses a broader analytical framework that incorporates cross-ecosystem linkages addressing upstream–downstream interactions.

SANREM learned many lessons from the five preceding decades of development work conducted by national and international institutions in developing countries. Local participation in the design of development efforts is necessary for effective project implementation. Therefore, in SANREM's 'research for development' approach, local farmers and other project participants, including government, research and non-government organization (NGO) partners, participated in the full spectrum of the research process that encompassed problem identification and diagnosis, identification and testing of potential solutions, and evaluation and adoption of practices or policies based on the research findings.

A second key tenet of the PLLA approach is interdisciplinary research teams.

The constraints on sustainability are not delimited by discipline. Addressing such constraints required the recognition of a complex array of processes within and between ecosystems, including the physical environment (soil, water, vegetation, climate), the human dimension (population patterns, social structure, cultural norms) and their interactions in the domain of agricultural and natural resource management (NRM) practices. Therefore, PLLAs require an interdisciplinary approach that transcends traditional boundaries among ecological, agricultural and social sciences, and that also encompasses a variety of conceptual and methodological tools to produce an integrated understanding of landscape ecology and the systemic relationships between its components.

A third key element of the PLLA process requires inter-institutional collaboration. By virtue of the cross-cutting nature and breadth of environmental and agricultural sustainability issues, PLLAs incorporate the diverse institutions implicated in them. Intersectoral partnerships and networks, including US and host country institutions, with a shared concern for sustainability include organizations such as universities, non-profit organizations, research centres, development agencies, farmers' organizations and women's organizations.

The fourth key PLLA element is the landscape/lifescape approach. It is based on the premise that in order to understand sustainability and thereby develop strategies to achieve it, the relationships and linkages within a landscape must be understood. Landscape refers to a complex mosaic of biological and physical processes. Lifescape refers to the social or human factors that influence that landscape. In crafting the PLLA approach, SANREM determined that the landscape/lifescape level of analysis was the most comprehensive for understanding as fully as possible the complex interactions between biophysical and social components for a given area.

Many of the tools used in a PLLA are familiar to development agents, field researchers, government employees and non-profit agents working in sustainable development. They include: interviews, transects, mapping (social or physical), timelines, problem trees, wealth ranking, etc. Every effort is made to be as inclusive as possible by accounting for any differentiation in a given community, be it religious, gender-based, wealth, ethnicity, caste, etc. PLLAs include incorporating any other data that may be available regarding the locale or issue. A complete description of the PLLA process is found in *Looking Back at the Landscape: Impacts and Lessons Learned from the SANREM CRSP*, 1999.

Pre-field Work Phase

Prior to undertaking the field component of the PLLA, the SANREM team conducted a West Africa wide literature review on the issues of conflict and NRM (Moore *et al.*, 1999). This synthesis was based on personal interviews with researchers and development experts across West Africa, grey literature and formally published documents. In the process, local partners were identified as key to understanding the rich and dynamic social and natural resource complexities found in Mali. The national Malian natural resource and social research organization, Institut de l'Economie Rurale (IER), was a natural choice as a research partner because of IER's long-standing history of applied, rural and participatory research in Mali. The SANREM project then invited Sahelian regional, Malian national and international NGOs as well as local consultants to participate in a conference (25–27 January 1999) at the Institut du Sahel (INSAH) in Bamako, Mali. The conference achieved its objectives of formally introducing the SANREM programme, identifying a research site, building relationships with local partners, and establishing working agreements.

Research Site Identification

The SANREM-West Africa team and conference participants needed to address a set of established criteria in its site selection process. The criteria included: the degree to which the site was representative of Mali and the broader Sahelian region; the presence of a variety of agricultural *or* livelihood activities (e.g. cultivation, animal husbandry, fishing, transhumance, etc.); the presence of conflict related to natural resources; year-round accessibility; receptivity and willingness to participate from rural partners; and availability of secondary data related to the site. An additional overarching element was the degree to which the site might be representative of other areas of the world, semi-arid or otherwise, characterized by increasing natural resource scarcity and competition or conflict for those resources, with respect to replication of lessons learned. In order to address these criteria, the assembled partners, through an intensive dialogue, selected the Madiama Commune (the commune being the smallest administrative unit in the then newly established decentralization rubric). Madiama is located in the Djenné *cercle*, Mopti Region. The conference participants then identified two primary communities, Madiama (town) and Tombonkan, for inclusion, pending their willingness to participate. Following its arrival at the site and preliminary investigation, the research team decided to include Nérékoro, a small village of Peul pastoralists with active social and economic interactions as well as shared NRM conflict issues in Madiama and Tombonkan.

Research Protocol Outline Established

Following the site selection process, the research team members worked together to design the research protocol. All members of the team were well-versed in the Participatory Rural Appraisal (PRA) or Méthode Accélérée de Recherche Participative (MARP) methodology. Therefore, they were well prepared to conduct a PLLA, an approach that draws heavily on PRA methods applied at the watershed level, comprehensively assessing human (or lifescape) and biophysical (or landscape) interactions. Once in Mopti (the regional capital of the Mopti Region), the research team fully developed the interview questions, identified which tools to use and divided into two smaller working groups. One team worked in Tombonkan and the other in Madiama and Nérékoro. (Nérékoro, the smallest of the three communities, is closest to Madiama and has a higher degree of interaction with Madiama dwellers than with more distant villages.)

Fieldwork in the Commune of Madiama

The PLLA Teams spent 2 weeks conducting field work, sharing results with community members, triangulating findings between the two community-based teams, posing follow up questions, conducting a village meeting, and synthesizing written results. Malian IER research staff (both local and national), SANREM scholars from Virginia Polytechnic Institute and State University (Virginia Tech) and Washington State University, a SANREM Management Entity Africanist, and a representative of the United States Agency for International Development (USAID) participated on the multidisciplinary research teams (soil scientists, animal scientists, agronomists, economists, sociologists, ecologists and rangeland specialists). The majority of researchers were Malians. Each team resided in the community for the duration of the research effort.

Upon arriving in the Commune of Madiama the research team was introduced to the village chiefs by the resident IER technician. Team members explained the research objectives of the PLLA, highlighting that the purpose of the work was to better understand NRM issues in the area and possibly work together towards their resolution. However, the team also made a deliberate point to clarify that SANREM was not the type of project that could provide a tangible physical benefit such as a schoolhouse, birthing centre, rice processor, fertilizer or the like. Given Madiama's history as an IER research site, the village elders were familiar both with research teams and participatory approaches. The team explained that it wanted to meet with women as well as men and older and younger members of the community. Since the research occurred during the cold season, a time of year when agricultural activities do not occur or are at a minimum, participation in research groups was not detrimental to completing key agricultural or resource garnering activities. The community representatives agreed to work together with the SANREM team.

The PLLA research team employed a variety of PRA-type tools that included: natural resource management maps, problem trees, activity and resource calendars, timelines of major environmental and historical events, access and control charts (regarding natural resource management use and conflicts), walking transects, pie charts, Venn diagrams and interviews.

Following a week of field work, the PLLA teams regrouped in Mopti to share preliminary results, triangulate findings and identify holes in data requiring further follow up. The teams then returned to Tombonkan, Nérékoro and Madiama in order to pose remaining research questions and to share their results at village-wide meetings. Information sharing is a key tenet of participatory research, because it demonstrates respect and equality between the outside researchers and local partners and provides an opportunity for local participants to respond to any key findings, clarify misunderstandings and ask the research team questions. Subsequent to these meetings, the teams reassembled once again in Mopti and began the written synthesis of their work.

Although in close physical proximity to each other, the ethnic composition of the three research sites varies. Madiama and Tombonkan are both composed primarily of sedentary agriculturalists dominated by the Marka ethnic group. Animal husbandry is practised at all three sites. Nérékoro, in contrast, is a sedentary village, whose composition is Peul (Fulani or Fulbe) herders and Rimaibé farmers. (The Rimaibé live in association with Peul and are culturally very similar. Historically however, the Rimaibé were Peul slaves.) Although Nérékoro is a permanent village, typically in Peul communities large numbers of men make seasonal migrations with their cattle, following the rains to verdant pastures.

The key research objectives identified for the Madiama PLLA included:

1. Seek to understand the landscape/lifescape interactions in the study area by focusing on:
- natural resource use, access and control
- food security
- poverty
- conflict related to natural resource use

2. Assess NRM practices and incentives that motivate these practices under varied circumstances.

3. Identify conflicts related to natural resource use.

4. Investigate natural resource conflict management strategies in historical, current and future scenarios.

5. Identify decisions made by natural resource decision makers at the communal level in historical, current and future contexts.

6. Prioritize research questions and information needs.

All three participating communities (Madiama, Tombonkan and Nérékoro) shared similar concerns and research priorities. The citizens of Madiama and Tombonkan (the two communities dominated by sedentary agriculturalists with some animal husbandry activities) expressed a desire: (i) to investigate how to improve agricultural and animal production; (ii) to examine how economic activities might be diversified in order to alleviate poverty; and (iii) to enhance food security. Tombonkan and Nérékoro (the two smallest communities) indicated a desire to investigate methods for mitigating inter-village conflict. Although the village elders in Madiama did not identify enhancing conflict resolution practices, the research team strongly concurred that this topic merited further investigation given the level of animosity between the communities, long smouldering disputes, severely constrained natural resources, and perceived ownership of the same resources by both parties. Additionally, the village chief at the time gave the impression of relative intolerance regarding resource use conflicts with his Peul neighbours.

In addition to these topics, each community highlighted its need for more water both for human consumption and animal husbandry. During the research period many of the wells in the village were already dry several months prior to the beginning of the rainy season. The research team recommended that SANREM possibly investigate area hydrology and groundwater availability.

Key PLLA Findings

The PLLA yielded several key findings. A summary of these results is presented in this chapter. However, the comprehensive report containing all of the data is presented in SANREM-CRSP West Africa Working Paper No. 99-02, *Participatory Landscape/ Lifescape Appraisal: Synthese du Diagnotic Participatif Realise a Madiama, Nérékoro et Tombonkan du 1er au 14 fevrier 1999*. The summarized PLLA results that follow are presented by village and research objective in the voice of the villagers.

Madiama (sedentary farming village)

Understanding landscape/lifescape interactions

Most of the available land within Madiama's village limits is used for agricultural production. Fallow is not practised. Extensive agriculture is conducted using animal traction and there is insufficient arable land. All available land is under cultivation regardless of soil fertility. Grazing land has diminished as field sizes have increased. Madiama farmers determine access and control issues regarding natural resource management in their territory.

During a poor rainfall year, 80% of the village population is food insecure. Various strategies are employed to meet their food needs, including selling small ruminants and engaging in small commercial activities. Twenty per cent of the population is food secure regardless of rainfall and harvest. According to the villagers' self-assessment, 20% of the population is rich; 60% is intermediate and 20% is poor. Eighty per cent of the village, or the rich and middle classes, possess animal traction equipment. The village infrastructure is well developed and includes a new school, a village health centre, a birthing centre, a pharmacy, water pumps and a grain mill. Revenue sources are quite diversified. Watermelon is a significant cash crop in the area.

NRM practices

The area is characterized by very poor soil fertility. Fallow is not practised; the same parcel is cultivated year after year without crop rotation. Intercropping is practised, most notably with cowpeas and cereals such as millet or sorghum. Organic fertilizer (manure) is used to as great an extent as possible in dryland farming. Chemical fertilizers are used in irrigated rice plots. Animal traction is used in 80% of dryland farming. The remaining 20% is cultivated using hand-held hoes. Fields are borrowed from neighbouring villages, particularly Kessedougou, Doukourani and Teguegne, for cultivation.

Factors that influence agricultural production decisions and NRM strategies include: soil type, anticipated rainfall (based on the duration and intensity of the cold season), family size, farming equipment, market prices for cash crops and social factors (e.g. bride price, funerals, etc.).

Natural resource conflicts

Conflicts between farmers and herders are the most frequent and contentious in relation to natural resources. These involve fields, pasture, water (seasonal wetland/ pond) and trees. Damage in Madiama farm fields caused by Nérékoro animal herds was the primary source of conflict identified. Passage routes for animals, either from neighbouring villages or further away, have diminished due to increased extensive agriculture that is the result of declining soil fertility and the availability of animal traction. Competition for a seasonal wetland/pond is intense between Madiama and Nérékoro.

There are some conflicts between farmers within Madiama due to land tenure, but they are settled amicably.

Conflict management strategies

Historically, the village chief and his counsellors managed NRM conflicts, and currently, they resolve most NRM conflicts at the village level. However, if the conflict is not settled at this level, it will continue to the Chef d'Arrondissement level (former pre-commune administrative unit), then the Justice Ministry at the *cercle* level (Djenné) or the regional authorities in Mopti. The Madiama village leaders currently see no areas for improvement in the conflict management strategies, despite a long-simmering conflict with the residents of nearby Nérékoro.

Decisions made by commune-level authorities

In historical and current contexts, supra-village decision making has not influenced natural resource management practices in Madiama. Madiama residents have heard of decentralization, but claim to be unaware of any of the details of it.

Nérékoro (sedentary herder/farmer village)

Understanding landscape/lifescape interactions

The majority of land within Nérékoro's village limits is used for agricultural production (i.e. cultivated fields). The size of pasture areas has decreased both within and outside of Nérékoro. The Nérékoro residents determine access and control issues in their immediate village territory. However, outside of Nérékoro during nomadic pastoral periods, sedentary farmers have greater access and control than transhumant herders.

Food insecurity is a common problem in Nérékoro. Strategies employed to respond to this problem include selling cattle, reducing the number of meals consumed per day, and increasing consumption of wild foods. Village self-analysis indicates high levels of poverty (80% of Peul and 95% of Rimaibé). However, it is possible that the indices used to establish wealth ranking were based on ideal circumstances and, therefore, not relative to current realities. Infrastructure in Nérékoro is poor or nonexistent. The villagers rely on the school and health centre in Madiama for education and health services. The primary revenue source in Nérékoro is animal sales.

NRM practices

Soil fertility is low, but fallow is practised, generally for 3-year periods. Organic fertilizer (manure) is used in fields, to the extent it is available. Zai (anti-erosion, moisture retaining rock structures) are employed in some farm fields. Pastures are not managed, and herders move their animals in order to maximize the health of the herd and not the state of the pasture. Factors that influence production strategies and NRM practices include maximum use of available space and maximal herd size.

Natural resource conflicts

The primary natural resource-related conflicts the Nérékoro inhabitants identified were between farmers outside of Nérékoro proper and Nérékoro herders. The reasons for these conflicts included: increased areas under cultivation, reduced pasture areas, reduced animal passage routes, damage caused in fields by cattle and competition for the seasonal wetland/pond.

Conflict management strategies

Historically, the village chief and his counsellors managed NRM conflicts. If a conflict could not be resolved at that level, it was taken to the Chef de Canton. Following independence unresolved conflicts were taken to the Chef d'Arrondissement. Currently, NRM conflicts are managed by a variety of structures including the village chief and counsellors, the Chef d'Arrondissement and the Justice Ministry at the *Cercle* level (Djenné) or regional authorities in Mopti.

A variety of practices are employed to

mitigate conflict. They include taxes for grazing rights, herder cooperatives, establishment of a pastoral calendar regarding herd movement, and animal corrals during the rainy season. Scenarios regarding more ideal management of natural resource conflicts include inter-village conflict management structures, adherence to the government-determined pastoral calendar, and a more just tax system.

Decisions made by commune-level authorities

Decision-makers historically included the village chief within the village. Tax systems for grazing rights in the *bourgou*[1] were implemented, but access to forest products and water was free. Supra-village level decisions now include establishment of the pastoral calendar by the Malian government, wood cutting permits issued by the administration, and tax systems established by other authorities (i.e. river crossing fees by the government, *bourgou* grazing rights by residential villages and field damage fees levied on implicated villages). Ideally NRM commune level decision-making should include recognition of Nérékoro land tenure rights, Nérékoro-authorized natural resource use in their territory and the creation of a fair tax system relative to animal passage and potential damage caused by herds in fields.

Tombonkan (sedentary agricultural community)

Understanding landscape/lifescape interactions

Tombonkan is experiencing severe land degradation. Nearly all land is under cultivation without the possibility of extension. Pastureland has been reduced to a minimum. There are insufficient permanent wetlands and only seven wells in the community. There is no forest and little wildlife. The founding ethnic group, the Marka (agropastoralists), have most access to and control of natural resources. Women have no control over natural resources. The Djoula ethnic group is the most prosperous. Eighty per cent of the population own agricultural equipment and there are more than 150 traction animals in Tombonkan.

Cereals compose the base of food production and consumption. The main sources of protein are fish and beans. The human hunger season lasts from August to the beginning of September, while that for livestock is from May to July. Agriculture is the principal source of revenue. Animal husbandry is a strategy for saving money. Tombonkan has poor socio-economic infrastructure, and the nearest school is 5 km away in Madiama.

NRM practices

All the inhabitants of Tombonkan farm, but they also fish, raise livestock, make crafts and engage in petty commerce. Women collect and process local fruits and participate in the more labour intensive planting and harvesting activities. Monoculture dominates dryland farming. Millet and sorghum are planted in fields based on soil type. Crop rotation and cereal–legume intercropping are infrequently practised. Extensive animal traction is the mode of cultivation. Tombonkan farmers use early-germinating varieties. Famine foods are cultivated. Rice is produced in the irrigated sector outside of the Commune, and in lowlands where water is retained longer. Fishing occurs in those same neighbouring areas.

Two types of livestock production, nomadic and sedentary, are practised. Sedentary livestock are kept in corrals due to the lack of local pasture. Crop residues and *bourgou* grass constitute the forage stocks. There is localized application of manure and compost. Participation in NRM is gender-dependent. The hunters' association oversees environmental protection.

Natural resource conflicts

Disputes between farmers occur over location of territorial boundaries, borrowed land, and use of wetland areas. Farmer–herder disputes (the most frequent form of

Fig. 6.1. Drawing a map of village resources.

conflict) involve livestock damage to crops, use of *bourgou* grass by the farmers (harvested and stored as fodder), and cultivation of former and current livestock passage routes. Women do not have secure tenure on the land they borrow.

There are also territorial disputes about land tenure between villages. Villagers from Madiama and Promani have borrowed land from Tombonkan households that have requested its return. The dispute with Promani has gone all the way to the Ministry of Justice. Disputes also include access to *bourgou* grass in the wetlands outside of the village territory.

Conflict management strategies

Historically, conflicts have been managed through the application of customary law. Colonial law regulated disputes as well, but much Islamic and traditional legislation of the *Dina* (established under Sekou Amadou) governs resource access and control. Currently, government extension agencies and the government administration apply modern legislation. The government-determined calendar of livestock passage to the *bourgou* pasture that is similar to the *Dina* legislation is a major controlling factor.

Ideally, it would be best if spaces were delimited between different territories and between different production systems. Agricultural intensification could free up more space for livestock production. Salt licks and manufactured animal feed could be used to intensify animal production. The former wetland areas could be managed in order to reintroduce *bourgou* cultivation and fish. Alternatively, soybeans could be produced for bouillon cube production to replace *sumbala*, a traditional bouillon made from Néré tree pods (*Parkia biglobosa*). Reforestation is also required. Improvements could be made through collaboration among different partners as well as education and information on land use management.

Decisions made by commune-level authorities

In the past, the heads of households, the village chief and his advisers, or the Chef de Canton in Promani made decisions. Today, the village counsel has more authority over heads of households and contested administrative decisions need to be referred to the justice system for conciliation. Ideally, responsible commune-level decision-makers should be chosen judiciously, because this would help avoid improper influence.

Fig. 6.2. Discussing women's nature resource management practices.

Primary Problems Identified by Villagers

As part of the participatory research process, the relevant villages were asked to identify and prioritize the issues discussed during the previous week's work. The main categories of discussion included natural resources, food security, conflict and poverty. Although the priorities varied slightly from village to village, the overall areas of concern overlapped and are complementary. Information regarding the specific problems they identified and their respective prioritization by village follows.

Madiama

The village participants in Madiama were quite clear and adamant that their primary problems are related to natural resources. The top two priority problem areas identified included: (i) lack of water (rainfall for cereal crops and river floodwater levels for irrigated rice); and (ii) decreased soil fertility in relation to dryland cropping.

Nérékoro

The residents of Nérékoro prioritized their problems in the following order (from highest to lowest): (i) food insecurity; (ii) natural resources (i.e. water shortages); and (iii) conflict. Nérékoro villagers identified a widespread lack of food self-sufficiency within their community as well as a lack of production methods adapted to local conditions. In reference to natural resources, they identified water shortages as the primary concern in terms of agricultural production, domestic use and animal husbandry. Conflict, the third priority, stems from the long-simmering disagreements between Nérékoro and neighbouring Madiama as well as from the presence of sedentary farmers in areas where the Nérékoro herders travel to graze their cattle.

Tombonkan

The types of natural resource and agricultural conflicts identified in Tombonkan were similar to those of Nérékoro and Madiama. In terms of natural resources, the following problems were identified: shortage of land, reduced soil fertility, water shortages, insufficient pasture, insufficient trees, and a shortage of fish. In reference to food security, the villagers cited several of the same natural resource limitations as well as crop pests and a lack of farming equipment. In reference to conflicts, they cited as problems a shortage of arable land and poor harvests. The problem of poverty was due to insufficient water, inadequate production, and natural resource degradation.

Summary

In general, although their top priorities varied, all three participating communities shared similar concerns. The research team did not attempt to reconcile inter-village priorities. Instead, it presented the common information needs and research questions the villages identified. The citizens of Madiama, Nérékoro and Tombonkan would like to investigate improved agricultural and animal production and examine how economic activities might be diversified in order to alleviate poverty and enhance food security. Nérékoro and Tombonkan, the two smaller villages, would like to investigate methods for mitigating inter-village conflict. Although the largest village in the commune, Madiama, did not identify enhancing conflict resolution practices as a pressing concern, the research team strongly concurred that this topic merited further investigation given the current level of animosity, long smouldering disputes, severely constrained natural resources and perceived ownership of the same resources by both parties.

In addition to these topics, it should be noted that at every site the participants highlighted water shortages as a concern both in reference to agricultural productivity and human consumption. Although SANREM is not likely to get involved in well digging activities, etc., it should be made aware of these concerns and participant eagerness to resolve these problems through an investigation of area hydrology and groundwater availability.

Although the topic of conflict has been highlighted at various points in this summary, it has not been discussed in any detail with respect to decentralization and the establishment of communes. A major point of concern from a number of interested parties, including villagers, NGO representatives and local United States Agency for International Development (USAID) mission staff, is how the establishment of communes will actually be operationalized, how natural resource management decisions will be made, and who will participate in those decisions. This is of particular relevance in the Madiama commune given the village of Madiama's dominance due to population, existing infrastructures, and land area under cultivation and its recalcitrance in reference to relations with Peuls in the nearby village of Nérékoro. This particular topic of farmer–herder conflict is relevant to Mali as a whole and is applicable in other Sahelian countries as well.

Finally, it should also be noted that a wide variety of development players, including UNICEF, CARE, government agencies and research organizations, have intervened in the Madiama commune. In particular, given recent government restructuring of its environmental ministries, government agents have become involved actively soliciting village problems, needs and solution identification. Government and NGO colleagues have suggested that SANREM first investigate what research has already been conducted in the study area prior to undertaking research that may be repetitive or complementary to previous or ongoing activities.

Research themes identified as a result of the PLLA included the following:

Agricultural production

- Rice production in both irrigated and rain-fed areas, particularly in the context of decreased availability of water.
- Soil fertility for enhanced harvests and related issues of wind and water erosion reduction.
- Intensification of agricultural production.
- Fertilizer management strategies (chemical and organic) and composting.
- Pest management strategies.
- Short season varieties of cereal grains.
- Potential for soybean introduction and useful tree species as well as investiga-

tion of appropriate techniques for introducing leguminous crops.
- Disease resistant species.

Water

- Availability of ground water and methods for determining the water table level.
- Water conservation and management methods, with reference to rice cultivation, dryland cropping, animal watering and human consumption.
- Techniques for improving water quality (related to human and animal health issues).

Economic activities

- Diversification of economic activities in order to enhance household incomes and food purchasing power.
- Support for existing or moribund village credit systems and/or cereal banks.

Animal husbandry

- Animal husbandry for production of milk and meat.
- Preservation of animal products (e.g. dried milk).
- Improvement of animal health.
- Forage varieties.
- Hay conservation.
- Techniques for equitable distribution of natural resources.

Conflict

- Conflict management techniques.
- Predictive tools to avoid conflicts.
- Information regarding use and improvement of the official government-determined pastoral calendar.
- Research efficient conflict management systems (consensus building among natural resource users before establishing rules, investigating methods for setting territorial boundaries).
- Information and training in NRM.

Poverty

- Women's market gardening options in light of unavailability of water (wells go dry and pumps do not provide adequate volume).
- Improvement of agricultural credit.

Building on the Participatory Relationship

Subsequent research was built on the initial relationship developed between SANREM and the participating Madiama Commune communities and has led to core long-term learning relationships developed between researchers and community members. As a follow-up to the participatory appraisal, and in conformance with the SANREM Annual Work Plan, researchers planned and implemented more detailed formal socioeconomic surveys at the household level (see Chapters 13 and 16). Also, biophysical baseline data were collected on weather, soils, crops and crop management practices (see Chapters 3 and 11), and a local interlocutor, the Natural Resource Management Advisory Committee (NRMAC), was established to represent and communicate between the full ten villages of the commune and SANREM on a sustained basis (see Chapter 7). Several partners worked together to establish the formal work plans that the SANREM CRSP Management Entity at the University of Georgia would review, and USAID-Washington, DC, would ultimately approved. These research-for-sustainable-development partners included a locally based NGO (CARE/Djenné), the IER's regional agricultural research centre in nearby Mopti (CRRA/Mopti), Virginia Tech and Washington State University.

The CRSP work plan process entails extensive review based on technical merit, resources available and USAID policy focus. The research priorities the villagers identified were paired with the scientific expertise of faculty at Virginia Tech and Washington State University. The research topics were in turn developed into annual work plans. The SANREM CRSP Management Entity at the University of Georgia, including a panel of neutral scientists composing the Technical Committee, and USAID-Washington, DC, reviewed the work plans. USAID made the final decision regarding what level of support to provide for which projects.

During a year-long process, the work plans were approved and funding initiated. In preparation for that work plan, the newly formed NRMAC took it upon itself to conduct meetings in each village and develop a list of priorities to focus SANREM research activities. These lists, and the committee's consolidation of them, were presented to IER and US scientists at a meeting on 1 February 2000 in Madiama. All NRMAC members, development agents, technicians, and local officials were present. An extended discussion ensued and two priority areas of biophysical focus were agreed upon: (i) increasing soil fertility in croplands (see Chapter 12); and (ii) increasing the quality of pastures (see Chapters 14 and 15). The NRMAC also signalled the importance of developing the institutional capacity of the committee to provide a commune-wide decision-making infrastructure for NRM in the commune (see Chapters 7, 8 and 9). In addition, a work plan was proposed to develop agronomic and socio-economic modelling based on the priorities the committee specified (see Chapter 11).

Note

1 *Bourgou* refers to the rich grass that grows in the Niger River and its flooded delta during the rainy season. It is a highly valued and nutritious source of animal fodder. Access to the *bourgou* may be contentious.

References

Moore, K., Kaboré, D., Gnoumou, B. and Bertelsen, M. (1999) Conflict and natural resource management in agricultural and pastoral systems of arid and semi-arid regions of West Africa (Sustainable Agriculture and Natural Resource Management Collaborative Research Support Program –West Africa Site, Working Paper No. 99-01–E). Office of International Research, Education, and Development, Virginia Tech, Blacksburg, Virginia.

Neely, C., Buenavista, G. and Earl, J. (eds) (1999) *Looking Back on the Landscape: Impacts and lessons from the SANREM CRSP.* University of Georgia, Sustainable Agriculture and Natural Resource Management Collaborative Research Support Program, Watkinsville, Georgia.

SANREM CRSP – West Africa (1999) Participatory Landscape/Lifescape Appraisal: Synthese du diagnostic participatif réalisé à Madiama, Nérékoro, et Tombonkan du 1er au 14 février 1999 (Sustainable Agriculture and Natural Resource Management Collaborative Research Support Program – West Africa Site, Working Paper No. 99-02). Office of International Research, Education, and Development, Virginia Tech, Blacksburg, Virginia.

7 Building Social Infrastructure for Sustainable Development

Keith M. Moore,[1] Salmana Cissé[2] and Abdoulaye Touré[3]

[1]Office of International Research, Education and Development, Virginia Tech, Blacksburg, VA 24061–0378, USA; [2]Centre Régionale de la Recherche Agronomique, Institut d'Economie Rurale, BP 205, Mopti, Mali; and [3]CARE/Tombouctou, Mali

---

Decentralization has changed the dynamics of natural resource management (NRM), but it has not yet yielded a methodology for effective local governance in the Sahel. In response to this insufficiency, the SANREM CRSP initiated a programme of action research to assist a local population in determining a new modality for management, in accordance with its environmental and socio-cultural context.

We began with the idea that a new platform for inclusive community action was required (Pretty and Buck, 2002; Vedeld, 1994) based on the premise that it was insufficient simply to revive old institutions and traditions or to work with one group of stakeholders at the expense of another. Recognizing that group decision makers needed to be targeted, we set out to build a social infrastructure (Flora, 1998) that could shape social capital for rural civil society in an image of modern Africa. The objective was to increase the horizontal linkages between stakeholders in the belief that such a common action platform can increase the chances that diverse ideas and people can: (i) increase open debate; (ii) increase the flow of information; and (iii) lead to more carefully considered management decisions and, as a consequence, increase trust between communities.

This chapter describes the establishment and operation of such an institutional innovation in the socio-political organization of a rural Sahelian population. This institutional innovation, the Natural Resource Management Advisory Committee (NRMAC), is an inter-village and pluriethnic civil society organization (CSO) in the 5th Region of Mali. The primary hypothesis of this research was that sustainable, decentralized management of natural resources can be achieved if a local population is equipped with a methodology for managing their natural resources and the conflicts associated with the exploitation of these resources, and a local organization to instigate inter-ethnic and multi-village dialogue.

Beginning with a description of the institutional context in which the population of Madiama is situated, we then present the constituent components of the NRMAC and the process of its establishment. The discussion highlights the significance of the organization's legitimacy, based as much in customary practice as in modern legal formalities. It shows that associational life is not only a matter of gathering people together – men and women, regardless of ethnic or socio-professional status, but also requires the development of individual capacities (training in functional literacy, association management, financial management, knowledge of NRM texts, laws and codes, management and reconciliation of the conflicts, etc.) in order for them to assume active roles in the development of their commune and to support the development of extra-communal relations and, in particular, the role played by training in conflict management, given its importance for the development of self-confidence and building of credibility for the committee in the development and the safeguarding of the commune's natural resources.

The Institutional Environment

This social infrastructure is designed to build linkages within the local institutional environment and bridges to external organizations and stakeholders (see Fig. 7.1). Two bases of institutional power in rural areas have vied for control of resource mobilization for development and environmental preservation. The infrastructure of state power has centred on the Commandant of the *Cercle* and the Chef de Arrondissement (*Sous-Prefet*), more closely associated with the local population. The village chiefs and other assorted resource chiefs, who control the immediate allocation of resources for household livelihoods, have held customary power. Overall, villagers have had little say in resource allocation or development initiatives.

The resources of the state have been mobilized by an array of development services that have attempted to assure environmental protection through threat of law and development initiatives designed to extract surpluses for the nation as a whole. For instance, the *Office Riz* has been a major state initiative (reformulated from time-to-time) to enhance the productive power of the peasantry in the production of rice. This effort has been implemented by organizing farmers into village rice growing associations based on traditional village associations. These associations work state lands, flooded annually by the *Office*-controlled

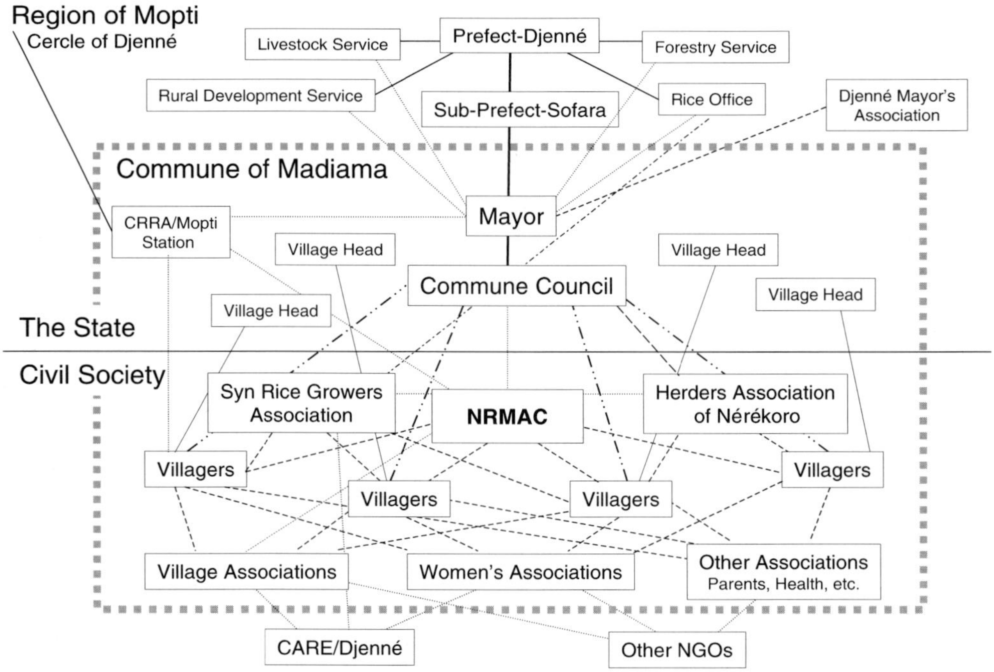

Fig. 7.1. Network of relationships between the population, government and civil society.

irrigation works. The livestock service monitors cattle health and the movement of herds within the zone. The forestry service polices the cutting of wood for fuel and timber to assure that over harvesting does not occur, frequently levying fines and collecting fees for wood cutting. The Centre Régional de Recherche Agronomique (CRRA)/Mopti is charged with developing new and adapted technologies to increase production on the farm and in the river waters.

Rural civil society has been restricted for the most part to the village level. Village associations are common among men and have provided a point of contact for state development efforts. Women's associations have been largely avoided until recently. The Herders' Association of Nérékoro was established as a local organization to protect the interests of migrant (transhumant) herders who have local bases within the region, but little or no administrative representation.

Since the end of the Traoré regime in 1991, the establishment of Rural Communes has brought government closer to the population while ostensibly empowering the people to improve their livelihoods. Supporting this movement, non-governmental organizations (NGOs) have provided an alternative source of resources that are stimulating new development initiatives and organizational skills development and providing incentives for self-improvement.

The communal elections in June of 1999 brought the first elected officials to live and work in the rural communes of Mali. Although these officials were elected locally, their names were drawn from national party lists and they are the government's voice at the local level. Being more closely in touch with the population, their representational experiences direct them to be more responsive to local concerns. Time will tell whether these elected officials will be able to be truly responsive to the expressed concerns of the populations they serve.

The most important services that the mayors now provide are simple administrative documentation (births, deaths, etc.) and the collection of local taxes. The mayor's office also provides a more direct connection to government services and advocacy for improving those services. The Commune mayors in the *Cercle* of Djenné also have their own association that helps them to better understand their new roles. Into this mixture, a new player, the NRMAC, has been introduced to support decentralized decision-making in NRM.

Establishment of the NRMAC

In September 1999, a SANREM CRSP-sponsored delegation of national agricultural researchers, the newly elected mayor of the Commune, and representatives of the local Mopti Rice Office and a World Bank NRM Project visited all ten villages in the Commune of Madiama. In each village, the chief and a group of his counsellors were informed about the objectives, value and role of participation in Village NRM User Groups and a commune level NRMAC.

The members of the delegation explained that the primary purpose of the NRMAC was to provide a forum for reflection on NRM in order to improve communal resource management. Village NRM User Groups would provide an essential link for communicating technological innovations to other groups in the Commune. The NRMAC would provide a network through which SANREM CRSP researchers could learn about commune priorities, technology, and information needs. In addition, it would help the local population improve the prevention, mitigation, and resolution of NRM conflicts and develop a plan for the rational use of natural resources within the commune. Committee participation would lead to improved leadership skills in working with technical assistance in planning and implementing NRM activities.

In each village, the chief held a gathering to select a diverse group of four or five delegates to represent the village in a commune-wide general assembly to elect the NRMAC. According to the relative importance of the activity to village livelihoods, either two herders, two farmers, or one of each were selected. In addition, two or three more villagers were selected to represent women, hunters and crafts/forest gatherers.

User groups had already been developed in three of the ten villages where an earlier participatory appraisal had been conducted. Three World Bank NRM Project (PGRN) villages had already developed their own Village NRM Committees that met and selected their representatives. Because the four additional villages were completely new to the idea of village NRM committees, their village meetings were more extensive. Two of these villages initially declined to participate, but one village later sent four representatives to the General Assembly. Each of the nine participating villages and the local irrigation management committee sent three to five representatives (45 in all, including seven women) to the General Assembly meeting held on 10 October 1999.

The Mayor of the Commune opened the General Assembly with the 45 village representatives and an additional 25 participants, including researchers and representatives from an NGO, a PGRN, the deconcentrated government services, Commune Council and the Sub-Prefect of Sofara in attendance. The anticipated objectives, role, and structure of the NRMAC were again described. Translations into the two local languages (Peular and Bambara) were provided. Participants were divided into four discussion groups (organizational and administrative issues; dryland farming, rice farming and fishing issues; livestock, hunting, gathering and craft issues; and the role of women) and enjoined to debate the concerns and priorities of the commune with respect to each topic. After these groups reported their conclusions at the plenary session, all research, technical service, NGO and elected officials withdrew and elections were held for positions on the NRMAC. This elected committee was made up of 11 men and two women. The village that had originally declined to participate for political reasons sent a representative to join the committee after it had been established.

Legitimacy of the NRMAC

Four factors contribute to the legitimacy of the NRMAC as a viable community

Fig. 7.2. The Natural Resource Management Advisory Committee of Madiama.

organization. They are all based on the foundation of the participatory approach developed between researchers and the community for establishing and operating the committee. The first involves gaining formal recognition of the committee as a legal entity. The second is the establishment of relationships with stakeholders and partners in the association's environment. The third is serving a valued purpose for the community. The fourth is the re-election of committee members after their first term had been served.

Formal recognition of the association

In order to become a legally recognized association, the NRMAC was required to be in conformity with national laws concerning associations and cooperative NGOs. Because legal recognition is based on an association's internal structure and governance, the General Assembly of the Village NRM User Group members drafted, discussed, and passed by-laws that the judicial authorities, in turn, formally approved. The NRMAC then informed the Prefect in Djenné of its intent to form the association and followed this with a formal request for legal recognition (including all official stamps). Initially, the Prefect refused the association's request, because the domain of the NRMAC activities fell within the range of authorities devolved to the Commune Council. Therefore, the NRMAC requested that the Mayor, who had assisted in the process of associational development, send the Prefect a letter in support of formal recognition. The Mayor obliged and, upon receipt of that letter, the Prefect approved their request. The NRMAC was formally registered as an association by the *Cercle* of Djenné on 23 October 2001.

Relations with other associations, technical services and villages

Partnerships have been developed to provide a framework for productive relationships. These partnerships are either formal ones, conforming to the standards of national civil society, or informal ones, based on shared understandings of customary practice at the village level.

The first step in developing formal agreements of co-operation/collaboration was the signing of a protocol with CARE/Djenné, the NGO providing the NRMAC with institutional development training and assistance. Once formally recognized as a registered civil society organization, the NRMAC also signed a partnership agreement with the Commune Council. This document, perhaps the most significant of the NRMAC's formal protocols, provides a framework for the NRMAC: (i) to influence NRM decisions within the commune; (ii) to be consulted concerning decisions of the Commune Council; (iii) to be recognized throughout the commune as a significant player in the resolution of conflicts linked to natural resources; and (iv) to actively participate in the economic development of the commune.

The NRMAC is pro-actively exploring other formal partnerships in order to better serve the population of Madiama. To develop collaborative relations, the NRMAC invited representatives of the technical services serving the *Cercle* of Djenné to recent meetings. For example, members of the NRMAC are interested in building a relationship with the Service Locale de la Réglementation et du Contrôle (SLRC), which is charged with protecting the forestry resources of the *Cercle*. The SLRC, in turn, is looking for partners to interface with and help it implement its activities at the local level.

The second type of partnership builds collaborative relationships with customary authorities and villagers. Although not formally documented as are those in modern civil society, these relationships have been formed while conducting activities that involve the key village actor, the village chief. This kind of partnership began with the establishment of Village NRM User Groups under the direction of the village chiefs. By sending representatives to the General Assembly to elect the NRMAC, the chiefs have in effect confirmed the legitimacy of the NRMAC. The reticence of certain chiefs to designate village members to participate in the initial General Assembly of the association bears witness to the validating role chiefs play. This form of legitimacy is fragile and ad hoc (or arbitrary) in nature. Unlike formal, documented recognition, it may be withdrawn at any moment.

Frequent communication with all partners is essential for effective organizational functioning. However, this communication is more than a matter of transferring information, because it involves continually renewing understandings between the NRMAC and the village chiefs. NRMAC members have regularly kept the village chiefs and Village NRM User Groups informed of their activities, the training programmes they participate in, and the research activities they monitor. In addition, NRMAC trainers have conducted several training sessions, first initiated as training-of-trainers by outside experts, at the village level.

A valued purpose

Unless the NRMAC does something valued by the community, community members have difficulty understanding why the association exists and why it should be of any concern to them. Built on the priority concerns of villagers, the NRMAC's mission has been to promote the management of natural resources in the commune by introducing and adapting technologies to local conditions so that the population can improve their standard of living. An essential element in this mission involves the management of conflicts generated in the use of natural resources by various community members. By providing such services (e.g. protecting and planting trees, resolving conflicts, and introducing new technologies), the NRMAC legitimizes itself in the eyes of local leaders and villagers.

Re-election of committee members

Validation of this legitimacy was recently witnessed with the re-election of the NRMAC for another 3-year mandate. Announcement of the process for re-electing NRMAC members was circulated throughout the commune, in each village, and on rural radio. As when it was first constituted, the process started in each village of the commune. The village chiefs re-established the Village NRM User Groups, and five representatives from each of the ten villages were sent to a General Assembly at the commune seat of Madiama. At this Assembly, presided over by the Mayor and the Sub-Prefect, the NRMAC President and the NRMAC Executive Bureau presented an activity and financial report of their accomplishments during the past 3 years. After a question and answer session, all the NRMAC members resigned. Following an open debate and consideration of their experience and training, all members of the committee were re-elected by acclamation, thereby renewing their mandate. As an additional outcome of the debate, a commission of peers that includes a representative from the Commune Council, the village chiefs, and other customary/religious leaders, the CRRA/Mopti, and NGOs working in the area will monitor this new term of office. With this renewed mandate, the NRMAC is ready for another 3 years of activities.

Training Received

The NRMAC has been provided with considerable training to accomplish its tasks. The SANREM CRSP has provided this training and the associated research activities through its partners in Mali, CARE/Djenné and the CRRA/Mopti. Even before receiving formal recognition as an association of the commune, the NRMAC visited each village to develop a list of priorities to help focus the SANREM research activities. These lists, and the committee's consolidation of them, were presented to CRRA/Mopti and USA researchers at a meeting on 1 February 2000 in Madiama. An extended discussion ensued during which the committee prioritized two biophysical themes, improved soil fertility in croplands and improved pasture management. The committee also stressed the importance of reinforcing its organizational capacities. These priorities formed the basis for the initial research and outreach relationship.

The training that the NRMAC has received can be grouped into three broad categories: (i) holistic management (HM) of natural resources; (ii) conflict management and resolution; and (iii) development of organizational and leadership skills. All training activities were developed and extended in a training-of-trainers format. In addition, a few key leaders of the NRMAC have profited from study tours. Four members went to Chad to observe an ongoing programme of HM of rotational grazing on open rangelands; two members travelled to Niger to observe a programme of conflict resolution through cattle trail delineation and maintenance; and one member attended a meeting of regional non-governmental organization representatives.

The first training introduced NRMAC members and their local technical assistance partners to the principles of HM. Six subsequent HM training workshops focused on applying these principles to evaluate on-farm research trials, establish wetlands management, and develop an HM grazing system for open rangelands. HM training has been coupled with the conflict management training from the outset. The five conflict management workshops held for the committee focused on consensus building, managing power, managing change, and adapting this training to the management of wetlands and open rangeland grazing. Since the initial workshops, the lessons learned at these training sessions have been routinely communicated at the village level, at first under the supervision of the SANREM trainers. However, the burden of this communication has increasingly shifted to NRMAC trainers. To date, NRMAC trainers have designed and implemented two HM and three conflict resolution workshops at the village level, in addition to hosting a workshop for Commune level representatives across the *Cercle* of Djenné.

CARE/Djenné has provided the committee with institutional reinforcement in the form of organizational and leadership training in two formats, formal training workshops and informal assistance and exchange visits. Institutional reinforcement training began with an institutional analysis to identify organizational strengths and weaknesses. Based on this analysis, an institutional development plan was drafted. Functional literacy training was a core training element and focused on concepts of democratic governance along with basic lessons in financial management and accounting. Subsequent training workshops addressed strategic planning, understanding NRM guiding texts, codes, and laws, decentralization codes and laws, and lobbying. The NRMAC benefited from the training, because a framework was created for exchange and dialogue between the NRMAC and other local actors involved directly or indirectly in NRM. The NRMAC also profited from the support of CARE in the promotion of the inter-village negotiations for establishing *bourgoutière's* (seasonal wetland basins) management agreements. Additional institutional coaching has included assistance in the drafting of by-laws and registration of the association, financial monitoring, seeking external funds, and building linkages with other local services and NGOs.

Services Provided

The NRMAC has served as an interface for the commune with government services. However, the NRMAC is not limited to this function alone. During its first 3 years of existence, the committee has initiated on its own many additional activities that are detailed below.

Monitoring research trials

Improved soil fertility

During the first year, local chiefs and village residents in collaboration with researchers set up user groups in three target villages. Each group included the village chief and at least two women. These Village NRM User Groups chose collaborating farmers for the field tests that included benchmark characterization studies and soybean and fertilizer trials. Contact between researchers and farmers and herders was monitored during the season, and formal meetings were held to discuss the achievements, expectations and difficulties of NR users at these sites.

Researchers had initially targeted the on-farm trials in a limited number of villages, but at the outset of the second year, the NRMAC made certain that each of the ten Village NRM User Groups participated. On-farm trials involved the application of rock phosphate (PNT), manure amendments, micro-dosing with inorganic fertilizer, field stabling, and the intercropping of millet and cowpeas or the rotation of millet or sorghum with cowpeas. Trial sites located in each village have allowed for participation by all community members. Although the farmer collaborators did not manage all plots successfully, committee members learned from this lesson and more closely monitored the quality of farmer participation in subsequent years. These trials have provided a focus for addressing issues of increasing soil fertility within the Commune.

Pasture improvement

Given the complexity of coordinating the management of communal pasturelands, pasture improvement research began more slowly, by building rapport and establishing common objectives within the community at both the commune and village levels. The NRMAC and collaborating researchers identified those management practices that led to a general regression of plant communities to the level of annual grasses and other species of low-forage quality that severely limit the forage potential. Biophysical research began with four rainy season exclosures (two in the village of Siragourou and two in the village of Nérékoro) to demonstrate pasture regeneration capacities and measure the available seed stock. This work was followed by the establishment of two open-range rotational grazing sites (in Torokoro and Siragourou) and ensilage trials to optimize forage resources for women based on *Cassia tora*, a prolific and noxious weed.

Early on the NRMAC identified the regeneration of *bourgou* (a wetland grass highly valued as forage) as a high-payoff investment. *Bourgoutière* regeneration is

Fig. 7.3. NRMAC members and researchers planning an on-farm trial.

more complex than communal dryland pasture regeneration, because the water resource on which it is founded is highly prized for agriculture and for livestock. Access to wetlands requires careful negotiation among all resource users. Of the four villages that embarked on this process only two have advanced to the stage of physical implementation and are attempting to master regeneration techniques (see 'Agreement development' below). Successful regeneration is also dependent on annual flooding, which did not occur last year.

Information exchange

Since the primary method of information exchange is through NRMAC meetings, the committee holds business meetings on the last Sunday of each month and less routinely animates training workshops at the village level. The Mayor or his representative routinely participates as an ex-officio member in these commune-level events. The village representative on the committee reports the information and/or issues discussed in these meetings to the village chief and the Village NRM User Groups. Village chiefs report being well informed of NRMAC activities.

A considerable amount of information filters through the NRMAC. Although much of it involves the routine administrative activities of the association, the NRMAC is also the interface between government services associated with NRM and the commune. This privileged position allows the NRMAC to learn about new techniques and technologies, gain improved understanding of the guiding documents, codes and laws concerning decentralization and NRM, and develop skills in the management of community affairs, including conflict prevention and management.

NRMAC members have been trained as trainers in HM and conflict management and resolution. They have appropriated this training and developed their own training modules that are often based on a local theatre format. This training programme has been conducted in four villages.

The NRMAC has also conducted four information and awareness building programmes on the local radio station (with three rebroadcasts to date). Five committee members, including two women, led each programme, with a total of ten committee members participating. These programmes have described the NRMAC mission as well as its objectives, activities, accomplishments and partners. Specific themes have included NRMAC leadership of the Mayor's campaign to protect the *Acacia albida* (Valanzan), local agreements for the promotion of *bougoutière* regeneration, and issues of decentralized administration that were conducted with the Mayor and a representative of the administration.

The NRMAC recognizes that communication is the key to its continued viability as a functioning commune level organization. NRMAC members are not equally proficient at reporting back to their villages, due in part to their educational levels, and consequently, the overall quality of these communications has suffered. Indeed, observers from the villages to the local administration in Djenné have noted that the poor flow of information has severely hampered the entire decentralization process. To improve their capacity to address this issue, the NRMAC, with the assistance of CARE, has submitted, and had accepted, a proposal to the *Fondation de France* for technical assistance in developing a communication strategy to better disseminate their messages. This strategy development was recently completed.

Forestry services

Early in the life of the NRMAC, the Mayor called upon it to assist him in the promotion of a national campaign to protect the *Acacia albida*. Drawing on the network of Village NRM User Groups, the NRMAC was able to disseminate the message quickly. This action resulted in a considerable decrease in damage to this nitrogen-fixing tree.

The NRMAC has also taken the lead in reforestation by purchasing and planting trees. Six villages had sufficient water at the time of planting to assure tree establishment and consequently requested a total of 149 trees of three species (neem, baobab and néré) to be planted in their *terroirs*. The Village NRM User Groups were responsible for tree planting and watering until these trees were fully established. The head of the Service de la Conservation de la Nature was impressed with the results of their independent action and assured them that any assistance they might need in the future for similar ventures would be available.

Agreement development

The NRMAC has initiated dialogues with selected villages in order to develop agreements for the regeneration of *bourgou* in certain seasonal basins in the commune of Madiama. These stakeholder negotiations were initiated in December 2000 in four villages, but due to a lack of consensus in one of the villages, only three villages were retained for the *bourgou* regeneration programme. Negotiations were also begun with neighbouring villages and stakeholders in order to establish local agreements governing the sustainable exploitation of these basins. These dialogues proceeded under the aegis of the NRMAC, which sees this activity as a core component of its mission, and the Mayor with the assistance of CARE/Djenné and CRRA/Mopti.

The objective of these agreements is to minimize conflicts between wetland users, improve the management of pastures, and develop the financial resources to maintain them. The agreements define the parameters of collaboration as well as the roles and responsibilities of each party. In the course of these discussions, it was determined that one of the *bourgoutière*, while constituting part of the customary *terroir* of the village, was in fact currently managed by the Casier de Bougoula (local water management association), which only allowed for the production of rice. Two multi-village agreements have been drafted, but due to the lack of flooding in 2002, no progress in implementation has been made.

Conflict resolution

Violent conflict in the Commune of Madiama has decreased over the past three years. Although difficult to verify objectively, community members claim that this is, at least in part, due to the awareness building of the NRMAC. The local population deeply appreciates the ability to resolve conflicts locally (i.e. without recourse to the administration).

On two occasions, the NRMAC has been called upon to intervene in local conflicts. The first occasion involved the early entry of cattle into the commune. This incident was the result of a need for water for the cattle and did not actually involve cattle entering unharvested fields. However, other herders in the commune were not happy about this breaking of the agreement concerning the date of entry, because their herds were still forced to remain outside of the commune. After informing the Mayor of the unauthorized entry into the commune, the other herders were ready to call the *gendarmerie* (of the *Cercle*). However, NRMAC members spoke with the principals in the conflict and negotiated a resolution, thereby avoiding the participation of the authorities.

Fig. 7.4. NRMAC President Sao clarifies a point during a meeting.

The second incident involved a fight between a Peul and a Marka in the village of Promani. When one of them was seriously wounded, the village chief called *gendarmerie* and the aggressor was taken to jail in Djenné. It was only after this incident that the NRMAC was called in. Although at this point there was still considerable animosity between the two families, the NRMAC was able to negotiate an entente between them. This base enabled the NRMAC to convince the family of the wounded participant to withdraw the charges against his assailant and get him released from jail.

Resource mobilization

The NRMAC has benefited from both technical and financial assistance through the SANREM CRSP project. However, the committee recognizes that this is really only start-up capital. It must be able to generate its own resources if it is to continue providing services to the community.

Internal

NRMAC membership cards have been designed and printed at the expense of the association to provide a credible presentation of the association, and to generate funds through a one-time membership fee. By the middle of 2003, over 250 persons had paid the 500 FCFA fee.

External

Learning of an opportunity to request funding from the Fondation de France, the NRMAC met to discuss what to propose. Initially a set of income generating activities, including animal vaccinations, soap production by women and techniques for seed multiplication, were suggested. After further discussion, however, it was concluded that its NRM mission would be best served through assistance in the development of a communication strategy for the association. With the assistance of CARE/Djenné, the NRMAC submitted a proposal. The Fondation de France tentatively accepted this proposal pending presentation of further information. The NRMAC identified a local consultant and resubmitted their application with the additional information requested and the consultant's CV. The *Fondation* then approved the proposal, and the consultancy has been completed.

Summary

The NRMAC has begun to mature as an organization at the service of civil society in the Commune of Madiama. It forms the core of a network linking socio-professional groups, village leaders, locally elected officials and government services. Built on a foundation based in both modern legal traditions and customary village law, the NRMAC is on the cutting edge of the transformation in rural civil society in Mali during this era of decentralization. It has provided space for dialogue between villages and ethnic groups and is building the confidence to address sensitive issues involving resource allocation during this period of socio-economic transformation.

NRMAC members see communication as key to NRMAC viability as a CSO. Linkages with CRRA/Mopti research trials, other service providers, NGOs, the village chiefs and the Commune Council place it in the centre of an important network of NRM decision makers. With its members trained not only in the management of their organization, but also in how to provide leadership for other community groups at the village and commune levels, the committee has assumed a leadership role in disseminating information concerning new technologies, innovative approaches to community based NRM, decentralization, tree planting and other issues of natural resource conservation. Members have dealt with conflict situations and facilitated their resolution. They have also coordinated the establishment of multi-village resource management agreements, although they have not as yet achieved the stage of full rule compliance. Nevertheless, from the villages to the

Commune Council, community members have been satisfied with NRMAC performance and have renewed their mandate.

In their article in *Science*, Dietz *et al* (2003) note that NRM governance is becoming more difficult since critical problems are at increasingly larger scales, involve non-local influences, and require adaptive capacity. They cite five capabilities necessary for adaptive governance: (i) providing information; (ii) dealing with conflicts; (iii) inducing rule compliance; (iv) providing infrastructure; and (v) being prepared for change. The NRMAC is well positioned to respond to these requirements. Chapter 16 evaluates the extent to which these expectations have been achieved.

References

Dietz, T., Ostrom, E. and Stern, P.C. (2003) The struggle to govern the commons. *Science* 302, 1907–1911.

Flora, J. (1998) Social capital and communities of place. *Rural Sociology* 63(4), 481–506.

Pretty, J. and Buck, L. (2002) Social capital and social learning in the process of natural resource management. In: Barrett, C.B., Place, F. and Aboud, A.A. (eds) *Natural Resources Management in African Agriculture.* CAB International, Wallingford, UK, pp. 23–34.

Vedeld, T. (1994) *L'Etat et la Gestion des pâturages: la création et l'érosion d'institutions pastorales au Mali* (Dossier No. 46). International Institute for Environment and Development, London, UK.

8 Conflict Management Training

Jeff Goebel,[1] Catherine Ampagoomian[2] and Abdoulaye Touré[3]

[1]Goebel & Associates, PO Box 2503, Redmond, OR 97756, USA; [2]Tulane School of Public Health and Tropical Medicine, Tulane University, New Orleans, Louisiana, USA; and [3]CARE/Tombouctou, Mali

Since conflict over West Africa's diminishing natural resources is impeding development efforts and perpetuating land degradation, conflict resolution is necessary to immediately diffuse disputes and avert violent clashes. Perhaps, more important in the long run, is ensuring that resource users have the skills to improve consensus building and build the social infrastructure necessary to manage conflict situations. Although current decentralization laws have allowed for natural resource management (NRM) decision-making and conflict resolution at the community level, the local populations have not received adequate training and are unprepared for these new responsibilities. Assessing the francophone civil society organizations (CSOs), Fox *et al.* (2001) state that few CSOs have had any previous experience in conflict resolution or in the related area of policy advocacy. Indeed, the successful management of conflicts requires re-empowering local institutions or creating new ones, based on the traditional practices that allow communities to address new challenges on their own (Pendzich, 1993).

In the Commune of Madiama, SANREM CRSP has introduced an Alternative Conflict Management (ACM) approach, 'Conflict Resolution and Consensus Building' (CRCB), to the newly formed CSO, the Natural Resource Management Advisory Committee (NRMAC). This experiential training addresses specific NRM issues of contention in Madiama through a three-step process designed to complement traditional conflict management and improve decision making. This modular training programme establishes skills necessary for short-term conflict resolution and for long-term management. Though conceived under the ACM umbrella, this approach differs from many other intervention programmes, because it goes beyond negotiating the resolution of a specific conflict. Rather, the training programme empowers local leaders by providing them with skills and a process for sustainable facilitation and management of a diversity of conflicts.

The SANREM Project did not implement ACM to resolve specific conflicts, but to build 'people' skills, training trainers and training locals in the necessary skills. Key stakeholders were farmers, herders and other natural resource users from the commune, members of the NRMAC, local NGOs and extension agents, traditional leaders and government administrators. Since the initial workshops, lessons from these training sessions have been routinely transmitted to the village level. Although initially led by SANREM trainers, NRMAC trainers themselves have become increasingly involved.

This chapter introduces the three commonly used approaches to conflict resolution in the rural communities of the Niger Inland Delta Area. In light of current conflict resolution and management theories, it explains how SANREM's approach and implementation of this conflict management training in Madiama reverses the cycle of conflict and natural resource degradation. It includes building confidence among local leaders in their ability to mitigate conflict in the commune of Madiama, and establishing a successful social infrastructure for long-term consensus building that can address the underlying causes of NRM conflicts, issues associated with diversity, scarcity, power and change.

Characterization of Conflicts

Messer *et al.* (1998) point out that food insecurity due to land degradation, scarcity and poverty is a major cause and consequence of conflict. In Madiama, this is certainly true, as the cycle of conflict and natural resource degradation perpetuates itself. In the Sahel, the most serious conflicts have generally involved resource access issues between and among farmers and herders (Cissé, 1999).

The Inland Delta of the Niger is attracting a growing concentration of herders, farmers and fisherman. Competition has come to replace more complementary and cooperative practices of resource use. The conversion of pastureland into cropland along with the expansion of cultivation into irrigated agriculture along water points has limited water and dry-season pasture access, thus disturbing traditional pastoralist movement. Meanwhile, the need for herder mobility instigates overt conflict involving the trampling and destruction of cultivated fields.

There is conflict-creating competition for access to wetlands, grazing areas and crop residues. This NRM conflict occurs not only between these different types of resource users (e.g. pastoralists and farmers) but also within user types as land degradation shrinks the resource base. Inter- and intra-community conflicts also occur as natural resources diminish. Conflicts are interstate as well, given that nomads' and transhumants' movements traverse national borders. In these cases, traditional conflict management has not been able to cope with the current conflicts, as conflicts and violence over immediate survival and long-term livelihood have escalated.

Conflict Resolution and Management

In response to escalating conflict around the world in recent years, there have been major developments in the management of competing interests of different natural resource stakeholders. Conflict resolution has merged with NRM to create a more specific field of conflict management, adapted to NRM issues. The methodology falls into this new field, generally referred to as ACM, which Pendzich *et al.* (1994) defines as 'a multidisciplinary field of research and action that seeks to address the question of how people can make better decisions together, particularly on difficult, contentious issues' (p. 5). Current literature tends to focus on forest conflicts, where communities are pitted against outside stakeholders such as the government or private sector companies. The Food and Agriculture Organization (FAO) Division of Forestry (2000), in reviewing this field of study, summarized three viable avenues of conflict management for local populations: Customary Conflict Management Systems, National Legal Systems, and ACM. These are described below.

Customary conflict resolution

Despite Mali having a National Domain Law and a judicial system, the Inland Delta region tends to follow traditional land allocation practices. If conflicting land claims arise after some years or in the next generation, village elders or the village head may arbitrate. Only if village authorities cannot resolve a dispute does litigation proceed to the Rural Council, and perhaps the Sous-

Prefet. Most NRM conflicts are currently resolved at the village level by the chief and his counsellors, or if they proceed further in the system the methods remain informal. However, if the conflict is not settled at this level it moves up to the Chef d'arrondissement level (pre-commune context), into the national legal system administered by the Justice Ministry at the *Cercle* level (Djenné) or the regional authorities in Mopti. Yet, at the heart of the perennial nature of most of these conflicts is the lack of a single accepted authority for the resolution of natural resource rights, and consequently tenure security (Touré, 1996; Lo *et al.*, 1996; Maïga and Diallo, 1998; Ngiado, 1996).

National legal system

The national legal system governing NRM is based on legislation and policy decisions, including regulatory and judicial documentations. Litigation and adjudication are the main strategies for addressing conflicts (FAO, 2000). These systems tend to disregard indigenous knowledge and long-term community needs, thereby creating a winner–loser situation (FAO, 2000). While access is universal to modern courts in Mali in principle, limits still exist due to costs of transportation and a lawyer to advocate a case. National codification of French laws has restricted the flexibility of agricultural and livestock producers to adjust their production systems as needed to assure subsistence (Mortimore, 1997). In recent years, Sahelian governments have been attempting to adapt village-level customary law to the need for standardized modern law (Bohrer and Hobbs, 1996).

Contemporary conflict management alternatives

Reviewing two theories that support recent conflict management approaches, Cousins (1996) notes that Burton and Dukes (1990) define conflict management as 'how to handle disagreements and arguments over choices and preferences that result from interactions between parties who have common interests and goals, and who differ only on the means of achieving them' (p. 47). Cousins states that their theory rests on a typology of conflicts which determines the most effective resolution/management approach. His classification system comprises three groups:

- *Management problems*, involving arguments or differences over the choice of alternatives among persons having the same goals and interests.
- *Disputes*, involving competing but negotiable interests, and issues of gain or loss.
- *Conflicts*, involving the development and autonomy of the individual or identity group that inherently involve non-negotiable human needs and questions of identity.

Cousins (1996) goes on to explain a second theory that deals with processes and procedures including fact finding, negotiation, facilitation, conciliation, collaborative planning, arbitration and adjudication as defined by Pendzich *et al.* (1994) and Anderson *et al.* (1996). These processes provide a primary alternative to the legal and traditional systems of resolution that commonly appear in the literature on Alternative Conflict/Dispute Management. This multidisciplinary field of research and action seeks to address the question of how people can make better decisions collaboratively, particularly on difficult, contentious issues (Pendzich *et al.*, 1994). ACM techniques are often used to resolve environmental disputes worldwide and encompass a range of lesser-known methodologies. Similar methods, based on these premises include the International Model Forest Network, Adaptive Management Areas – Western USA, Zimbabwe's CAMPFIRE, India's Joint Forest Management programme and Stakeholder Analysis.

ACM refers to a variety of collaborative approaches that seek to reach a mutually acceptable resolution of conflict issues through a voluntary process that has been developed as an alternative to adversarial or non-consensual methods, such as judicial or legal recourse, etc. (FAO, 2000). ACM's three main premises are: (i) it is assumed that conflict is a normal process in society; (ii) successful alternative conflict management relies on constructive communication between all legitimate parties or 'stakeholders' in a dispute; and (iii) power imbalances are nearly always an issue in negotiations, and problems that result from negotiating in situations of unequal power can seriously undermine efforts at reaching a lasting accord (Pendzich, 1995). Within this general ACM approach lies the SANREM methodology used in training the Madiama Commune stakeholders. This methodology framework is based on the fundamental elements of Alternative Conflict/Dispute Management, including fact finding, facilitation and collaborative planning, and negotiation.

The available literature on AMC and its approaches indicates a lack of case studies that articulate the actual methods used. The SANREM ACM approach emphasizes long-term consensus building and addresses the conflict resolution process itself, instead of facilitating resolution of a particular conflict. The following method incorporates these basic premises and fundamental elements of ACM, and was adapted to fit the dynamic context of natural resource conflicts in the Commune of Madiama.

Methodology

Alternative conflict resolution/management training – creating facilitators

This section describes a method for training facilitators in ACM. The SANREM project did not implement ACM to resolve particular conflicts; rather the team trained members of the NRMAC in the basic elements of ACM, including communication, confidence-building, and other 'people' skills in order for them to engage in ACM, and in facilitation, thereby creating trainers to implement ACM and further disseminate the necessary skills. The conflict management process engages all conflicting parties in a secure, power-equalizing atmosphere of respectful listening, while allowing opportunities for confrontation and expression of concerns. It is the process itself that fosters the desired outcomes. The training workshops were designed around conflict management within the context of Madiama's NRM disputes and were associated with various situations such as resource scarcity and issues associated with power, control, change and diversity. Key stakeholders were farmers, herders and other natural resource users from the commune, local NGOs and extension agents, traditional leaders and government administrators.

This training model consists of a three step process that fosters conflict management and successful resolution. The first two steps incorporate different sets of exercise tools to help the NRMAC become successful in resolving internal and external conflicts. This approach complements traditional conflict management and is adapted to local circumstances. In Madiama, the tools used in the first step set the stage for the workshop and provide the lead facilitator the information needed to choose a specific path of training. The second step's tool set includes the basic conflict resolution process. The third step is a series of modules chosen specifically to address the participating community's issues and conflicts. After explaining each technique, the technique's rationale and the exercises used to teach it are discussed. Training sessions follow a set pattern consisting of an overview of the training modules and their objectives, the teaching/training techniques and finally, a review of the exercises comprising the consensus building and conflict resolution training modules.

Overview of workshop training objectives

Since SANREM's conflict resolution training workshops were considered a training of trainers (TOT) for ACM, the objectives were participatory in nature and focused on introducing new skills and strategies to Madiama that would complement traditional conflict management. The primary objectives were to:

1. Increase each participant's capacity to communicate effectively regarding emotionally charged issues, and demonstrate new communication skills.
2. Increase participants' confidence in dealing with confrontation and develop a sense of empowerment.
3. Train participants in the art of facilitation and recognition of its value in conflict resolution.
4. Become aware and understand the physiological and behavioural responses in various conflict situations.

The participatory training process aims to increase participants' knowledge of ACM, and to raise their self-confidence in addressing difficult issues (FAO, 2000).

Teaching techniques – training of trainers

These TOT workshops used real life situations to achieve the goal of transferring skills directly to participants, while simultaneously building their capacity for future community dissemination and guidance of the conflict management process. Training participants to implement conflict management means training *facilitators* to guide the process. Facilitation training is more than an exercise in learning 'people skills' and conflict management practice itself, because it also teaches vital teaching techniques. Good facilitation skills are best learned through observation of other trained facilitators and through experience. During the first workshop in Madiama, participants observed the lead facilitator and rotated into the facilitator position themselves, developing their empathy and building functional skills.

Lead facilitator

The lead facilitator's role is to guide the process towards the desired consensual outcomes of the group. Three levels of facilitation occur during these workshops. The lead facilitator, who is conducting the training workshops, carries out the highest level of facilitation. The other two levels are discussed below as part of the teaching techniques and workshop exercises.

The lead facilitator is the external/outside implementer of the workshop and the training sessions. Early on in the workshop programme this facilitator hands over the task of routine daily facilitation, and acts as a 'guide on the side, not a sage on the stage'. Nevertheless, he or she opens the training session each day, determines which modules are needed to tailor the workshops to the local context, oversees the sessions, and ensures that the workshops run smoothly and remain focused.

The lead facilitator uses a training module called 'The Interview Process' in order to gather information and contextualize the workshop. Usually done on the first day, this process helps the lead facilitator determine vital facts about the situation, such as what the issues are, who the players are, and what behaviours, powers, resources and stakes are involved. A key responsibility of the lead facilitator is to ask effective questions and observe the group processing *their* understanding of the questions and answers. The lead facilitator ensures there is the opportunity for 100% participation.

Facilitation: roles of the successful facilitator and recorder

Within the first day of workshops, the other two levels of facilitation are initiated. Once the workshop commences each day, a facilitator (not a mediator, but often one of low positional power in the group) is chosen to take on traditional facilitation roles and lead activities. Facilitators are also chosen for all small group conflict resolution exercises.

At the beginning of a workshop, the participants are requested to define successful roles. Staying in a circle, one at a time, the participants answer the following questions: (i) What is the role of a successful facilitator? and (ii) What is the role of a successful recorder? The recorder, who was the facilitator for the previous activity, records their responses on newsprint. Throughout the workshop, all small group activities begin by selecting a group facilitator who then becomes the recorder for the next activity.

This process ingrains a sense of empowerment within the group as the facilitators model power-sharing behaviours throughout the workshop. Facilitator behaviours, such as allowing the participants to direct themselves, selecting participants to co-facilitate with the facilitator, and facilitating while not directly participating (from outside the circle), create a sense of equity within the group and further demonstrate the importance of empowerment. By defining the roles of facilitator and recorder, the participants gain a sense of self-direction. This importance of balance is illustrated through shifting between the roles of powerful facilitator and submissive recorder. The recorder's role is to develop the trust of groups in conflict by accurately listening and recording what each individual says.

Balance of power is encouraged throughout the workshop and is modelled in behaviours such as having a representative from one ethnic group begin the workshop and a representative from another ethnic group close the workshop, or having a farmer, a herder, Mayor or NRMAC President serve on a panel. Panels are formed to begin a dialogue about specific conflict issues. Selecting powerful representatives or spokespeople from a variety of perspectives allows for an open sharing of perceptions. These activities develop listening skills and foster the importance of respectful listening and allow participants to have a sense of being heard. Grounding, Adaptive Learning, Defining Successful Roles and Defining the Worst/Best Possible Outcomes provide ample opportunities to practice listening exercises.

Repetition and imitation

Since repetition and imitation are fundamental training tools used throughout training workshops, this participatory instructional model also uses these tools as primary pedagogic strategies for teaching techniques and allowing natural, successful learning. Imitation ensures that skills are highly transferable. The process of imitation is most effectively learned through an *experiential process* and through *repetition*. People will observe others in order to learn what is and is not effective. Consequently, the facilitator must model the desired behaviour. A great deal of thought goes into the actions and interactions of the facilitator with the group, because the participants are keenly observing the actions and interactions. The purpose of every action should focus on fostering the desired outcomes of the workshop. Repetition allows participants to observe the facilitator's behaviour and thus experience a consistent pattern of conflict resolution and consensus building.

Small group organization

People have a tendency to coalesce with others similar to themselves. The old adage 'birds of a feather flock together', describes this behaviour. One challenge with 'like thinking' or 'group think' is that new ideas are seldom created. Small group activities must include a diverse cross-section of participants. New ideas to facilitate old problems can frequently come from the introduction of new perspectives and new ways of seeing things.

When working with groups in conflict, it is important for them to exchange perspectives. A stalemate indicates that current thinking, or perception, is not providing the group with workable solutions for unresolved conflicts. The opportunity for people to express their assessment of a present situation, their concerns and fears about addressing the situation, and their best hopes for the situation in a setting, where respectful listening occurs, allows new perspectives to be formed. This group organization gives people the opportunity to change their 'point of view' to a 'viewing point'. It is more difficult to create new solutions to gridlock if groups are using the same base of information, and if the groups are made up of people who think alike. For this reason, diversity tends to foster a level of insights not produced by homogeneous groups.

Alternative conflict management training exercises

Step one: Tools for empowerment

Since the first step is designed to introduce the workshops, gather information for the direction the workshop will take and develop a safe venue for conflict resolution, step one tools model effective behaviour when striving for conflict resolution by creating a sense of potential equity and respectful listening. The process includes but is not limited to fact finding, balancing power and facilitation. This step involves participants in a generic set of activities in which self-awareness and participant interaction create an open-mindedness that makes them receptive to the workshop's next steps. In step one, they neither process issues, nor are they in the process of conflict resolution *per se*; rather they are gaining confidence, mutual trust and facilitation skills to engage fully in the workshop.

THE 'GROUNDING' In this opening tool, participants sit in a circle and answer three questions, one person at a time, going around the circle. The three questions are: (i) Who are you and what is your relationship to the topic of the session?; (ii) What are your expectations of this workshop? and (iii) How do you feel about being here? Participants are free to say as much as or as little as they want, but usually follow the lead of the first person to speak.

The rationale is that this participatory process models listening with respect and establishes 'verbal territory' for all participants. It engages the whole brain through thinking and feelings questions, gets people rooted in the 'here and now', allows the expression of hidden agendas, shared hopes and apprehensions for the meeting, and provides initial information to the facilitator. Someone besides the lead trainer usually facilitates this, demonstrating a willingness to share power as a facilitator and allowing participants to realize they can facilitate the process themselves.

THE 'GREETING CIRCLE' The greeting circle, an American Indian adaptation exercise, allows participants to introduce themselves individually to each other. The participants stand and the greeting circle activity begins with a designated leader. The

Fig. 8.1. NRMAC trainers lead session on conflict resolution.

lead person moves into the centre of the circle, turns to the person next to him or her and greets them. The lead person then moves to the next person and so on around the circle. Meanwhile, the person first greeted also steps into the centre of the circle, and, following the lead person, greets the next person in the circular line and continues greeting people around the circle. This continues until each person has gone around the inside of the circle greeting others, and the outside of the circle, being greeted.

The rationale is that the circle is a design of shared power, a theme continually reinforced throughout the conflict resolution process. It is a turn-taking exercise and again, is usually led by someone in the community. This activity breaks down intimidations and levels power. Greeting is one of the oldest ways of connecting for humans. From past experience, many conflicts, particularly ones dealing with power and intimidation, are resolved through the greeting. The perspective of village participants was that the greeting circle allows conflicting parties to build a friendship and that it allows them to put the conflict in the centre of the circle where it can be resolved. Malian custom has its own greeting circle, usually done at the end of a conflict resolution process.

AN ADAPTIVE LEARNING PROCESS Adaptive learning occurs at the end of any enriching experience such as the greeting circle, or at the end of the day. Again going around the circle, one person at a time speaks, answering two questions: (i) How do you feel about the experience or situation? and (ii) What did you learn from it that will make you successful, resolve the conflict, overcome the impasse, etc.?

The *rationale* is that there are three main reasons for asking these adaptive learning process questions. First, asking reinforces topics covered during the session and brings out individual insights concerning these topics. Having others introduce what was experienced during the session is a strong method of reinforcing learning. Second, asking provides feedback to the group and the facilitator about what did and what did not work in the session. This feedback allows participants to improve their abilities as facilitators and provides the lead facilitator with ideas for continual improvement. Finally, asking provides direction for upcoming experiences. The adaptive learning process at the session's end provides direction for future sessions because of the emphasis on addressing fears and focusing on best outcomes. The human mind wants to find a solution.

'HONOURING' Honouring is a process of formally recognizing individuals who have benefited the group in some way, such as people who facilitated, prepared meals, made contributions to the group in other venues, etc. The group is asked to stand in a circle and the individual(s) being honoured are asked to come to the centre of the circle. Someone in the outside circle states why the people in the centre are being honoured. The group applauds once the words are spoken.

The *rationale* is that honouring people builds community. After modelling the process of honouring, the facilitator often uses it to resolve conflicts between parties by asking one side to speak to the other by saying something respectful or explaining why one party appreciates the presence of the other in its community. The people being honoured in the centre of the circle are often very self-conscious. Again, conflict resolution is about awareness; self-consciousness, or self-awareness, is an important first step towards change. When people are aware of their internal feelings in a given situation, they can then make a conscious decision to take action and change behaviour to what is consistent with their desired outcomes.

Step two: A process for coping with conflict

This is where the actual process of alternative conflict resolution begins. Step two tools continue to allow participants to explore and understand the natural human response to potentially threatening situations and issues. The progression of questions noted below moves from developing a common understanding of the situation and the associated feelings to putting the situation in an emotional context. The second question explores the 'worst possible outcomes' imagined of confronting the situation. These imagined results are why confronting the issue is 'dangerous', because they often paralyse individuals and groups into inaction. Should this occur, it is important to ask the other side of the question: What is the worst possible outcome of *not* confronting the situation?, because this question allows groups to recognize the hopelessness of either side of the question.

Step two tools allow participants to successfully move from a focus on worst possible outcomes to a focus on best possible outcomes as a distinct possibility. Often, allowing the group to explore conflict resolution at a generic level, before confronting real and potentially threatening issues, provides the group with an understanding of the human behaviour surrounding conflict resolution. As a result, they are more capable of successfully focusing on the issue. These tools also help in developing a more complete understanding of the issue, particularly if diverse perspectives are involved. With this understanding and clarity of the best possible outcomes, the human mind works to solve the problem of moving to the desired outcomes of the group.

THE BASIC PROCESS QUESTIONS The following set of questions are asked of the participants in order to develop an understanding of the conflict, gather concerns the participants have about confronting the situation, define their best outcomes and foster the best outcomes of the situation. These questions are directed to the individuals or groups.

- What is the situation? (Define the conflict. What is the evidence of this conflict in your environment?) How do you feel about it?
- What are the worst outcomes of confronting/not confronting unresolved conflict?
- What are the best outcomes of confronting unresolved conflict?
- What beliefs / behaviours / strategies / actions will foster the best outcomes?

WORST/BEST POSSIBILITY The worst/best possibility exercise allows the participants to explore and understand the importance of worst and best outcomes as well as possibilities. The following two questions are asked: (i) What are the worst possible outcomes of the workshop? and (ii) What are the best possible outcomes of the workshop? Understanding these possibilities is important, particularly if the group is confronting a serious conflict. Also, allowing the group to express the worst/best outcomes at the workshop level helps the group begin to understand the physiological patterns of individuals who confront potentially threatening situations or issues.

Worst outcomes are feared *future* outcomes, often based on *past* experience, with a *presently* experienced emotion and physical reaction. When believed, they affect people's perceptions, beliefs, values, and strategies and, consequently, tend to be self-fulfilling prophecies. (Outcomes are automatic and based in the lower brain stem).

In contrast, *best outcomes* are hoped for *future* outcomes, sometimes not *previously* experienced, but intensely imagined, with a *presently* experienced emotion and physical response. When believed, they affect people's perceptions, beliefs, values and strategies. And, consequently, tend to be self-fulfilling prophecies. (Unlike the fear of worst outcomes, the hope for best outcomes is not automatic and originates in the neocortex part of the brain that is seeking the good of community.)

Possibility thinking is the acknowledgement that both worst and best outcomes are inherently present in each moment, up to, and often after the event. This *balanced* view allows movement towards desired outcomes, because it is based both on negative and positive thinking. A simple analogy can help with this explanation. It is a fact that the sun will rise (or the earth will turn towards it). However, there is a possibility of not seeing the sun rise because of cloud cover in the sky or some other impediment such as a windowless room, even though it

Fig. 8.2. Conflict resolution through the village theatre.

has risen. The worst possible outcome is believing the sun has failed to rise, even though it always rises. Possibility thinking in conflict resolution approach acknowledges the balance of a focus on both worst and best possible outcomes.

An example of this possibility thinking occurred at a workshop in the village of Tatia Nouna when the villagers and participants in a group were asked to identify the worst possible outcomes that could result from confronting the conflicts between farmers and herders. The answer was violent confrontation and death. If the mind imagined this as *the* future, then why would anyone take any action to resolve the situation if any attempt to resolve the issue would result in death? In a second worst possible outcomes question, the participants were asked to name the worst possible outcome of *not* confronting the conflicts between farmers and herders. Again, the answer was death. This created a difficult problem for the mind to solve. If confronting the problem led to death, and not confronting it also led to death, and no one wants to die, then what is to be done? At this point, there is a third possibility that asks the participants to identify the best possible outcome of confronting the conflicts between herders and farmers. The answer was that people would learn to live in harmony with each other and foster actions and behaviours that are consistent with that outcome. The group from Tatia Nouna liked this alternative and decided to focus their actions towards this outcome. Throughout the process of moving towards peaceful coexistence, any of the three outcomes are possible, and by acknowledging these possibilities, the community is able to choose to move towards their desired outcome.

The process also changes behaviour concerning creating desired outcomes because beliefs change as people learn from each other in a respectful environment. Because the human brain is a problem solving organ, when it is presented with a problem, as when the space between a present situation and best possible outcome are created, the mind goes to work to seek resolution to the problem. Sometimes, because the answer is difficult to determine, the way the mind finds resolution is to say 'I don't know' or 'It is impossible to solve'. Once a person acknowledges the possibility of not knowing or the impossibility of finding a resolution, restating the question in terms of possibility (i.e. 'if you did know?' or 'if it were possible, what would you say?') recharges the mind to examine the problem again, but with a different emphasis. The mind has a very difficult time leaving an unresolved issue alone and, therefore, continues to search for a solution so that it can be at peace.

Because of the power of the human brain to focus and the limitations of the brain to gather all relevant information for problem solving, the mind often confuses fact with possibility. The mind tends to project potential outcomes as real due to past experience, when, in fact, the future has not happened yet. This sets the body up to take appropriate action based on the 'perceived' outcome. Human beings are often quite different if there is a perceived positive outcome instead of a negative outcome. These actions and behaviours can have the effect of creating the outcome most focused on. This reaction is based on the automatically 'hard-wired' system for self-preservation generated from the lower brainstem that protects us from the physical and emotional damage created from worst outcomes. A less powerful force in the mind is the ability to imagine best outcomes generated by the neo-cortex part of the brain.

The 'worst possible outcomes' response allows the exploration of the possibilities of best possible outcomes in confronting a situation or issue. Because most people are paralysed with the fear of confronting an issue, the potential for fostering best possible outcomes of confronting a situation never arises. When people are allowed the opportunity to explore best possible outcomes, they begin to imagine ways to encourage these outcomes. Two questions arising from the determination of the best possible outcome engage the problem solving nature of the human mind: (i) What beliefs and behaviours will foster the best possible outcomes? and (ii) What strategies and actions will foster the best possible outcomes? Beliefs and behaviours focus on a fundamental transformational change within people. Strategies and actions are focused on modification as change. Usually in conflict resolution, transformational change is required to move people to foster the best possible outcomes of confronting conflict. This process is about modifications and transformations. The participants were shown the Change model diagram (Fig. 8.3) with results being either the Worst Possible Outcomes or Best Possible Outcomes. The diagram demonstrates that if change at a basic level is desired, someone must modify strategies and actions. If change at a fundamental level is desired, someone must transform beliefs and behaviours. If natural resource users want to change the Sahel, that change will have to be at the fundamental level, because the underlying cause of these conflicts is the perception of a scarcity of degrading resources. Hence, in order to resolve these conflicts, it is necessary to have more than conflict resolution strategies and good communication; rather, it is necessary to change the management of these resources. This is where conflict resolution methodology becomes linked to the underlying issues of these disputes and brings in holistic management (HM).

Step three: Module tools for contextual adaptations

The third step in the conflict resolution workshop is tailored to the particular local context. A series of modules are chosen and adapted to the specific issues and conflicts facing the participating community. Each module uses a set of questions, adapted to the issues at stake, to guide people through specific conflicts. The modules are: one-on-one conflicts, managing scarcity, managing power, managing change, managing diversity and conducting interviews.

Discussion

The CRCB training workshops conducted in Madiama have demonstrated the success of an ACM programme in Mali. The methodology was tailored to the Commune of Madiama and specific NRM conflicts that are occurring across West Africa. As both a

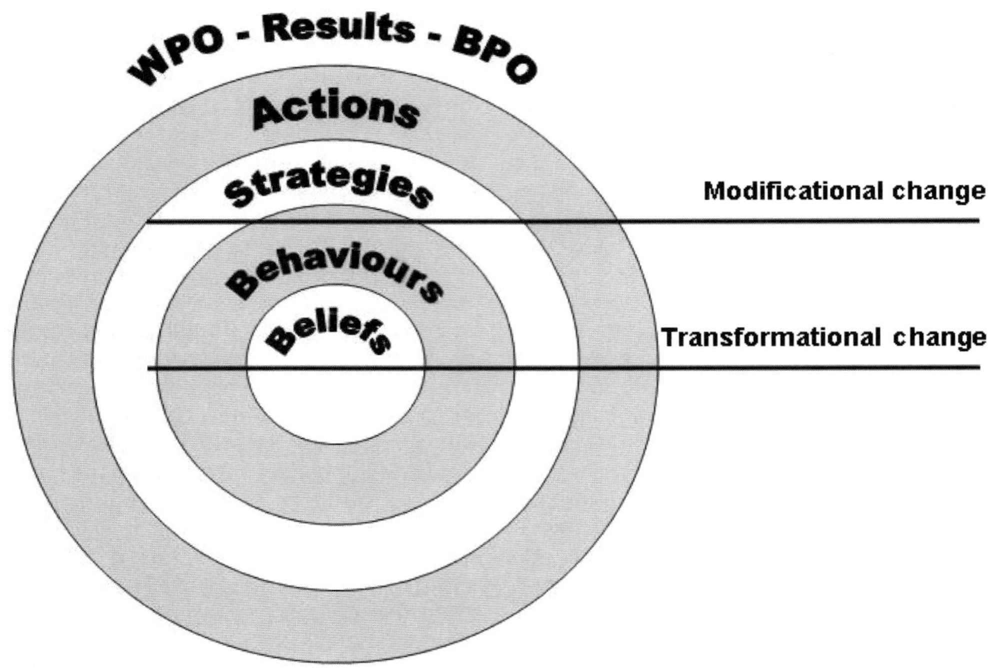

Fig. 8.3. Change model.

short- and long-term approach to manage conflict, the SANREM programme has led to considerable changes in the ways that conflicts and NRM decision-making are handled within the community. In this sense, its method has developed into a more participatory approach to ACM by engaging participants themselves to lead the conflict management process and ultimately resolve their own conflicts and build consensus in management decisions. In training community members and the NRMAC in sustainable ACM methods of conflict management, SANREM has built the communities' capacity to manage conflict more successfully, independent of outside intervention.

Conflict resolution rests on a handful of premises about conflict, change and power (Pendzich, 1993). The training method is flexible enough to be adapted to different cultural contexts in order to build sustainable community capacities to resolve conflicts and reach consensus. The process of conflict management engages all conflicting parties in a secure, power-equalizing atmosphere of respectful listening that allows for the confrontation and expression of concerns to foster desired outcomes. The workshops in Madiama were designed around conflict within the context of NRM disputes and included modules covering concepts such as scarcity, power, control, change and diversity. SANREM's approach to CRCB has led to changes in the community's perceptions of conflict and local power struggles.

Equalizing power

Creating an atmosphere of respect and equality is essential to this process and promotes potential power equalizing. Exercises like 'greeting circles' enable each individual of differing status and power to interact on a level playing field, creating an increased level of confidence and a sense of empowerment for participating farmers and herders. Power tends to equalize itself, especially when using such techniques as rotating facilitators, balancing gender and power in group settings, and having the lead facilitator on the side rather than centre stage. When each person is expected to engage in the same exercises and interact as stakeholders during workshop exercises, individuals feel less intimidated and more empowered to collaborate as equal stakeholders and resource users with diverse interests. These transferable techniques allow for a more open forum and an awareness of how to best cope with issues of intimidation and power.

New skills, perspectives and behaviours

Based on general principles of conventional conflict resolution and ACM, the participatory workshops helped to develop a sense of trust vital to conflict management and the consensus building that is fundamental to effectively applying newly formed human capital. Facilitation skills empowered every participant to become a part of the process. Each workshop had a purpose, and each exercise was done with a purpose. Through repetition and imitation, the participants come to understand the purpose that the 'adaptive learning process', completed after the workshop exercises, reinforces. The skills demonstrated in each exercise are transferred directly to the participants; consequently, while the group was creating, individual consensus-building *skills* were learned during each workshop. Ultimately participants learned to alter the ways in which they communicate and engage in social relationships. The conflict management process created new perspectives and brought diverse views to the table. By creating willingness to explore, communal solutions became a more viable option. Consequently, the learned and applied CRCB skills have allowed best possible outcomes to emerge from NRM disputes in the commune.

A local observer noted that the conflict management training workshops demonstrated that conflict resolution was 'everyone's business'; someone did not need to be a village chief to have such facilitative skills and capacities. The workshops have shown the community that any individual or group of individuals can create a secure forum to discuss conflicts and their reasons, and serve as a mechanism for change. The training workshops enabled individuals to rethink their views of resolution processes by using skills such as listening, recording and facilitating. Success was not measured in negotiated agreements and resolution of specific conflicts, but was defined by enthusiastic participants engaging in training sessions and sharing their experience with the villagers. Most of the participants continue to use facilitation and conflict resolution techniques in various roles and settings beyond major NRM conflicts. These settings include a women's cooperative and individual households, where behaviour has been modified. This illustrates the adaptability and sustainability of the approach, as well as the skill-building dimension of the workshops. Participating community members have learned to use ACM for larger community conflicts, and they have used individual ACM skills and approaches in other settings, as well.

From conflict to consensus

SANREM's CRCB approach to helping communities cope with conflicts over NRM and extend their learned skills beyond present conflict is an important addition to the field of ACM. This ACM process complements other aspects of the SANREM project, thereby increasing the success of current research designed to transform the cumulative underlying biophysical and social causes of conflict. As previously mentioned, there is a void in the literature concerning consensus building approaches and ACM training methods, where the purpose is to develop the community's capacity to manage conflicts as they arise, not merely to solve one specific conflict. Disseminating ACM approaches and different experiences will change community-level capacity to manage natural resource conflicts in the region and beyond. To bring change to their land and community, NRMAC members are changing their views on confronting and managing conflicts, and consequently are adapting their traditional methods of conflict resolution via alternative processes and newly learned individual 'people' skills. Beyond the cases in which empowered participants have applied new skills, there is a broader recognition in the community of behaviours that lead to and increase conflict and, as a result, behaviour is becoming more positive and respectful. This shift in behaviour creates a more peaceful tone in communities when confrontations between disputing land users occur.

Movement is occurring in the region to expand conflict management training to more communes in surrounding areas. As the process proves adaptable and effective with the transhumant population, the opportunity exists for this process to extend across national borders. The initiation of activities to regenerate natural resources (i.e. holistically managed *bourgoutière* and rangelands) exemplifies the consensus-building aspect of these conflict management workshops aimed at natural resource decision-making. NRMAC members have shared their CRCB workshop experiences with villagers. Using alternative methods to confront disputes and resolve conflicts is leading to a more open community with beliefs that confronting NRM disputes can potentially create positive outcomes. There is a greater sense of security regarding negotiations and agreements. Compared to customary and legal mechanisms of conflict resolution, this ACM approach transcends important gate-keeping issues found in both. CRCB creates a forum empowering all NRM actors, including herders and women, which tend to become marginalized during traditional conflict resolution. CRCB builds relationships across traditionally divided ethnic and production system communities working towards mutually beneficial outcomes and avoiding winner-take-all legal system resolutions.

National and regional governmental support is needed to continue the use of ACM strategies to manage widespread NRM conflicts. The cyclical nature of conflict and natural resource use requires a multi-levelled approach in order to mitigate both short- and long-term sources and perpetrators of conflict. One-time negotiated resolutions and static legal agreements are not always a viable or sustainable option for rural communities given the dynamic, ever-changing nature of NRM. Initial facilitation along with consensus building/decision-making skills and mechanisms, in sync with local traditions, will allow local Malian and West African communities to successfully manage their resources within the context of decentralization. Consensus building skills are an essential form of human capital for reaching agreements on NRM. Capacity building for conflict resolution creates a critical mass of human capital necessary to change social behaviours and relationships that can lead to more effective decision-making and management of natural resources. Efforts should be made from governments and NGOs to support ACM interventions, using consensus building skills to improve decentralized, community-based NRM.

References

Anderson, J., Gauthier, M., Thomas, G. and Wondelleck, J. (1996) Addressing natural resource conflicts through community forestry: setting the stage. E-conference on Addressing Natural Resource Conflicts through Community Forestry. Food and Agriculture Organization. Rome, Italy.

Bohrer, K. and Hobbs, M. (1996) *Post-Praia Progress towards Tenure Security and Decentralization: Review of CILSS-member County Legislative Reforms.* University of Wisconsin, Land Tenure Center, Madison, Wisconsin, pp. 1–41.

Burton, J. and Dukes, F. (1990) *Conflict: Practices in Management, Settlement and Resolution.* St Martin's Press, New York.

Cissé, S. (1999) Décentralisation et développement dans le delta interieur du Niger (Mopti Republique du Mali). In: Bourgeot, A. (ed.) *Horizons Nomades en Afrique Sahélienne: Sociétés, Développement et Démocratie.* Edition Karthala, Paris, pp. 137–154.

Cousins, B. (1996) Conflict management for multiple resource users in pastoralist and agro-pastoralist contexts. *Institute of Development Studies Bulletin* 27(3), 41–54.

FAO (2000) *Conflict and Natural Resource Mana-*

gement. Food and Agriculture Organization, Rome, Italy, pp. 1–20.

Fox, L., Ouattara, R., Boly, A., Ba, M.B., Konaté, Y., and Charlick, R. (2001) *Francophone Civil Society Organization Assessment: Mali and Côte d'Ivorie Synthesis Report.* West Africa: Civil Society Strengthening for Conflict Preventions Study. Associates in Rural Development (ARD), US Agency for International Development, Broadening Access and Strengthening Input Market Systems (BASIS), Burlington, Vermont.

Lo, H., Sylla, C., Ndione, P. and Agne, A. (1996) Ressources forestières et conflits au Sénégal: quelles approches pour une typologie? *Bulletin Arbres, Forêts et Communautés Rurales* 8, 26–36.

Maïga, I. and Diallo, G. (1998) *Les conflits fanciers et leurs gestion dans la 5ème Région du Mali* (Dossier No. 76). GRAD. International Institute for Environment and Development, Programme Zones Arides, London, UK.

Messer, E., Cohen, M.J. and D'Costa, J.D. (1998) *Food from Peace: Breaking the Links between Conflict and Hunger* (Food, Agriculture and the Environment Discussion Paper 24). International Food Policy Research Institute, Washington, DC, pp. 1–44.

Mortimore, M. (1997) *History and Evolution of Land Tenure and Administration in West Africa* (Drylands Programme Issue Paper No. 71). International Institute for Environment and Development, London, UK, pp. 1–35.

Ngaido, T. (1996) Redefining the boundaries of control: post-colonial tenure policies and dynamics of social and tenure change in Western Niger. Ph.D. dissertation, University of Wisconsin, Madison, Wisconsin.

Pendzich, C. (1993) Conflict management and forest disputes: A path out of the woods? *Forests, Trees, and People Newsletter,* 20, 4–9.

Pendzich, C., Thomas, G., and Wohigent, T. (1994) The role of alternative conflict management in community forestry (Community Forestry/FTPP Working Paper No. 1). Food and Agriculture Organization, Rome, Italy. http://www.fao.org/ DOCREP/005/X2102E/ X2102E00.htm (accessed 4 May 2004).

Pendzich, C. (1995) Gestion des conflits sur les ressources forestières : est-on sorti de l'auberge? *Arbres, Forêts et Communautés Rurales* 7, 31–37.

Touré, D. (1996) *Prévention et gestion des conflits.* Centre Canadien d'Etude et de Coopération Internationale, Ouagadougou, Burkina Faso, pp. 1–50.

9 Holistic Management Applied to SANREM Research and Development in Madiama

Sam Bingham

Allan Savory Center for Holistic Management, 394 Vanderbilt Road, Asheville, NC 28803, USA

The length of title indicated the breadth of the challenge. Each word in 'Sustainable Agriculture and Natural Resource Management Collaborative Research Support Program – West Africa (SANREM/CRSP – WA)' coded one of the Big Ideas of turn-of-millennium thinking about development and progress. If the dominance of northern, developed-world, science had in the past promoted destructive exploitation of nature, it would now exploit the synergy of local and regional knowledge to discover, or rediscover, the science of *managing* a return to stable productivity of one of the most devastated ecosystems of the world.

SANREM reflected a widespread academic interest in studying traditional practices for previously overlooked environmental strengths and in conducting useful dialogue with traditional practitioners about ways to preserve these in modern conditions. Indigenous practitioners working in respectful partnership with American and African scientists would discover, or rediscover, a way to reverse the degradation of the Sahelian landscape and its agropastoral economy. The programme would honour the scientific traditions of the USA and Europe as well as the conventions and practicalities of multi-ethnic West African communities while assuring inclusion of gender, ethnic and caste diversity.

Although no categorically successful model of such a programme had yet been seen in West Africa, new techniques for improving communication among parties of varied background and point of view gave grounds for optimism. The traditional anthropologist's questionnaire, for example, has been replaced by much more open ended and often graphically and symbolically enhanced dialogues structured according to methodologies bearing names such as Rapid Rural Appraisal, Participatory Landscape/Lifescape Appraisal (PLLA), etc. Mali, like many other countries in Africa today, now enjoys a strong contingent of home-grown scientists and technicians who can call on deep cultural knowledge and communicate directly in local languages.

Despite such assets, however, the grail of managing resources in sustaining and sustainable ways spontaneously embraced by the local population remains elusive. Holistic Management® (HM) was one packet of seeds that SANREM broadcast on the real and virtual soils of the Commune of Madiama in its attempt to find answers. As the copyright symbol suggests, Holistic Management, capitalized, refers to a rather strictly defined management model, in this case one developed through the non-profit Allan Savory Center for Holistic

 Conflict, Social Capital and Managing Natural Resources (ed. K.M. Moore)

Management and largely embodying the thinking of Allan Savory himself (Savory, 1999).

Savory, a trained wildlife biologist born in southern Africa, grew up amid the economic, cultural, and environmental challenges facing his native continent and has spent half a century personally involved in the ever-evolving succession of European and North American policies to address them. He is known among scientists first for synthesizing from existing research and his own observation the radical scientific hypothesis that the ecological vigour of certain environments *requires* large herds of large herbivores behaving in the manner induced in 'wild' conditions by the presence of pack-hunting predators.

Savory's efforts to incorporate this idea into a practical model for managing what he calls 'brittle' environments, of which the Sahel offers a prime example, led him to a second, more generalized conclusion. Holistic Management reflects the view that the ecosystem, including human economic activity, cannot be studied or managed piecemeal. Research, policy, and management risks miss the point by attempting to isolate parts of the system – as in laws to protect a particular endangered species or research on the productivity of a particular plant.

Over the years, Savory and others of similar persuasion have criticized much research, policy, and management for failure to account for externalities of the social and environmental context. However, they have also contended that overemphasis on producing clean, quantifiable data often produces irrelevant data, or at best, data of little use for informing practice.

The Holistic Management Model® presumes to help the researcher, politician, or manager avoid these pitfalls, but the stridency of arguments among scientists over the charge reflects significant differences in perception of how science should operate and the relationship between science and practice. As the project began, preconceptions about what should be done were shaped in different disciplinary domains and intellectual cultures of American and West African scientists and development professionals. Madiama itself brought together three ethnic groups, each interpreting economic and social activity and natural phenomena according to its own culture.

For this reason, bringing the Holistic Management point of view into SANREM's Mali project represented a rather courageous choice on the part of the project leaders. The holistic approach took its place as one more player in a marketplace of ideas governed by a creative tension. Nevertheless, the Holistic Management Model's recognition of cultural priorities and its emphasis on practical management choices accord well with SANREM's mandate to involve local people in the scientific discussion. Its concern for only pushing action when social organization could support it mapped perfectly on to the conflict resolution component of the project and the development of the Natural Resource Management Advisory Committee (NRMAC) within the commune.

In retrospect, the relevant questions are:

1. What ideas did Holistic Management contribute to a general *scientific* understanding of the problems of Madiama?
2. How did it affect perceptions among the various constituencies within the commune?
3. How did it affect communications between people in the commune and the various researchers who worked there?
4. Will it account for any significant and lasting change in decision-making or management at any level?

To answer all four questions requires some understanding of how the Holistic Management Model evolved, and how it may be applied.

The Gestation of Holistic Management

Allan Savory (1999) credits the revolutionary insight that overgrazing is a function of time management and animal behaviour, not herd size, to the French scholar of pastures, André Voisin. Savory himself worked in the rather different environment of Southern Rhodesia, now Zimbabwe, where he sought to explain why the 'unmanaged' grassland he knew from his youth supported enormous herds of wild ungulates and recovered from even severe droughts without loss of biodiversity but degraded rapidly when grazed by much smaller numbers of domestic stock under human management.

Inspired by Voisin to test the common wisdom that the damage resulted from simply overstocking the range, he eventually found four keys to unlock the riddle.

- In so-called 'brittle' environments, characterized by low humidity, a prolonged dry season and erratic precipitation, plant communities stagnate when rested, and bare ground increases, whereas rest in stable, humid environments (such as France) results in covered soil and rapid development of diverse communities, typically culminating in forest.
- Rest allows succession to advance in non-brittle environments, because microbes and other small organisms efficiently recycle the carbon accumulated by plants. In brittle environments, however, carbon is better recycled by grazing animals through both trampling and digestion. Large herbivores are thus *necessary* to maintaining diversity, productivity and stability in brittle areas.
- Overgrazing and overtrampling are principally functions of time, not numbers, as Voisin had determined. Dense herds of great size that move frequently and allow plants to recover before regrazing them recycle nutrients without weakening plants. Even single animals that do not migrate overgraze the most palatable plants in the areas where they linger.
- On the enormous unfenced ranges of pre-colonial Africa (and similar brittle environments elsewhere), pack-hunting predators assured beneficial herd behaviour, and nomadic herders generally developed similar patterns. Limited lands, however, demand management that is 'holistic' to the extent that it responds to ever-shifting conditions of weather, economics, culture and environmental conditions.

Savory figured that the conjunction of these principles would bring a new dawn to livestock production, game management and most efforts to reverse desertification in areas impacted by livestock. Sheep, goats, cattle and horses would become instruments of restoration. Reducing livestock stock numbers, a policy that no herding society has ever accepted, would fade as a policy imperative. The economic potential of long-lost grassland would salvage the balance sheet where irrigation, fertilizer and supplementary feed had failed. The documented success of individual managers over the past 30 years leaves no doubt that Savory had got something right, though the revolution is taking its time.

Anthropologists, meanwhile, followed a parallel path to similar conclusions. In their seminal paper 'Rethinking range ecology: implications for rangeland management in Africa' (1991), British-based scholars Scoones and Behnke (1993) argued the genius of traditional nomadic societies in managing land and animals for maximum efficiency. They debunked the stereotype that traditional people acquired large herds for prestige only and attacked a brace of standard development policies such as veterinary programmes, fencing, water development, genetic improvement, confined feeding, sedentarization and matching stocking rate to 'carrying capacity' – this concept being meaningless where conditions fluctuated and herds moved so dramatically.

Their findings rested on a decade of research issuing from the same institutions that did the original development of Rapid Rural Appraisal techniques – the International Institute for Environment and Development in London, the Institute of Developmental Studies at the University of Sussex, and the British Government's Overseas Development Institute. A stream of studies in the same vein gave academic support to recent policy initiatives to establish (re-establish) the grazing rights of transient herders in the brittle environments of Africa (Behnke *et al.*, 1993).

Kerven and Alimaev (1998) published similar findings from an extensive study of Central Asia following the Soviet collapse. They also concluded that productive, environmentally sound, husbandry of brittle lands depended on large, moving, herds and the enormous flexibility of herding societies, including their willingness to take up other professions in time of drought.

This refrain of extreme flexibility, common to Savory and the British anthropologists, has made the claims of both difficult to document through methodologies based on controlled studies of short duration (less than 20 years, depending on prevailing weather cycles). If there is a 'science' of management, it is the study of how organizations and their leaders respond to change in order to achieve goals. Good *management* responds quickly and properly to new opportunities, including conditions changed by management itself (feedback).

A Schematic Rendition

A good model predicts. You tweak it here and see what wobbles there. It interprets. You can look at a wobble and deduce the tweak that produced it. It instructs. You can tweak it this way and that until you learn what makes the wobble you want. And best of all, you can do all this cheaply without wrecking anything important. Computers allow you to poke around complicated models, but the most powerful are simple, and a participatory research/development programme requires one that equally amplifies the brainpower of the Peul shepherd trying to supply goat milk for her baby sister and the scientist sipping coffee at the stop light on the way to her institute.

The Holistic Management Model rests on four axioms:

- The condition of the resources always reflects the management of them. No excuses.
- Management means management of 'wholes' which have characteristics not present in, or predictable from, their constituent parts studied in isolation. (Knowledge of hydrogen and oxygen does not give the whole story on water. Knowing the alphabet does not imply understanding of words, sentences or Shakespeare.)
- In complex, self-organizing systems, processes and the relationship patterns of wholes are the only constants and therefore become the focus of research and management.
- Management, being a human endeavour, must have a goal, an assumption of a better state.

Therefore, the shepherd/scientist-friendly Holistic Model looks like this:

The whole is at the top, followed by a three-part goal describing the quality of life you expect, the production needed to support it, and the condition of your resources required to produce that. Next are the processes that define the ecosystem – community dynamics (succession), water cycle, mineral cycle and energy flow. Therein lies the genius of nature – the interplay of untold variables boils down to measurable effect on only four basic processes (Fig. 9.1).

Since holistic managers posit that conditions reflect management, they can assess past practices and future propositions by looking at the effect of management tools on the four processes. Fortunately there are only six direct tools to consider, plus money, labour and creativity. If this seems oversimplified, consider that virtually all past policies and management practices have not included living organisms, grazing, or animal impact in the tool kit and do not recognize the differential effects of long-term rest in brittle and non-brittle environments.

When planning future actions, ethical, economic, ecological and social considerations weight the value of a given tool, and these appear in seven testing questions that are augmented by certain guidelines that determine how the tools are applied. Finally, a feedback loop joins planning, monitoring (the four processes), controlling and replanning into one process.

Over the course of the project, a number of workshops and training sessions were

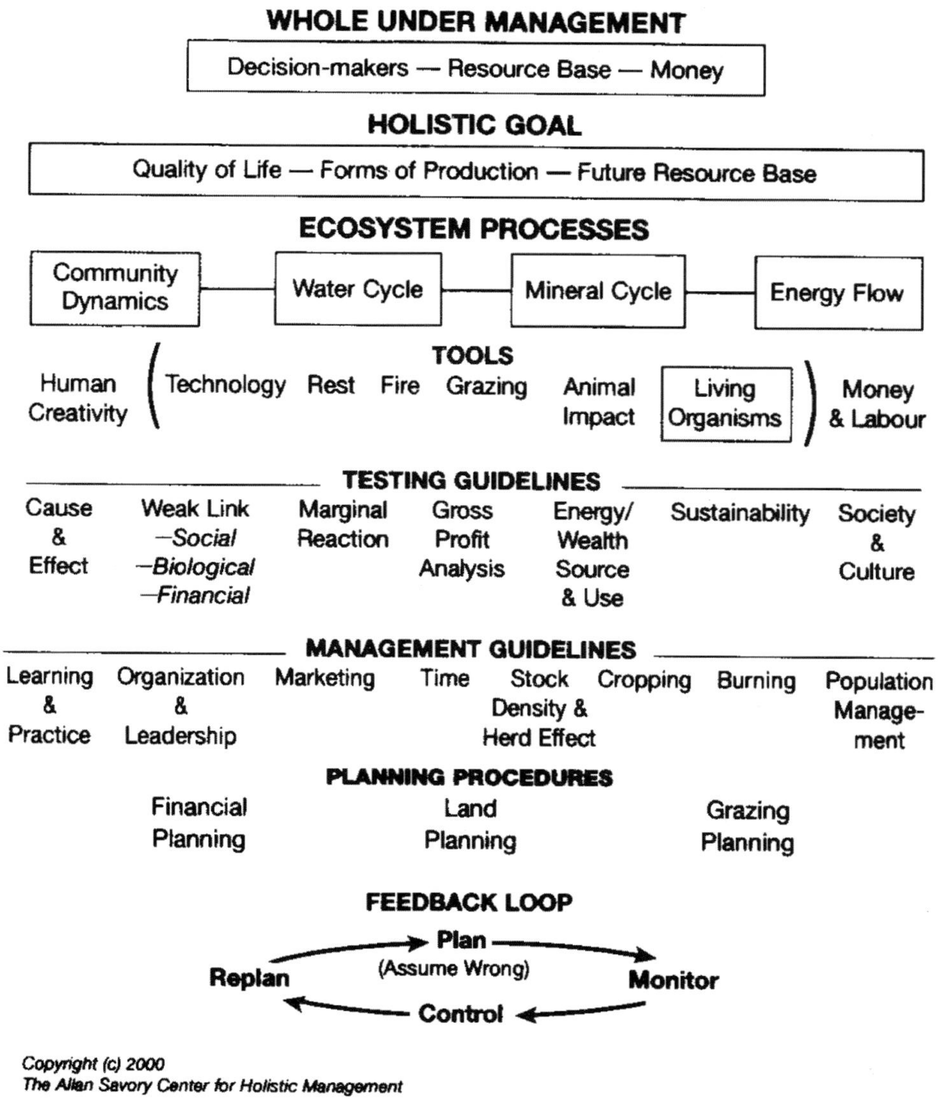

Fig. 9.1. Schematic of the Holistic Management Model.

conducted to promote this line of reasoning. Generally these events were carried out in conjunction with conflict resolution training and focused primarily on members of the NRMAC composed of representatives from all the villages in the commune. Institut d'Economie Rurale (IER) personnel attended some of these and others were exposed to the Holistic Management Model in separate sessions.

Holistic Management Training in Madiama

The NRMAC selected two women and three men participants to join ten IER scientists, three US scientists, seven community service technicians, the mayor of Madiama and his assistant in a 5-day Holistic Resource Management Workshop on Consensus Building and Conflict Resolution

(1–5 November 1999). After the workshop, NRMAC members (under the guidance of the HM trainers) reviewed the workshop for the full NRMAC and then led the partners in a transect tour and an HM-focused discussion of natural resource features of the Commune. The committee expressed an interest in training for all NRMAC members in the national language.

From 2 to 17 May 2000, three consecutive 3-day workshops were conducted with all members of the NRMAC, two in Holistic Management and one in Conflict Resolution. These sessions included the new CARE/ Djenné SANREM Facilitator. Researchers of IER participated in some sessions. Committee members learned to evaluate different options for improving resource management in the Commune. Group discussions were encouraged and the committee generated several proposals to improve soil fertility and the productivity of wetlands. The conflict resolution workshop included training in group facilitation and consensus building, including many techniques that were immediately applied and continue to be used. NRMAC members also roughed out their holistic goal reflecting universal values – a desire for food security, health, safety, education, respect, free time and opportunity; a stable income from small enterprise, crops, livestock, fish and increased skills; and a resource base of good water, fertile soil, lush grazing and vigorous people.

The autumn 2000 training session in Holistic Management began with a review of the research trials on soil fertility and pasture management. Visits to farmer field sites facilitated these discussions. Small group exercises reinforced the Holistic Management themes and the application of tools and management techniques presented. The NRMAC analysed the conditions for *bourgou*[1] management (*bourgou* being an important source of fodder that grows in flooded areas during the rainy season), identifying stakeholders and their perspectives, as well as technical issues of *bourgou* restoration. They developed an action plan for implementing *bourgou* restoration.

SANREM brought in an HM trainer/ consultant from Chad, who had in-depth experience in introducing Holistic Management in his country through the World Bank's West African Pastoral Perimeter Project. Consultations followed with the NRMAC in March and again in July 2001 to assist in setting up managed grazing trials in two sites (Siragourou and Torokoro) and training local environmental monitors to track compliance. In September, a delegation from the NRMAC had an opportunity to visit the HM site in Chad and speak with villagers who had applied it, albeit in circumstances much less complicated than prevail in Madiama.

Three additional training programmes in October 2001 were conducted by the Holistic Management and Conflict Resolution trainers focusing on managing power conflicts, the management of conflicts associated with wetlands, and the management of pasturelands in general. The overriding objective of this training was to provide a structured opportunity for NRMAC leaders to gain experience and become confident as trainers.

In March 2002, an HM trainer provided training on how to develop a management plan for timed grazing in open-range pasturelands. This workshop involved introducing the NRMAC to the timed grazing of open-range pasturelands, identifying villages ready to participate in the experiment, evaluating the pastures to be managed, dividing the pastures into parcels, and preparing village level environmental supervisors to monitor the activity. This was followed up in July 2002 by a second HM training session to assist in the establishment of the timed grazing programme in two sites (Siragourou and Torokoro). The primary objectives of this training were to familiarize the environmental monitors with the terrain and the routines they must follow, examine the rotational parcels to determine the initial grazing periods, establish a schedule of tasks and designate responsibilities for implementation.

A conflict resolution workshop on managing change in October 2002 built on the previous training in managing the grazing trials and was designed to prepare NRMAC members and the environmental supervisors for the inevitable conflicts that would arise in making such a management change. The primary objectives of this training were to strengthen local capacity to face and manage conflict situations, develop techniques to manage change, and develop and refine techniques for the mediation of conflict. This provided an important opportunity for NRMAC trainers to take the lead in the training exercise.

Holistic Management for open-range pasturelands was reinforced in May 2003 with a review of principles and discussion of the results of the previous year, which were influenced by extremely poor rainfall. This also led to the final Conflict Resolution and Consensus Building Workshop (October 2003) in which NRMAC trainers led the training activity in the twinned village of Tatia Nouna where the build-up of herds during the descent into the bourgoutière routinely leads to confrontations and conflicts. The NRMAC brought together opposed interests within the village to confront and discuss the natural resource management (NRM) issues related to these conflicts.

Holistic Management Contributions to Scientific Understanding

Beginning with the initial Participatory Landscape/Lifescape Appraisal (PLLA), groups from Madiama Commune produced lists of 'constraints' that isolated problems from their holistic context making them irresolvable. The issues included: wells are drying up; the parasitic *Striga* plant destroys the grain; the cattle are starving; the soil does not produce like it used to; the all-sustaining *bourgou* plant is vanishing from the lakes; firewood has disappeared; and the transhumant herds come through before the harvest and ravage the crops. This constellation of sorrows is typical of Sahelian communities and well-documented across semi-arid lands worldwide. Since the typical response is a demand for technical remedies for each problem, the introduction of a different line of reasoning into discussions in Madiama Commune is an important legacy of the project.

Training in Holistic Management began at the end of the first year of project activities in Madiama. At an intellectual level members of the NRMAC and other villagers participating in the training quickly mastered the principles of the Holistic Management Model. They had no difficulty in seeing erosion, drought and falling water tables as manifestations of a damaged water cycle, weed infestations and pest outbreaks in terms of succession, and fertility loss as a failure of the mineral cycle. They frequently returned comments on the order of, 'Thank you for explaining what is going on. We see things in our own way, but up to now outside advisors just told us what to do but never why they thought it was right.'

Faith in theory has a long gestation period, however. The final step of turning faith into a community consensus strong enough to risk complex changes in organization and practice did not develop during the life of the project. Nevertheless, an important change in perception did occur.

Acceptance of the Model did not come as easily to many of the researchers as it did to the average villager, given that the former tended to have invested enormous effort in mastering a particular speciality. Quite a number found the Model's insistence on distinguishing between cause and symptom confusing, because they had been trained to seek remedies. Observing that the soil fertility and tilth that previous generations enjoyed did not depend on periodic doses of rock phosphate does not automatically cause a fertilizer specialist to reflect on why it might now, or to seek alternatives if chemical fertilizer proves economically unprofitable or ecologically unsustainable. A real dialogue from one of the training sessions follows:

Question: 'Seriously, why are Madiama's soils so much worse today'?
Response: 'They lack phosphorous and nitrogen!'
Question: 'Why?'
Response: 'Because they are degraded!'
Question: 'Why are they degraded?'
Response: 'Because they lack necessary nutrients!'

Nevertheless, at least one IER researcher did become intrigued by the suggestion that direct importation of soil nutrients and organic matter by transient livestock could play a large role in restoring and maintaining soil fertility, and preliminary experiments he conducted with village collaborators in stabling herds for short periods on fields after the harvest produced very promising results.

For input, the Holistic Management Model asks:

- *What happens to the water that falls on Madiama and the surrounding area (the Water Cycle)?* Historical data on well levels and rainfall may be interesting but are mostly useful to substantiate the unanimous opinion that enormous amounts of water that once entered the water table through percolation into friable, protected soil, now leave the system through runoff and evaporation.
- *Where do soil nutrients come from in Madiama, and where do they go* (*Mineral Cycle*)? The process of loss is of more concern than the fact of it.
- *What changes in plant and animal communities have occurred* (*Community Dynamics*)? Concern for biodiversity loss is a step in the direction of holism as opposed to the customary fixation on controlling pest and weed outbreaks.
- *What happens to the available solar energy* (*Energy Flow*)?

Every one of these questions could form the basis of several research projects designed to quantify the state of basic ecosystem processes, but neither NRMAC members nor even outside observers had much difficulty reaching consensus on general answers. In Madiama, the almost complete loss of perennial grasses and the expanding extent of bare ground where nothing grows at all mean a high percentage of rainfall never enters the productive cycle. Consequently, chronic ‘virtual drought’ coexists with persistent flooding.

Historically, nutrients in surface soils had to have been brought from lower levels by deep rooted perennial plants or imported through the manure of transient livestock, and they would have been held in topsoil by organic material. Now, evidently one or more of these processes had broken down. The cycle no longer functioned as it should, and nutrients had moved somewhere else. Plant communities had shifted towards non-palatable species and annuals. Perennial grasses had almost entirely disappeared except for vestiges of the least desirable. Larger game species had all but vanished, and domestic stock had shifted towards small ruminants. A visiting biologist noted that biodiversity is now actually greater in cultivated areas than in grazing areas, although in the former *Striga* and weed infestations have increased markedly.

Energy too is an issue. If someone conceived of agriculture as the harvesting of solar energy, Madiama abounded in signs of decline. Crop production per hectare had fallen. Herd fertility and milk production had fallen. Ponds yielded fewer fish and less *bourgou*. Trees like the *Acacia albida* that produced greenery in the dry season and fixed nitrogen in the soil were all in steep decline.

What tools had been used? How had the use of tools changed? Going tool by tool, what impact on the four processes could be expected?

- *Technology:* Steel ploughs pulled by bullocks have largely replaced hand cultivation, enabling exploitation of much larger fields, and this has practically eliminated the old practice of a 6–10 year fallow. Older inhabitants note that today even land that *is* taken out of crop production does not recover, but in fact degrades further.
- *Rest:* Cultivated ground gets little rest now; nor does grazing land, as increasing numbers of small ruminants are kept year round. Overrest is not a serious issue.
- *Fire:* Burning has not been used as a tool much in recent years, due to suppression by the Forestry Service, although annual grassland is fire prone.
- *Grazing:* Although transhumant herders with cattle continue to pass through the commune, more small ruminants and

draft animals graze continuously year round than in the past, thereby assuring severe overgrazing of plants.

- *Animal impact:* Passing herds continue to create significant trampling events and deposit large amounts of manure, but much of the latter does not enhance crop land. Wide areas around ponds and other water points are severely degraded despite enormous deposits of manure and moisture in the root zone of perennial plants, if any remained.
- *Living organisms*: Little has been done to promote habitat for beneficial creatures.

Such an analysis, crude as it may appear, allowed both villagers and researchers to focus on probable causes of degradation that they might be able to address, thus immediately raising the question of whether to attempt that or simply mitigate the symptoms. In the case of Madiama, this discussion resulted in a conclusion that the soil fertility question and the grazing land biodiversity loss are linked and that livestock and crop issues cannot be solved separately. This conclusion is reinforced by the fact that, in an erratic rainfall environment, the food security envisioned in the holistic goal will depend on maintaining multiple forms of production.

A shift from looking for discrete solutions to individual problems to correcting basic flaws in a complex system compels an admission that real progress might depend on changes in social organization, personal relationships, and improved management rather than a new technology. Unfortunately hugely complicated and contentious land tenure issues and jurisdictional ambiguities arise. Longstanding family and inter-ethnic conventions come into question.

Even though no major changes in management were achieved during the course of the programme, the general consensus that the problems in crop and livestock production cannot be solved separately probably stands as the Holistic Management Model's most significant contribution to the future of Madiama. If the social, political and research implications of this observation overwhelmed the capacity of the NRMAC, IER and American partners to effect significant change in 5 years, the challenge to the future has been fairly communicated.

The holistic analysis can also take some credit for planting the notion among villagers, researchers and project leaders that lasting solutions to Madiama's woes had to involve people and land outside the commune. Most of the cattle and much of the smaller livestock actually spent more time outside the commune than within it, and much of the stock that passed through belonged to people living elsewhere. Exploitation of other resources such as firewood, fish and thatch also involved outsiders. Certainly much of the conflict has had roots in this fact.

The experience in Chad highlighted for the Madiama delegation the importance of including external stakeholders in planning and negotiation. Unfortunately the logistical and administrative difficulties of this challenge limited what could be done within the project period, but a start was made.

Impact of Holistic Management

The Holistic Management approach probably did not precipitate much change in the way IER and expatriate researchers worked in Madiama, given the fact that most of the research proposals were generated in response to the PLLA and the list of constraints provided by the NRMAC before they or participating researchers had been exposed to its principles. The NRMAC did generate a proposal to attempt restoration of a previously productive *bourgou* zone around a wetland basin. The project failed for multiple reasons, though the genesis of the idea is a fair illustration of the potential of a holistic discussion.

The idea arose when a discussion of the Holistic Management Model's 'marginal reaction' guideline focused on rehabilitation of a *bourgou* site as the action that the NRMAC felt promised the greatest return for the effort. Other tests in the Model raised questions about the difficulty of controlling grazing access to a rehabilitation site and social and cultural resistance to fences as a means to control access. This, in turn, led one of the women members of the NRMAC to propose that resistance to fencing by the several customary users of the *bourgou* might be overcome if the area were consigned to a committee of women who would use the forage to produce milk for malnourished children and elderly.

This was tempered by the memory of a discussion that took place following the first training session when members of the NRMAC reported to a broader selection from the community. After they made a presentation of the four ecosystem processes, when asked to identify the causes for the decline of the *bourgou*, an elderly man opined that a lot of it succumbed to drowning, a counter-intuitive diagnosis for a plant that grows in water. But the man pointed out that the water cycle had changed. He argued that now, since the land around had lost its cover, the runoff came in fierce sheets that filled the lakes faster than even *bourgou* could grow, and silt from the erosion blocked sunlight. The *bourgou* would not come back, he predicted, as long as the adjacent land remained bare. He also remarked that *bourgou* was not the first good plant to vanish from Madiama's ponds.

Unfortunately, neither American nor IER scientists were able to incorporate any of these ideas into a proposal for rehabilitating *bourgou* within a timeframe that fit SANREM's conditions. The weak link of commitment and understanding had yet to support concrete community action. Lacking volunteer labour to plant *bourgou,* a broadcast seeding was done without a grazing management plan, but it was washed out altogether as the old man and the Model warned it might be.

Nevertheless, the experience may yet bear fruit. In the summer of 2002, one of the NRMAC members suggested that the commune ought to grant temporary rights to someone to fence an area of potential *bourgou* production, plant it as a crop and sell it. The profit motive would assure that it was planted properly and protected. The re-established plants would provide a seed source for the rest of the area, and the community would see the benefit of managing the restored common land for the benefit of all. As this would be a revolutionary initiative on a number of counts, it is hard to say how realistic it might be, but it does illustrate the variety of ideas that issue from the holistic process.

Perceptions among constituencies within the commune

The very notion of grouping villages into communes came to Mali only shortly before SANREM did. It arrived in Madiama as part of a nationwide administrative reorganization aimed at decentralizing power, but many questions of jurisdiction and relationships with residual layers of authority going back to pre-colonial times remain open to this day. Madiama's Natural Resource Management Advisory Committee was the first and only one in the country. It had no statutory or traditional authority, but it did open communications among ten villages that had not had any institutional relationship before. That in itself was a major achievement, not directly the result of the Holistic Management, but certainly in the spirit of the approach.

Much of the conflict within the commune has an inter-ethnic character, but most of it stems from competition over resources. Different groups specialize in farming, herding and fishing. The conflict resolution component of the programme generated a great deal of healthy discussion of how human nature reacts under conditions of scarcity, power imbalance and change. The Holistic Management Model, particularly its emphasis on the contribution of animals to soil fertility and the importance of watershed health to the productivity of wetlands and rivers opened at least the possibility of win–win solutions. Bambara farmers and Peul stock raisers on the NRMAC at least began to discuss the possibility of mutual benefit in a way that could not have happened at the beginning of the project.

Lasting influence of Holistic Management?

Since 1993, the Center for Holistic Management has operated a demonstration site in Zimbabwe near Victoria Falls and trained and supported local field workers in neighbouring villages. After 10 years several do have grazing programmes, but that only happened after years of patient discussion and organization-building. Success in other projects, including water development and women's microbanks, and a training programme for game guides came first, as community capacity grew.

It is unreasonable to lament the fact that Madiama's ad hoc aggregation of villages has not reorganized communal grazing, integrated its crop and livestock production, negotiated grazing and parking plans with transhumant stock raisers, or sorted out the property rights around water points and ponds. However, the idea that a different approach to management could restore a devastated landscape where many technical solutions had failed was new and will be remembered.

Note

[1] *Bourgou, Echinochloa stagnina,* is a highly productive plant that grows underwater with the seasonal rise of the Niger River and the associated system of lakes and marshes. As the water falls the *bourgou* becomes the major source of fodder for a wide region.

References

Behnke, R. and Scoones, I. (eds) (1991) *Management and Sustainability of African Communal Rangelands* (ODI/P59). Commonwealth Secretariat Conference, Overseas Development Institute, London, UK.

Behnke, R., Scoones, I. and Kerven, C. (eds) (1993) *Range Ecology at Disequilibrium: New Models of Natural Variability and Pastoral Adaptation in African Savannas*; Overseas Development Institute and International Institute for Environment and Development, London, UK.

Kerven, C. and Alimaev, I.I. (1998) *Mobility and the Market: Economic and Environmental Impacts of Privatisation on Pastoralists in Kazakstan*, Paper presented at the conference on Strategic Considerations on the Development of Central Asia, CODOCA and Chinese Academy of Science, Urumchi, China.

Savory, A. (with Butterfield, J.) (1999) *Holistic Management: a New Framework for Decision-Making*, 2nd edn. Island Press, Covalo, California.

Scoones, I. and Behnke, R. (1993) Rethinking range ecology: implications for rangeland management in Africa. In Behnke, R., Scoones, I. and Kerven, C. (eds) *Range Ecology at Disequilibrium. New Models of Natural Variability and Pastoral Adaptation in African Savannas*. Overseas Development Institute, London.

10 Farmers' Knowledge and Perceptions of Soil Fertility

Todd A. Crane[1] and Boureima Traoré[2]

[1]*Department of Anthropology, University of Georgia, Athens, Georgia, USA; and*
[2]*Centre Régional de la Recherche Agronomique, Institut d'Economie Rurale, BP 205, Mopti, Mali*

According to participants in the Participatory Landscape/Lifescape Appraisal (PLLA) (see Chapter 6), soil fertility is a major constraint for villagers of Madiama. Scientists (Powell *et al.*, 1996; Brouwer and Powell, 1998; Scoones and Toulmin, 1998; Gobin *et al.*, 2000) likewise cite soil fertility decline as the major limit to agricultural production in the Sahel. However, despite this similarity of problem identification, it is surprising how often farmers or scientists question the usefulness of each other's soil fertility science and practice. This problem raises several questions: What are the different assumptions and beliefs that they bring to the problem? Are they really talking about the same thing? Do they really have the same perception of the problem?

Unsurprisingly, scientists and farmers often approach the problem of soil fertility from vastly different theoretical and practical perspectives. Due to their unique approaches, both scientists and farmers have valuable insights to contribute to each other. Consequently, explicit efforts to integrate scientific and folk knowledge of soils through participatory research methods have several benefits. Among these benefits is that folk knowledge can guide aspects of the scientists' research programme, scientific research findings can be better communicated to farmers, and farmers can improve production practices.

Based on generations of practice and observation, farmers have developed their own knowledge system and theories of soil fertility maintenance that might or might not match up with scientific knowledge and theory. For example, while the scientific approach to soil classification is typically based on the geological origin and chemical composition of soil, farmer classification tends to be based on more utilitarian factors such as texture, water retention and cropping uses. Regardless of scientific validation, it is these folk models that guide management behaviours at the local level. Therefore, efforts to provide scientific research in support of local decision-making need to be integrated with folk knowledge systems for optimal efficacy.

This chapter examines the subject of linking local knowledge with scientific research on soil fertility, in both theoretical and applied terms. The end goal of this effort is to collaboratively develop improved soil-management techniques via epistemological integration of local and scientific knowledge. The chapter begins with a brief examination of some of the historical problems and failings of introduced agricultural technologies. In an effort to explicate the shared and divergent perspectives, local knowledge systems and institutional scientific research are compared according to classificatory systems and fertility management regimes. The scientific bases for soil classification and soil fertility management are based on a review of the literature. The folk model of soil classification and soil fertility management developed from field interviews with producers. This parallel presentation highlights where scientific and folk knowledge overlap and where they diverge regarding the subject of soils. The theoretical aspect of this relationship is, however, only part of the picture. The final part of this chapter examines how cultivators in the Commune of Madiama have received and perceived scientific research. The chapter then concludes with a discussion of ways that local knowledge and practice has related with the agronomic experiments on the ground in Madiama.

Problem Statement

Historical weaknesses in top-down agricultural research

Historically, development efforts in the agricultural sector have often relied on the introduction of technologies and techniques by the scientific research community and government bodies. However, while institutional science is a powerful way to understand soil nutrient cycles and develop technologies to influence these cycles favourably, its weakness has often been that the techniques and technologies developed do not translate well into the realities of agriculture as practised by the farmers themselves. There have been reasons for this weak translation, including lack of consideration for the socioeconomic realities of peasant life (such as financial or labour limitations), lack of consideration for the long-term environmental or social impacts of introduced technologies, and differences in conceptualizations of how and why agriculture works the way it does (Rhoades, 1984, 1989; Chambers *et al.*, 1989; Scoones and Thompson, 1994). By involving the end-users from the beginning, participatory approaches to technological development go a long way towards avoiding these pitfalls and contribute to the ultimate goal of developing feasible, sustainable and culturally appropriate production systems.

Participatory approach in soil management research

In the past 15 or so years, participatory approaches to natural resource management (NRM) research have become more prevalent. These research approaches incorporate local knowledge and local practice into the development of improved technologies for peasant producers. The broader field of soil science literature (Arrouays, 1987; Tabor, 1993; Gobin *et al.*, 2000) provides examples of integrating local knowledge and scientific research of soils, and there are specific examples that focus on various topics of relevance to Mali, including seed varietal development (Adesina, 1992), striga control (Debrah *et al.*, 1998) and soil fertility management (Defoer *et al.*, 1998). Many similar cases can be found throughout the world. As described in Chapter 1, the SANREM approach stresses participatory research, and it was within this framework that soil fertility trials were undertaken

Following the PLLA and subsequent follow-up by the NRM Advisory Committee (NRMAC) (see Chapter 6), several meetings focused around the theme of participatory technological development were held between researchers, NRMAC members and other peasant collaborators in an attempt to broadly identify technologies adapted to the local context. At the end of these meetings, the peasants had chosen technologies to test as well as collaborators whose fields would be used to test them. In an effort to understand how they would evaluate the tests in their own terms, local collaborators mentioned four criteria: the vigour of the seedlings, the colour of the leaves, the condition of the grain heads, and most importantly the quantity of grain harvested.

Visits to the fields were organized each year under the direction of the NRMAC accompanied by senior researchers from CRRA/Mopti in order to assess and discuss the season and the performance of the experiments. During these visits, the peasant collaborator would discuss the performance of the field as well as the difficulties and the advantages related to technology. Following the harvest, a researcher's presentation of the quantitative experimental findings at a meeting with the NRMAC, the peasant collaborators, the field technicians and local dignitaries augmented the informal field assessments. This approach allowed for the qualitative and quantitative evaluation of the experiments by the local collaborators and researchers together.

Linking scientific and farmer knowledge

One of the challenges in participatory research is linking the sometimes disparate folk and scientific knowledge systems. The contrasting of 'local' and 'scientific' is not to suggest that local knowledge is not, at some level, scientific. In as much as local knowledge is based on systematic observation and experimentation through time, it can be said to be scientific. However, for the purposes of this chapter, 'scientific' is used to indicate knowledge that is institutionally supported, informed primarily by formal education, and inclined to universalistic scientific laws. The scientific approach to improving soil fertility is based on the body of knowledge developed by institutional scientific research through the years, broadly referred to as 'the literature'. Peasants have, however, rarely read 'the literature'. Consequently, their body of knowledge of soil fertility is based on insights gained through practice, observation and informal information flows. This includes knowledge developed and passed across generations as well as knowledge based on personal experience.

Over the last 15–20 years, the study of local knowledge of soils, ethnopedology, has become increasingly prevalent in an effort to make scientific research on soils more accessible to farmers and to make farmers' soil management practices accessible to scientists (see Talawar and Rhoades, 1998 for summary of ethnopedology; and Furbee, 1989; Pawluk *et al.*, 1992; Dialla, 1993; Tabor, 1993; Zimmerer, 1994; Sillitoe, 1996 for case studies). Both local and scientific systems frequently recognize that soil is a foundational and unifying element in any landscape, domesticated or otherwise (see Chapter 3). This, however, is not to say that they always approach or even understand soils in the same way. Because ethnopedological research provides a clear and strong means by which the ideologies of research scientists and rural producers can be linked, a study of local knowledge of soils was conducted in association with the soil fertility trials.

Methods

Characterization of scientific soil classification schemes are presented based on a literature review. The local knowledge aspects of this chapter are based on semi-structured interviews with peasants across the Commune of Madiama. Interviews were conducted in 2001 and 2003–2004 and centred on knowledge of soils and soil management, specifically principles of soil classification, management practices associated with various soil types, and indicators of quality. The findings here do not represent a random sample, but this was an acknowledged concession to the participatory approach to research. Since NRMAC members served as contacts and gatekeepers within each village, they are disproportionately represented in the sample and they usually decided who else in their villages would make a good informant. This sampling method was chosen because it is more likely to actually represent channels of information flow of SANREM research, which is typically presented to NRMAC members first. They then take information to their villages and share it with members of village-level NRM User Committees, after which the information circulates informally among other villagers. This model of information flow was the basis for presentation of findings from the soil fertility field trials.

Scientific Model of Soils

In order to contrast scientific and local knowledge, a brief review of the literature for scientific perspectives on principles of soil classification and prescribed management practices follows. Despite the universalistic nature of science, there are numerous distinct systems of soil classification used by soil scientists around the world, each based on different principles. In the late 19th century, the USA and Russia independently began to develop soil classification systems in efforts to assess soil resources and rationalize their exploitation. By the mid 20th century, many industrialized and agrarian nations had undertaken similar endeavours, the result being a large variety of classification systems, some very similar and some vastly different.

However, most scientific soil classification systems have been constructed to deal specifically with types of soils found within the relevant national boundaries. The literature provides an American system, a French system, a Russian system, a Brazilian system, a Chinese system, a German system, a Japanese system, a Canadian system, an Australian system, a New Zealand system, a South African system, and so forth (Finkl, 1982; Guillet, 1989; FAO, 1998; Spaargaren, 2000). Inasmuch as they were all constructed to address a need that exists within the geographic boundaries of a nation, each system could be said to be a local form of knowledge in its own right. In the latter half of the 20th century, serious efforts were undertaken to create a universally applicable system of soil classification under the auspices of the FAO and the United Nations, but this is by no means complete or universally accepted or used even to this day (FAO, 1998).

Some systems are based on perceptible properties of the soils themselves, whereas others are based on geological origins that can only be inferred from extensive geological survey and laboratory analyses, and still others are based on soil–climate interactions. An extremely brief review of some of the major systems is presented below to illustrate the diversity within the field. It should be noted that the SANREM soil survey combines elements of the French and American classification systems, which are subsequently informed by local knowledge and practice (see Chapter 3).

Principles of classification

The first step of any classificatory system is defining the domain of things to be classified. Soil is defined as a natural formation of variable thickness resulting from the combined action of atmospheric agents, plants and animals on stone, which constitutes the surface of the continents (Bockman, 1990). A strength of this definition is that it underlines one of the essential characters of soil: it is a complex medium at the crossroads of mineral matter and living organisms (Vannier, 1987; van Breemen, 1993). Indeed, soil is comprised of:

- Mineral components (generally the majority by weight); characterized according to the mineral size: boulders, stones, gravels, sands, silts, clays.
- Organic components (normally a minor component). One can find remains of this organic matter in humus. It is estimated by the quantity of carbon in the soil.
- Living organisms, microorganisms and other animals.
- Organo-mineral complexes; organic and mineral matter which are chemically bonded to each other.

American classification

There have been several different attempts over the last 150 years at classifying soils within the USA. The most recent American soil classification system, originally published in 1975, is hierarchically structured, with processes of soil genesis forming the primary point of distinction. Below this, within each grouping, soils are classified by observable ongoing processes that control soil formation. Below this level are found distinctions based on properties of the soils themselves (Ahrens and Arnold, 2000). The

Brazilian system originated from an early American system, but was significantly modified in order to accommodate the types of soils found in Brazil that did not appear in the American taxonomy (Spaargaren, 2000).

French or ecological classification

An early French system, the Commission of Pedology and Mapping of Soils was a hierarchically based system of classification. The first level was distinguished by processes of soil formation, and then subsequently divided by ecological conditions and context, followed by soil properties. This system, however, was replaced in the early 1990s by the Pedological Reference Base (PRB), a non-hierarchical classification that is instead intended as a general 'reference system'. The PRB typology is three dimensional in character and recognizes 90 different major groups that are defined by soil properties, soil origins and position in the mantle. The French system is similar to what is known as the Ecological classification system because it considers a soil's ecological context and processes in its categorization (Spaargaren, 2000).

Russian classification

The Russian classification system, published in 1977, is characterized by heavy reliance on soil-forming processes, such as weathering and biochemical processes, as the primary point of distinction. It recognizes 71 different types of soils. At lower levels, it distinguishes soils based on their respective bedrocks, geographical position and degree of soil development. In the Russian classification system, soils are characterized according to their processes more than they are by their properties (Spaargaren, 2000).

FAO-UNESCO Legend of the Soil Map of the World

The contemporary FAO-UNESCO classification, finalized in 1988, is in some ways similar to the American system, except that it is not a hierarchically structured method of analysis. Instead, it simply lists 28 major soil groups based on parent materials and geographical zones, followed by 153 discrete soil types within those groups. The contemporary system is a revision of an earlier effort that listed 24 major soil groups and 106 types of soils. It should be noted that one of the goals of the FAO-UNESCO effort was to 'promote the establishment of a generally accepted soil classification and nomenclature' (Spaargaren, 2000, p. E-138), a first step towards a completely universal system.

World Reference Base for Soil Resources

Despite the fact that it was co-developed by FAO in conjunction with the International Society for Soil Science (ISSS), the World Reference Base (WRB) was explicitly intended to provide a universally accepted system of soil classification that could be integrated with national systems. This system focused on easily observable soil properties as the primary points of classification, including the horizon in which a soil is found. The geological processes that led to the formation of the soil are secondary in significance. The WRB identified 20 major soil groups before merging with the Legend of the Soil Map of the World effort, wherein it settled on 30 major soil groups (Spaargaren, 2000).

In summary, American and FAO classifications particularly take into account the degree of evolution of the soil profiles. Western European classifications take into account a soil's milieu, processes and the characteristics. Dominant modern classifications (Russian, American and FAO, French) retain the following criteria that depend on the degree of development of the soil profile in relation to the degree of evolution:

- *Weathering*: Weathering is related to the evolution of the mineral fraction of the soil, particularly with the formation of clays.

- *Movement of material*: This criterion is used to speak about eroded or alluvial soils.
- *Pedoclimate*: The pedoclimate makes it possible to distinguish hydromorphic soils.
- *Time*: There are soils whose formation can be rapid and others whose formation is slow, which means soils can be spoken of as having short and long cycles.
- *Organic matter*: This plays a significant role in the formation of the organo-metallic complexes upon which many properties of the soil depend.

The development of any classification system makes it possible for soil researchers to have shared reference points. However, as anyone can see, even university trained research scientists approach the problem of soil classification from a number of angles using appreciably different logics, depending on their training. Is a soil best characterized by its geological origins and developmental history, or is it best characterized by its component parts and physical characteristics? To what extent is the ecological context important in characterizing a soil unit? All are equally valid methods in terms of their ability to describe a soil accurately. However, the primary question in the applied research context is their utility. Which way of approaching soils is most valuable to the end user, whether researcher or farmer? This is a legitimate concern of the local beneficiaries of research that makes the incorporation of local knowledge into soil research even more important. Acknowledging that science has many ways of approaching a single topic makes it easier to see local classificatory schemes as just another way of approaching soils, a useful way that can be integrated with scientific schemes.

Description of soil fertility management

One of the factors that any soil classificatory system will address is the question of soil fertility. From an agricultural point of view, the fertility of a soil represents, in a given climate, its aptitude to ensure crop growth and harvests in a regular and repeated fashion (Vilain, 1989). The problems of soil fertility maintenance, while universal among agriculturalists, varies with climate, type of soil, cultivation practices and techniques and environmental change (Sebillotte, 1989).

From a scientific point of view, in order to give an opinion on a soil's fertility, it is necessary to evaluate not only its capacity to provide crops with the conditions and factors they need, including light, water, mineral elements, but also the costs and risks associated with cultivation (Sebillotte, 1989). Indeed, the same soil would not allow the same levels of production if it were in a different climate or water regime. Similarly, a soil can be fertile for one species and unsuitable for another (Bockman, 1990). Generally speaking, in order to manage soil fertility, it is necessary to use techniques that enable soil to produce abundantly in the short term, but also maintain or improve its long-term fertility.

To fertilize soil is to improve its physical, chemical and biological properties so that crops flourish. Four major processes can be identified as parts of soil fertility maintenance in an agriculturally productive system. First, the soil must be managed so that the plants gain the nutritive elements they need. Second, the soil needs to be managed in such a way that its structure supports the rooting and the circulation of air and water. Third, it must be managed so that its physicochemical properties support the biological activity necessary for good decomposition of organic matter. Fourth, the soil must be managed so that its biological processes limit the presence of pathogenic organisms and encourage advantageous ones.

Depending on the context, management of soil fertility can entail any number of techniques, but typical among them are the addition of organic matter (in the form of manure, compost, etc.), the addition of minerals necessary for plant health (nitrogen, phosphorus, potassium, etc.), and the management of abiotic factors such as water, either by draining or irrigating. In this last case, the fertility of the soil itself (if measured by organic matter or mineral content) is not improved, but the principal constraint, water excess or shortage is changed so as to improve agricultural productivity.

Folk Model of Soils

Principles of classification

Where scientific approaches, such as those used in SANREM, often employ geological origins and processes of transformation to classify soil types, the folk system of soil classification found in Madiama is based primarily on the salient component materials of the top horizon, which in this case are sand, clay and rock. The primary difference between these three broad categories that three categories generally correlate with basic field management strategies is the degree to which they contain and retain organic matter and water. Millet and groundnuts do better in sandier soils while sorghum and okra favour clayey soils.

While sand, clay and rock provide the fundamental reference points when referring to soils, subcategories are divided out based on colour. As Fig. 10.1 shows, black is the most salient colour for agricultural soils due to its association with organic fertilizers and thus soil fertility. It should be noted that red and white clay and white sand are not regarded as agricultural soils, but are subsoils identified by only a few informants who had experience digging wells. The fine and very fine clay types are exclusively located in the river floodplain. These are characterized by their stickiness and the degree to which a person can sink into them when they are inundated. Although not all informants presented the typology below in all of its detail, no one presented any contradiction to it.

Even though peasants discuss soil in terms of the categories presented above, they also quickly acknowledge that these categories are not defined in absolute terms. Instead, soil is understood as a constantly shifting gradient of these three component materials. For example, most sandy soil in the commune has at least some measure of clay in it and most clayey soil had some measure of sand. Soil categorized as rocky soil always has either a sand or clay base, but sand is more common. Further questioning found that black sand is actually acknowledged to be a blend of sand and clay, thereby revealing that these categories are not mutually exclusive or entirely distinct. The widespread acknowledgement of mixed types illustrates that although there are specific named categories for soil

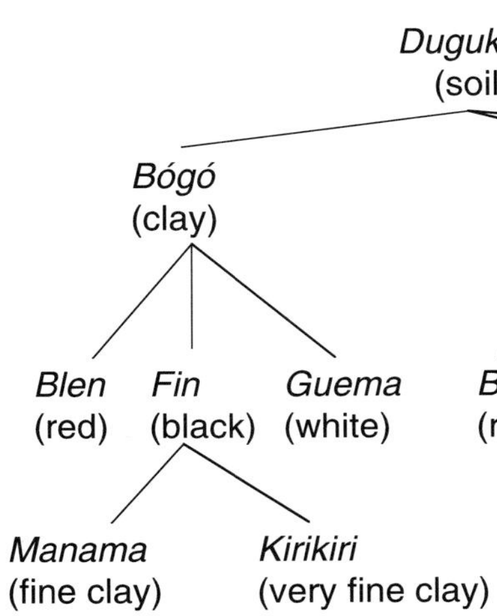

Fig. 10.1. Folk soil typology resulting from semi-structured interviews.

types, farmers understand soil as a three-component continuum that is constantly varying through space rather than discrete plots of one type or another.

It is important to point out that focusing exclusively on soils themselves misses a key aspect of local knowledge. Local knowledge, being situated in the context of an active production system, does not entirely distinguish between soil and landscape features. Distinct from, though related to soil types, hydrology and elevation changes within fields also play into management strategies. Clay is often associated with low spots and sand with high spots, but several informants speak of how elevation can play a role in their field management regardless of soil type. Although conventional wisdom is that millet is grown in sandy soils and sorghum is grown in clayey soils, these 'rules' are frequently broken according to the presence of elevation changes that affect microhydrology. For example, a low spot in a sandy field might be planted in sorghum due to the higher accumulation of water there during the course of the growing season. This is all to say that soil typology alone is insufficient to understand what guides management behaviour. Landscape level features also play a role, as does cognition of how the component parts of the management system relate to one another.

Description of soil fertility management

When asked what shows that soils have 'strength' (*fanga*), peasants uniformly say that crops grow well there, because they are large, green, and produce lots of grains. This is to say that agricultural productivity is the primary indicator of soil fertility. When asked specifically if looking at the soil itself revealed if it was strong, the primary indicator offered was consistently its colour. A strong soil is a dark soil, because darkness indicates that there is *nógó*, organic matter, in it. Water retention is a secondary indicator and is recognized to be linked to quantity of *nógó* in a soil. Water retention is typically tested by picking up a handful of inundated soil and squeezing it in the fist. The degree to which it sticks together exhibits the relative quantity of water it contains, which is, of course, linked to quantity of organic matter, particularly in sandier soils. In as much as clayey soils tend to retain more water, they are also thought to contain more *nógó* and so are considered stronger by nature than sandy soils. It is of note that even though black clay is thought to be the strongest soil, it is not necessarily the most preferred. Black sand is widely agreed to be preferable because it is more friable and therefore easier to work when wet.

The scientific analysis of soil fertility is typically expressed in terms of processes that affect chemical composition and physical structure. The farmer's point of view generally agrees with this, though its expression differs. In Madiama, soil fertility is a function of several processes, some socioeconomic and some ecological. Briefly stated, the basic process in maintaining soil fertility is the application of organic fertilizer, *nógó*. Although fallowing is known to rejuvenate soil strength, it is very rarely practised due to perceived inability to leave fields out of production, because there is a shortage of land. As the application of *nógó* is the primary means of maintaining soil fertility, farmers' conceptualizations of *nógó* are essential to understanding their beliefs regarding the management of soil fertility. Farmers describe *nógó* as being food for plants, food that is placed on the surface of the fields and then ploughed in at the beginning of the agricultural season, where it begins to break down until it is consumed by the plants.

According to informants, all sorts of organic matter, including leaves, grasses, manure and so forth, fall into the category of *nógó*. But as with human foods, some plant foods give more strength than others. Animal manure is the preferred fertilizer and its acquisition, a socioeconomic process, is of central importance to soil fertility maintenance. However, not all animal manures are the same. Informants consistently agree that small ruminant manure is 'stronger' than cow manure, because it endures in the soil for up to 7 years whereas cow manure only lasts for 2 or 3 at best.

There are three general means of acquiring manure for the fields: corralling animals in the actual agricultural fields at night, corralling animals at night in household pens and transferring the deposited manure to the fields, and gathering manure from open pasture areas to haul to fields. Manure acquisition strategies vary across households in correlation with wealth, which often manifests itself in terms of animal ownership.

For all farmers interviewed, there is a clear and significant difference between *nógó* and chemical fertilizer. The first indication of this is that two primary categories are commonly used in informal conversation about fertilizer: *nógó* (*farafin*) (fertilizer [African]) and *tubab nógó* (European fertilizer). When asked to list all forms of *nógó*, without using the modifiers *farafin* or *tubab*, farmers would frequently not mention chemical fertilizer. When subsequently asked if chemical fertilizer is a form of *nógó*, they would reluctantly tack it on to the list as an afterthought. This might be partially attributed to the fact that it is too expensive for most farmers to buy and therefore not in their regular cognitive repertoire of fertilizer options. However, the fact that chemical fertilizer is widely used on watermelons, which many farmers grow, would contradict this explanation.

When asked how chemical fertilizer and *nógó farafin* are similar, farmers all responded that they both help crops grow, which is the quality that appears to unify the overall category of *nógó*. When asked how chemical fertilizer and *nógó farafin* are different, four major factors arise. The first and universally expressed difference is that chemical fertilizer, while powerful, is only good for one season. Its efficacy, its *fanga*, does not endure in the soil. Conversely, *nógó farafin* is seen as effective for anywhere from 2 to 8 years after application, depending on the type and intensity of application. This is connected to the second observation that while chemical fertilizer helps plants grow, it does nothing to improve soil quality itself. This observation is significant because although farmers all say healthy-looking crops are the best way to tell the quality of soil, they still recognize that there is a difference between feeding crops and improving soils.

A third quality which farmers use to distinguish chemical fertilizer from *nógó farafin* is that chemical fertilizer requires sufficient rain to be effective and can actually kill crops if there is not enough rain. In contrast, *nógó farafin* works regardless of the amount of rain, and actually increases water retention in the soil. The fourth distinction made by a few informants is that chemical fertilizer is made in factories and sold in markets whereas *nógó farafin* is produced by households and is not available for purchase.

Overall, the folk model of soils in Madiama is not unlike a scientific model in any significant way. Motivation is provided by the need to manage agricultural soils and the principle of categorization is based on observable soil properties such as texture, colour, component materials, and water retention. The application of manure increases the quantity of organic matter in the soil and fosters beneficial biological processes in the soil. Nuanced knowledge of abiotic aspects of soils is illustrated by planting strategies that are based in landscape features and microhydrology, as well as by soil preferences based on physical traits of soil such as friability and water retention. Finally, the distinction between organic fertilizer and chemical fertilizer indicates that there is a difference between biological processes of decomposition and the addition of chemical nutrients needed for plant growth. While chemical fertilizers may make plants grow better, they do not improve the quality of the soil itself in the same way organic fertilizers do. These findings have contributed to the integration of local and scientific knowledge in SANREM research.

Linking Local and Scientific Knowledge

Presentation of findings to NRMAC and others

This chapter has described how, in theoretical terms, local knowledge of soils in

Madiama can be linked with scientific research on soils. However, it is more important that they are well linked in the applied context. Consequently, after each harvest a debriefing is organized for members of the NRMAC, collaborating peasants, and other persons in charge of the commune such as the mayor. The peasants describe to the group the difficulties they have encountered during the growing season, the state of the crops during the vegetative cycle and the amounts harvested. The financial and labour requirements for each experimental technique are presented to the group along with the quantified totals of the harvest. After the third season of field tests, a final debriefing was presented to the NRMAC members and participating peasants. As a part of this, NRMAC and other participants were able to ask questions regarding the tests, keying in on important questions of labour inputs for the various techniques, scheduling problems and diminishing returns on financial and labour investments.

Discussion

There is a wide range of soil fertility maintenance strategies already employed in Madiama, and most correlate with techniques tested in SANREM field trials. In fact, farmers in Madiama know and use all of the introduced techniques, though to variable degrees, intensities and situations. However, to say that the experimental techniques were already known among local participants does not devalue the experiments themselves. Rather than introducing new techniques, the experiments served to scientifically measure and illustrate to the community the variable utility of known techniques and their long-term impacts. The restitution has already caused several participants to revisit their soil fertility management regimes and reconsider the application of certain techniques.

Prior to SANREM, microdosing of chemical fertilizer and manure was known and practised. The cultivation of watermelons as a cash crop has been done with microdoses of chemical fertilizer for at least two generations. Chemical fertilizer was not used for grain crops. As explained by Madiamans, the primary reason it is not used with grain crops is that chemical fertilizer is expensive and grains, being subsistence crops, do not generate income. Consequently, it is difficult to justify spending that much money on chemical fertilizer. Additionally, they explained that the technique requires too much time considering the size and number of grain fields. Although microdosing with manure has been seen in both grain and calabash fields, it is relatively rare.

Field observations indicate that the frequency and intensity of intercropping is low. Even when intercropping and rotation of grains with legumes are practised, they are not always done as a means of improving soils. One explanation offered for this by two different farmers was that groundnuts and cowpeas improve the soil, because they are low-growing and have lots of foliage which is good for the soil. Although intercropping can be widely found in fields, the benefit of intercropping cowpeas with millet and sorghum was most often expressed in terms of labour and space efficiency rather than soil management. By mixing beans in with grain seeds, the planting and harvesting of beans requires little extra time and energy, as well as no additional space.

The use of cattle enclosures can be found in Madiama, but it occurs only at low intensity. Small ruminants are typically penned adjacent to the household and their manure is collected and then spread on fields. In the short season during which transhumant herds are passing through the commune, some herds pass the nights in herd owners' field (if he is local) or in a local host's field, a practice that provided the basis for the corralling experiments. Field proprietors already recognize the benefits of keeping animals in the fields, but there is low systematicity of this practice, partly due to the short season and partly due to lack of sufficient grazing material in the fields.

In some instances even before the completion of the soil fertility experiments, practices could be seen changing among NRMAC members and participants in the experiments. Preliminary interviews with peasant collaborators who attended the debriefing indicate particular interest in two of the methods presented: microdosing and corralling. Although they found the rotation and association experiments useful, the peasant collaborators did not express as much interest in the impact of these technologies as they did in the microdosing experiments that demonstrated even a relatively small amount of fertilizer, when applied intensively, can make a marked difference in crop productivity. One research participant had already begun utilizing this technique in his other fields prior to the completion of the soil fertility experiments, because he saw the efficacy after the first year and that was all the convincing he needed.

In discussions with NRMAC members and test collaborators, three caveats accompanied the enthusiasm for microdosing: (i) even microdosing requires what for the peasants is a significant amount of cash and therefore is not as useful for the large segment of seriously cash-strapped households; (ii) chemical fertilizer is best used *in addition* to traditional organic fertilizers, which aid in water retention, but not as a replacement; (iii) chemical fertilizer, due to its great and fast strength, can actually kill sprouting millet if the early rainfall is weak and therefore carries some risk with it, even though it is perceived as promising under the proper conditions.

Corralling has also gained the interest of several peasant collaborators, who see two distinct benefits of this method. First of all, it minimizes labour, because it requires no gathering, hauling or spreading of manure. Instead, manure is deposited directly in the fields by the cattle themselves. Furthermore, the field gets the urine, which gathered manure lacks. A second benefit is that the manure is more evenly spread within the field than it is with the haul-and-spread method, which tends to create clumps of manure. The primary constraint cited for this method is that the period of availability of large and numerous transhumant herds of cattle is relatively short in Madiama. Immediately after completion of the experiments, one peasant collaborator had already sought out a passing Fulani herder to ask him to let his herd pass the nights in his fields for a couple of weeks after the harvest was finished. No money was paid, though food was delivered to the herder and his family in exchange for this service. This mirrors traditional arrangements found elsewhere in areas of overlap between farming and pastoral systems of production (Painter, 1994; Van den Brink, 1995).

Conclusions

Before the field tests were even finished, a certain interlinking of local knowledge and scientific practice was underway. Local knowledge contributed to the soil typology developed by SANREM through the inclusion of a locally salient soil property, water retention, which is not part of a typical scientific soil survey (see Chapter 3). Furthermore, local knowledge of soil management also shaped one of the field experiments. As previously mentioned, the results of the ethnopedological research indicated that peasants widely agreed that small ruminant manure was significantly more powerful than cow manure in terms of maintaining soil fertility. This hypothesis was directly incorporated into the field tests the following 2 years (see Chapters 11 and 12). Not only did scientific research validate the local knowledge, but it was also able to elaborate the efficacy of variable quantities of manure, showing the point of diminishing returns on the application of small ruminant manure and quantifying its efficacy relative to cow manure, a point which greatly interested the local participants. This demonstrates how local knowledge can inform scientific research that then bolsters it through experimentation.

Within the scientific literature, there are numerous systems for characterizing and classifying soils. There are undoubtedly vastly more folk systems. When conducting participatory, *in situ* soil fertility experiments and land management research, it is useful to elicit local knowledge on soils and incorporate it into the research as well as the extension of the findings. Scientists can benefit from farmers' observations and insights. In the SANREM context, such knowledge has guided experimentation, enhancing the sense of participation and ownership among local collaborators. Scientific experimentation can bolster and expand local knowledge by systematically quantifying and describing the results of various management practices, both introduced techniques as well as those already known. While there is no magic bullet solution to soil fertility maintenance, participatory experimentation and exchange of ideas between peasants and researchers has been found to stimulate new approaches within scientific research as well as the adoption and adaptation of scientifically developed techniques by peasant farmers themselves.

References

Adesina, A. (1992) Village-level studies and sorghum technology development in West Africa: case study in Mali. In: Moock, J.L. and Rhoades, R. (eds) *Diversity, Farmer Knowledge and Sustainability.* Cornell University Press, Ithaca, New York, pp. 147–168.

Ahrens, R.J. and Arnold, R. (2000) Soil taxonomy. In: Sumner, M.E. (ed.) *Handbook of Soil Science.* CRC Press, Boca Raton, Florida, pp. E117–E135.

Arrouays, D. (1987) Farmers' knowledge and soil survey: a multivariate treatment of agropedologial data. *Soil Survey and Land Evaluation* 7, 81–86.

Bockman, O. (1990) *Agriculture et fertilisation: Les engrais – leur avenir.* Norsk Hydro, Oslo, Norway.

Brouwer, J. and Powell, J.M. (1998) Increasing nutrient use efficiency in West African agriculture: the impact of micro-topography on nutrient leaching from cattle and sheep manure. *Agriculture, Ecosystems and Environment* 71, 229–239.

Chambers, R., Pacey, A. and Thrupp, L.A. (1989) *Farmer First: Farmer Innovation and Agricultural Research.* Intermediate Technology Publications, London, UK.

Debrah, S.K., Defoer, T. and Bengaly, M. (1998) Integrating farmers' knowledge, attitude and practice in the development of sustainable striga control interventions in southern Mali. *Netherlands Journal of Agricultural Science* 46, 65–75.

Defoer, T., De Groote, H., Hilhorst, T., Kante, S. and Budelman, A. (1998) Participatory action research and quantitative analysis for nutrient management in southern Mali: a fruitful marriage? *Agriculture, Ecosystems and Environment* 71, 215–228.

Dialla, B.E. (1993) The Mossi indigenous soil classification in Burkina Faso. *Indigenous Knowledge and Development Monitor* 1, 17–18.

FAO (1998) *World Reference Base for Soil Resources.* FAO, Rome, Italy.

Finkl, C.W. (1982) *Soil Classification.* Hutchinson Ross, Stroudsburg, Pennsylvania.

Furbee, L. (1989) A folk expert system: soils classification in the Colca Valley, Peru. *Anthropological Quarterly* 62, 83–102.

Gobin, A., Campling, P., Deckers, J. and Feyen, J. (2000) Integrated toposequence analyses to combine local and scientific knowledge systems. *Geoderma* 97, 103–123.

Guillet, D. (1989) A knowledge-based-systems model of native soil management. *Anthropological Quarterly* 62, 59–68.

Painter, T., Sumberg, J. and Price, T. (1994) Your terroir and my 'action space': Implications of differentiation, mobility and diversification for the approche terroir in Sahelian West Africa. *Africa* 64, 447–463.

Pawluk, R.R., Sandor, J.A. and Tabor, J.A. (1992) The role of indigenous soil knowledge in agricultural development. *Journal of Soils and Water Conservation* 47, 298–302.

Powell, J.M., Fernandez-Rivera, S., Hiernaux, P. and Turner, M.D. (1996) Nutrient cycling in integrated rangeland/cropland systems of the Sahel. *Agricultural Systems* 52, 143–170.

Rhoades, R. (1984) *Breaking New Ground: Agricultural Anthropology.* International Potato Center, Lima, Peru.

Rhoades, R. (1989) The role of farmers in the creation of agricultural technology. In: Chambers, R., Pacey, A. and Thrupp, L.A. (eds) *Farmer First: Farmer Innovation and Agricultural Research.* Intermediate Technology Publications, London, pp. 3–9.

Scoones, I. and Thompson, J. (1994) *Beyond Farmer First: Rural People's Knowledge, Agricultural Research and Extension Practice.* Intermediate Technologies Publications, London, UK.

Scoones, I. and Toulmin, C. (1998) Soil nutrient balances: what use for policy? *Agriculture, Ecosystems and Environment* 71, 255–267.

Sebillotte, M. (1989) *Fertilité et système de production 'ecologie et aménagement rural'.* INRA, Paris.

Sillitoe, P. (1996) *A Place Against Time: Land and Environment in the Papua New Guinea Highlands.* Harwood, Amsterdam.

Spaargaren, O.C. (2000) Other systems of soil classification. In: Sumner, M.E. (ed.) *Handbook of Soil Science.* CRC Press, Boca Raton, Florida, pp. E137–E174.

Tabor, J.A. (1993) Soil survey and indigenous soil classification. *Indigenous Knowledge and Development Monitor* 1, 28–29.

Talawar, S. and Rhoades, R. (1998) Scientific and local classification and management of soils. *Agriculture and Human Values* 15, 3–14.

van Breemen, N. (1993) Soils as biotic constructs favouring net primary productivity. *Geoderma* 57, 183–211.

Van den Brink, R., Bromley, D. and Chavas, J.P. (1995) The economics of Cain and Abel: agro-pastoral property rights in the Sahel. *The Journal of Development Studies* 31, 373–399.

Vannier, G. (1987) The porosphere as an ecological medium emphasized in professor Ghilarov's work on soil animal adaptations. *Biological Fertility of Soil* 3, 39–44.

Vilain, M. (1989) *La production végétale: Agriculture d'aujourd'hui.* Lavosier TEC et DOC, Paris, France.

Zimmerer, K. (1994) Local soil knowledge: answering basic questions in highland Bolivia. *Journal of Soil and Water Conservation* 49, 29–34.

11 Biophysical Assessment of Alternative Soil Fertility Technologies Through Modelling

Oumarou Badini,[1] Boureima Traoré[2] and Mohamed Sidi Mohamed Touré[2]

[1]International Programs/Development Cooperation, Washington State University, Pullman, WA 99164–5121, USA; [2]Centre Régional de la Recherche Agronomique, Institut d'Economie Rurale (IER), BP 205, Mopti, Mali

One of the objectives of the SANREM biophysical research in West Africa has been the assessment of the effects of land use patterns and agricultural practices on the landscape and lifescape systems of smallholder farmers. The overall goal of the biophysical tools and methods development activity conducted in the last 3 years in Mali is to support researchers and natural resource decision-makers at local and regional scales with appropriate data, tools and methods to identify and evaluate alternative technologies. This has led to the creation of a natural resource baseline information and modelling system as a tool to support decision making, and as a platform for more focused research. To reach such a goal, the strategy on the biophysical side has been to integrate baseline database development, soil and land use cartography, geographical information systems (GIS), and cropping systems simulation modelling tools. The baseline data, soil and land use maps will increase natural resource stakeholders' awareness and knowledge about existing natural resource sustainable production potentials and limitations.

The challenge faced by farmers in the semi-arid region of West Africa in general, and in the study region in particular, is to find sustainable ways to improve crop productivity and reduce production risks due to irregular and low rainfall as well as low fertility of soils. Technical options to improve the productivity of crops exist. However, the identification of sustainable agricultural options requires analyses that can make assessments over long timeframes of uncertain and changing biophysical factors, and variable social and economic settings (Stöckle, 1989). The selection of alternative technologies suitable to a given environment requires a long time series (30 years or more) of crop responses to weather, soil and management. Therefore, it is impractical to rely solely upon empirical observations to demonstrate that a particular set of cropping systems is sustainable (or not sustainable) over a 40–50 year timeframe.

New tools such as simulation models that portray the response of crops to environmental, site and management variables offer the best opportunity, other than seemingly endless experimentation, to evaluate the long-term effects of farming practices on the landscape and lifescape (Loomis and Connor, 1992). They can provide the information necessary to assess the long-term biophysical viability of different crops and alternative

 Conflict, Social Capital and Managing Natural Resources (ed. K.M. Moore)

technologies suited to a given environment. They can also provide coefficients and stochastic crop yield outputs to be used for risk and socioeconomic analyses of different management practices (see Chapter 12).

The main objective of the present chapter is to share the experience in the assessment of the viability of existing and alternative technologies adaptable to the study landscape and similar areas, using cropping systems simulation modelling. Cropping Systems Simulation Model (CropSyst) (Stöckle and Nelson, 1996) was developed with a focus on crop growth simulation in response to soil, weather and management, has been adapted to conditions in the Madiama region, and used to simulate the long-term effects of selected soil fertility management practices. Simulation modelling is used here as a complement to field-testing to allow for long-term evaluation and comparison of existing and alternative soil fertility management practices. The present assessment of the simulation results allowed for the identification of soil fertility technologies that are best suited to the region.

An approach integrating biophysical indicators of productivity and sustainability generated from field monitoring and simulation is used for this assessment. Yield variability and production risk as well as soil losses due to water erosion were analysed over multiple seasons to assess the viability of existing and alternative cropping systems and management practices. This approach provides a simple and more complete method for evaluating a set of complex interacting variables in an easy-to-understand manner applicable to most cropping systems. It allows yield efficiency, variability and environmental sustainability to be advanced on an equal basis.

Methodology

Scale considerations

Individual fields are the fundamental geographical units in the modelling and analysis of cropping systems in the present study. A cropping system here is defined as a crop (or pasture) community, together with management practices such as tillage methods and rotation used in crop production (Loomis and Connor, 1992). At this field scale it is possible to examine the impact of management on crop productivity, sustainability, and overall biophysical viability. By observing and/or simulating field conditions over years, effects of crop rotation, tillage practices, soil conservation, soil amendments and subsequent yields can be seen, and analyses made of the use of resources such as nutrients and water. Economic analyses of costs and benefits can also be conducted at the field scale.

At a higher level, fields are components of farms under the management of particular farmers. The principal crops and management practices employed on a particular farm constitute a farming system. Ruthenberg (1980) states that because farms are organized to produce a net economic return they are the fundamental units for economic and sociological analyses. The regional organization of farming systems is the landscape level. It offers an additional venue for analysis, including matters such as drainage and water quality analyses. At this level, economic and sociological studies might include service roles of and impacts on towns and villages within the region.

Details on biophysical landscape characteristics and farming systems of the study site are given in Chapter 3.

CropSyst: cropping systems simulation model

Brief overview of model

CropSyst is a multi-year, multi-crop, daily time-step generic crop growth simulator. The model is developed to help decision-makers analyse current and alternative cropping systems and predict their environmental and socioeconomic sustainability. It is a comprehensive cropping system model to be used as an analytical tool to assess the effects of weather, soils, crop rotations and management practices on complex cropping systems and the environment. The modelling framework has a capacity to reflect site specificity and integrates several different components and management options. This holistic approach permits users to assess the sustainability of cropping systems from the point of view of constraints imposed by productivity, resource conservation, protection of water quality and socioeconomic considerations.

Details on management options and model components can be found in the model's user's manual (Stöckle and Nelson, 1996) and elsewhere (Badini *et al.*, 1997). The model simulates the soil water budget, soil-plant nitrogen budget and crop growth and development. The amount of water used for crop growth can be determined by modelling the components of the soil water balance. The water budget components considered are: irrigation, precipitation, rainfall intercepted by the crop and surface residue, surface runoff, soil and residue evaporation, infiltration through soil layers, transpiration, deep percolation and water storage in the soil profile.

Nitrogen is an important nutrient whose availability is primordial for crops to reach their growth potential. The components of the nitrogen budget in CropSyst include transport, transformation, ammonium sorption, crop nitrogen uptake and residue mineralization. Nitrogen types (nitrate and ammonium), form (organic or inorganic, solid or liquid), and application methods can be specified in each simulation. CropSyst simulates crop growth and development using a generic crop simulator. Species and cultivars are characterized by a set of parameters that determine crop response to the environment. The crop parameter file is structured in the following sections: phenology, morphology, growth, residue, plant nitrogen, harvest index and salinity tolerance.

Model inputs used in the study

In order to run CropSyst, input data describing the location, weather, soils, crops and management from the study site are needed. Separation of files allows for an easier link of CropSyst simulations with GIS allowing the extrapolation of research results to other sites. A simulation control file combines the input files as desired to produce specific simulation runs. In addition, the control file determines the start and ending day for simulation, defines the crop rotations to be simulated, and sets the values of all parameters requiring initialization. The CropSyst model has been adapted for West African conditions. Database and model parameter files (previously non-existent for the study area) were established for weather, soils, crops and different management schemes in the present study.

LOCATION AND WEATHER DATABASE Parameters characterizing the site of interest are name (Madiama), latitude, longitude, elevation and daily rainfall as well as minimum and maximum temperatures. Actual weather data from 1970 to 2002 for Djenné (located at about 25 kilometres from the study site Madiama) and Madiama were used with the weather generator CLIMGEN to generate 50 years of daily data including solar radiation, rainfall, air temperature and relative humidity. These compiled weather data were used for the simulation modelling.

SOIL DATABASE Generally, the soils of the commune are poor in nutrient elements. Baseline values have been derived that show very low values for total nitrogen (less than 0.1% in all soils), very low values for organic carbon content (0.08 to 0.5%) and deficient phosphorus content (2.5 to 8 ppm). The soils range from strongly acid (pH 4.2) in the hydromorphic floodplains to slightly acid (pH 6.2) in the loamy and sandy plains.

Using data provided by the Soil Survey of Madiama Commune (Badini and Dioni, 2001) and seasonal biophysical monitoring data from the site, a soil parameter file was constructed for each of the eight land units identified in the commune (Chapter 3). The plains of sandy to loamy materials (unit t2 – tropical ferruginous soils or USDA *Ultic Haplustalfs*), which cover about 25.5% of the commune and on which millet is grown, were used for the present simulation, and the comparative biophysical assessment of alternative and unimproved soil fertility management technologies.

CROP DATABASE From the existing database, a 90-day millet cultivar (local name: *sagnori*) was used in the present simulation. Calibration for the millet cultivar parameters and yield estimate was performed using four years of field data on phenological events (emergence, flowering and maturity dates) and crop yields from the biophysical monitoring and field trials conducted in Madiama since 1999. These were adjusted in accordance with results compiled from similar tests in the region.

MANAGEMENT DATABASE In the course of the SANREM West Africa project, a number of soil fertility management technologies (Traoré, 2002) were field tested to determine which were the most viable and adaptable to the research area and region. Simulation modelling was used as a complement to field-testing to allow for long-term evaluation and comparison of existing and alternative farming and NRM technologies without long-term field tests to identify crops and technologies that are best suited to the region. Technologies evaluated include crop rotations and organic (manure) and inorganic (chemical) fertilization (Table 11.1). Information and data for the management database were derived from a review of soil fertility literature (Traoré, 2002) and soil fertility trials conducted in Madiama.

After calibration of millet cultivars, the databases for soils, weather, crops and management practices were combined through a simulation rotation table in CropSyst to run a 30-year simulation of crop growth, yield and soil water erosion rates of selected soil fertility management practices. The outputs from the simulation were used to assess the biophysical viability of selected soil fertility management practices (Table 11.1).

Table 11.1. Soil fertility management simulated.

Fertilization practices	Treatment	N fertilizing unit (kg/ha)
Traditional management	Control (no N input)	0
Crop rotations		
• Continuous cropping	Millet–millet	0
• Continuous cropping	Cowpea–cowpea	6.03 kg N every year
• 2-year rotation	Cowpea–millet	6.03 kg N every 2 years
• 3-year rotation	Cowpea–millet–millet	6.03 kg N every 3 years
Organic fertilization		
• Cow manure	2 t/ha every year	19.6 kg N every year
	5 t/ha every 3 years	49 kg N every 3 years
• Small ruminant manure	2 t/ha every year	25 kg N every year
	5 t/ha every 3 years	62.5 kg N every 3 years
Inorganic fertilization		
• Recommended	100 kg diammonium phosphate (18-46-0) at planting, 50 kg urea (46-0-0) each at 10 and 50 days after emergence	15 kg N at planting, 23 kg N at 10 days, and 23 kg N at 50 days
• Microdose complex 1 (6 g/poquet)	60 kg/ha Complex cereal NPK (15-15-15) at planting	9 kg N at planting
• Microdose complex 2 (6 g/poquet)	60 kg/ha Complex cereal NPK (15-15-15) at 10 days after emergence	9 kg N at 10 days
• Microdose DAP and urea (2 g/poquet DAP and 1 g/poquet urea)	20 kg/ha of diammonium phosphate (18-46-0) at planting, and 10 kg/ha of urea (46-0-0) at 30 days after emergence	3.6 kg N at planting and 4.6 kg N at 30 days after emergence

Assessment of crop rotation, organic and inorganic soil fertility management practices

From the many outputs CropSyst allowed, the present study used yield and erosion rates as the key indicators of productivity and sustainability. Crop yields are the dominant factors affecting, for example, farmer adoption of alternative practices such as soil fertility and conservation practices. Erosion rates are determinants of soil fertility status and water holding capacity, both of which are critical to plant production. The amount and quality of soil eroded or retained on a plot of cropped land is a primary determinant of sustainability for either food or cash crops. Quantification of yields and expected erosion rates over time is important in deciding on the productivity and sustainability of a given natural resource base.

To determine the overall biophysical viability of each technology, the variables of productivity and sustainability were combined into a unique viability factor determined as a weighted value of yield efficiency, stability and soil sustainability. The best-adapted technologies are then determined as the ones with the best yield efficiency, stability and sustainability. Details on each of these indicators are given below.

Indicator of productivity: yield

Productivity is explicitly defined by yield, the amount of useful product per unit land area. The concepts of efficiency, stability and stochastic dominance (riskiness) are used here to analyse and compare the yields obtained from the different management practices simulated in the present study.

EFFICIENCY Because yield is the resultant of various natural and human factors such as radiation, water, nutrients, and labour, it has the property of also measuring efficiency relative to such inputs (Loomis and Connor, 1992). Observed yields may fall anywhere on a continuum between crop failure and potential yield. In the Food and Agriculture Organization (FAO) terminology, *potential yield* is considered as that obtained under the ideal conditions with no limitation set for water, nutrients and technology. *Attainable* or *optimum yield* corresponds to the best yields achieved through skilful use of the best technology. High yields reported by experimental stations and the best farmers serve as measures of attainable yield. The concepts of actual, attainable and potential yields assist in assessments of farming systems and help to identify opportunities for improvement. They also serve to define the intensity of farming. The average yield obtained from each management scenario will be compared to the attainable (optimum fertilization) yields to determine their level of efficiency. In this study, optimum fertilization corresponds to the fertilization levels recommended by research stations in Mali (Table 11.1).

STABILITY Stability is defined here as the variation in yield over time, i.e. as relating to repeatability and predictability in farming. Yields may vary over years with weather and other causes. Marten (1988) used the term stability in reference to the degree of such variation. Standard deviation (SD) and coefficient of variation (CV%) are used as measures of degree of variation or stability of yield. Comparisons are made between the different management practices to determine the most stable in terms of productivity.

YIELD PROBABILITIES Yield probabilities of exceeding a given level of yield were computed and plotted to compare different management practices, and to determine the most risk-efficient and dominant management systems. For example, if the plots of yield probability level of different management practices do not intersect, the most risk-efficient management is the one whose probability at all outcomes (e.g. yields) is more than, or occasionally equal to, the probabilities of all other managements. This is known as first-degree stochastic dominance (FSD) and is based on the fact that decision makers prefer more income to less (Anderson *et al.*, 1977).

In the common situation where more than one probability function first degree dominates the others or if no yield function dominates, then, the second-degree stochastic dominance (SSD) principle is applied to produce a ranking. SSD can be established by measuring the cumulative area below two functions (Anderson *et al.*, 1977; Johnson *et al.*, 1994). For example, a yield probability distribution, I, dominates another, J, by SSD if the cumulative area under I exceeds the area under J. SSD is appropriate for a risk averse decision maker, but is inappropriate for risk-neutral and risk-preferring individuals. SSD is appropriate in the Sahel, because most farmers and most government policy makers cannot afford to be other than risk averse.

Indicator of sustainability

Sustainability is concerned with whether the production level can be maintained over the years at the site. An overlap evidently exists between concepts of stability and sustainability. Large yield variations and persistence of crops both affect sustainability and stability, but maintenance of soil resources is the key issue, because soil degradation has the effect of gradually lowering the potential of site until farming is no longer possible. Proper management of soil resources is the key to sustaining agriculture, because soil properties determine the type of farming that can be practised. Soil characteristics important to sustainability include the ability to supply essential plant nutrients and water and the ease of tillage. Therefore, rate of soil loss and nutrient exhaustion serves to quantify sustainability. The soil water erosion rates (SWER) are used to compare the management practices and technologies in the present assessment.

Overall biophysical viability

To determine the overall biophysical viability level of each technology, the variables of yield efficiency, stability and soil sustainability were combined into a unique viability factor. In reference to each of these biophysical indicators (efficiency, stability, SWER), each technology was given a dominance rank (from 1st to *n*th) that is then combined into a final biophysical viability rank to determine the overall dominating technologies. The best-adapted technologies are then determined as the ones with the best yield efficiency, stability and sustainability.

Results and Discussions

Productivity assessment of soil fertility management practices

Crop rotations

It is assumed that farmers can avoid or reduce problems such as weeds, soil-borne diseases, low fertility and soil hardpans through rotations. Rotations are also used to manage water availability by sequencing crops that have different patterns and amounts of water and nutrient use. In this way, the system can fine tune to rainfall variations over the years and give higher or more efficient, stable, risk-efficient yields, and allow a more sustainable production environment than the existing systems. The results from the field tests and the long-term simulation confirm this assertion.

In the simulation runs, monoculture (continuous cultivation) for millet was compared with a 2-year cowpea–millet rotation and a 3-year cowpea–millet–millet rotation. Figure 11.1 shows that long-term average millet yield was 583 kg/ha for the low input traditional practice compared with 650 kg/ha for the cowpea–millet–millet rotation (n-m-m), and 726 kg/ha for cowpea–millet rotation (n-m), a difference of 67 kg and 143 kg/ha, respectively. Overall yield efficiency (Table 11.2) compared to the attainable optimum production conditions is, respectively, 52% for the 2-year rotation and 47% for the 3-year rotation. There is no difference in stability between the two rotations with a CV of 56% for both the 2-year and 3-year rotations.

In addition, the long-term simulation has allowed for the determination of yield probabilities and risks linked to production and the choice of the best technology among the compared systems.

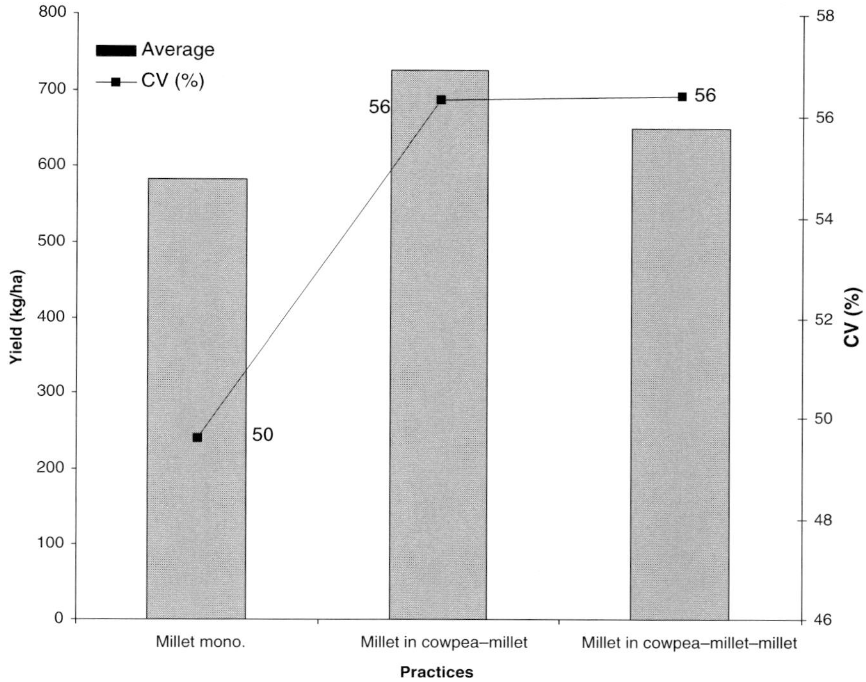

Fig. 11.1. Millet long-term average yield and variation (CV%) in rotation practices (monoculture, cowpea–millet, and cowpea–millet–millet).

Table 11.2. Long-term millet yields (mean, stability and efficiency) for soil fertility management practices.

Fertilization practices	Mean	Stability		Efficiency recommended yield (%)
		SD	CV (%)	
Potential productivity	1949	694	38	140
Recommended (attainable) fertilization	1392	458	33	100
Traditional low input monoculture	583	298	51	42
Cow manure (2 t/year)	1011	277	27	73
Cow manure (5 t/3 years)	928	271	29	67
Small ruminants manure (2 t/year)	1078	293	27	77
Small ruminants manure (5 t/3 years)	993	279	28	71
Microdose complex 1 – NPK (6 g/hill at planting)	910	287	32	65
Microdose complex 2 – NPK (6 g/hill at 10 days after emergence)	924	280	30	66
Microdose DAP and urea (2 g/hill DAP at planting and 1 g/hill urea 30 days after emergence)	908	279	31	65
Millet in cowpea–millet (n-m)	726	409	56	52
Millet in cowpea–millet–millet (n-m-m)	650	367	56	47

Given the erratic nature of the weather in arid and semiarid Africa and the subsequent stochastic yields, yield probability levels represent another way of looking at production stability and the risk involved in growing a given crop in a particular environment and management scheme. Figure 11.2, which shows probability distributions for the two rotations and the traditional continuous monoculture practice, indicates that in 1 year out of 10, the yield for millet in the cowpea–millet rotation will exceed 1217 kg/ha compared with 1154 kg/ha and 950 kg/ha for millet in cowpea–millet–millet rotation and continuous millet, respectively. In 5 years out of 10, the millet yield in a cowpea–millet rotation is expected to exceed 628 kg/ha compared with 550 kg/ha and 496 kg/ha for millet yield in cowpea–millet–millet and continuous millet, respectively. Since Fig. 11.2 shows the dominance of the 2-year cowpea–millet rotation over the 3-year cowpea–millet–millet and the continuous cropping of millet, this 2-year rotation appears to be the most risk-efficient practice among the three in the present comparison.

Considering yield efficiency level, stability and stochastic dominance, the 2-year cowpea–millet rotation is the most productive, stable, and risk-efficient of all three rotations under comparison. The continuous millet system is the least efficient. This analysis confirms the hypothesis that crop rotations will give more risk-efficient yield than mono-cropping systems.

Organic (manure) fertilization

Organic fertilization from manure transported to the field or corralling animals directly in the field is a known traditional practice in the Sahel. But given the difficulties in producing the required amount for optimum crop growth, SANREM proposed a number of alternative improved technologies that basically bring in lower quantities in manure. Manure from small ruminants (sheep and goats) and large ruminants (cattle) at a rate of 2 t every year or 5 t every 3 years was simulated and their corresponding yield outputs compared to existing unimproved management systems (Table 11.1).

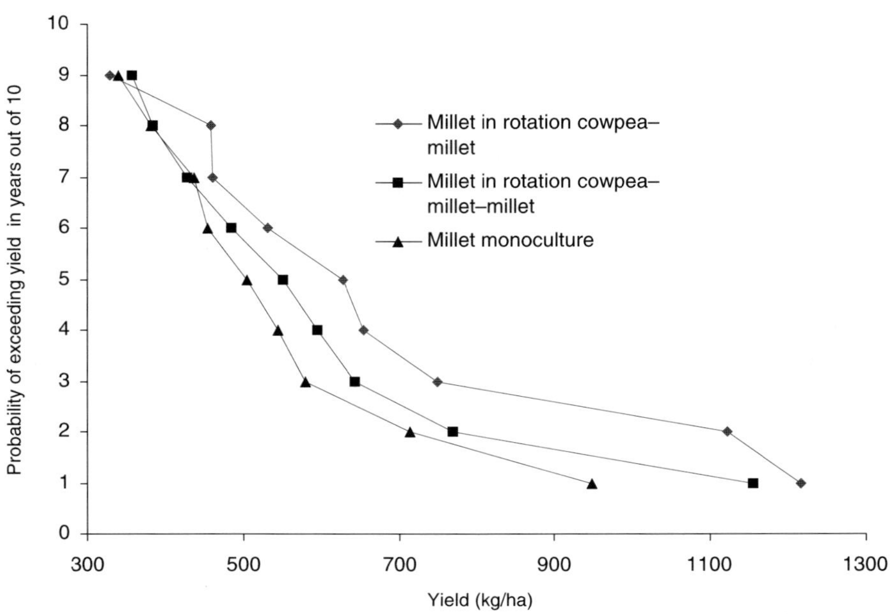

Fig. 11.2. Yield probability functions for rotation practices.

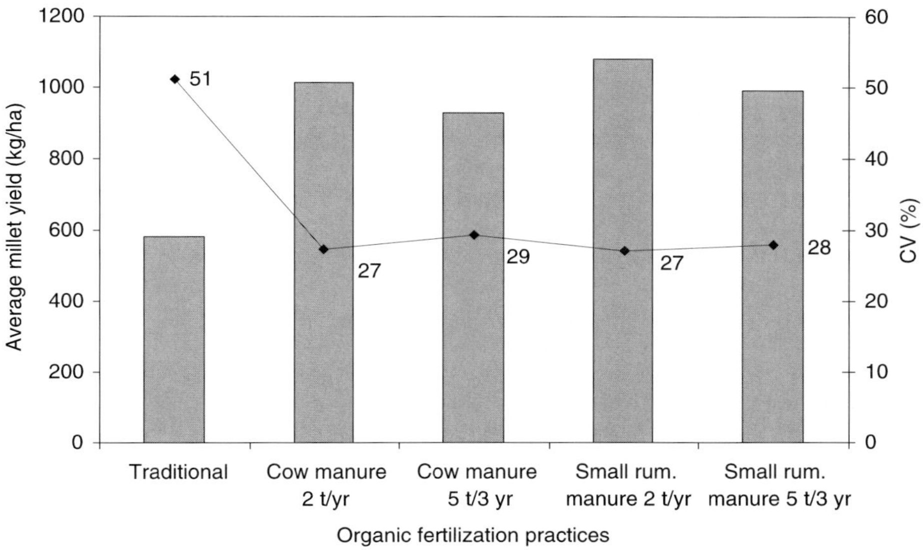

Fig. 11.3. Millet long-term average yield and variation (CV%) for organic fertilization practices.

Figure 11.3 shows that among the four manure type and rate combinations, small ruminant manure at 2 t every year (PR 2t) has the highest long-term average (1078 kg/ha) compared with yields between 1011 kg/ha and 928 kg/ha for the other organic fertilization practices. Yield efficiency for PR 2t is 77% of recommended attainable yield compared to 73% or less for the other manure practices (Table 11.2). Although not statistically different from each other, all manure practices showed a significant difference in yield from the traditional practice (only 583 kg/ha) as well as lower variability (27–29% of CV) compared to the traditional practice with 51% of variability. Also, organic fertilization has exhibited the lowest yield variability of all management practices simulated.

Overall, Fig. 11.4 shows that the rate of 2 t of small ruminant manure applied each year stochastically dominated all other organic fertilization simulated in the analysis. In 1 year out of 10, yield from PR 2t will exceed 1424 kg/ha compared to 1279 kg/ha for cattle manure at a rate of 5 tonnes every 3 years (Bov 5t). Similarly, in 5 years out of 10, PR 2t allows a productivity of 1061 kg/ha compared to 890 kg/ha for Bov 5t, and 496 kg/ha for the traditional low input practice. It shows that from a biophysical standpoint, PR 2t is the dominant and most risk-efficient practice among the four organic fertilization technologies screened in the present work.

Inorganic fertilization

It is assumed that crops that are not limited by the availability of nutrients will grow best and will have the highest yields. But given the fact that local farmers often cannot afford the recommended doses, a number of improved technologies that basically bring in lower levels of fertilizers (microdosing) and different dates of placement are proposed and compared to existing (unimproved) and recommended fertilization systems (Table 11.1).

The long-term 30-year average yield outputs from the three fertilizer microdosing management practices (Fig. 11.5) show no significant yield difference. But the average yields from all microdosing practices were significantly higher than the control (traditional practice with low input) with a difference in yield between 325 and 341 kg/ha. Microdose complex 2 (with 60 kg/ha

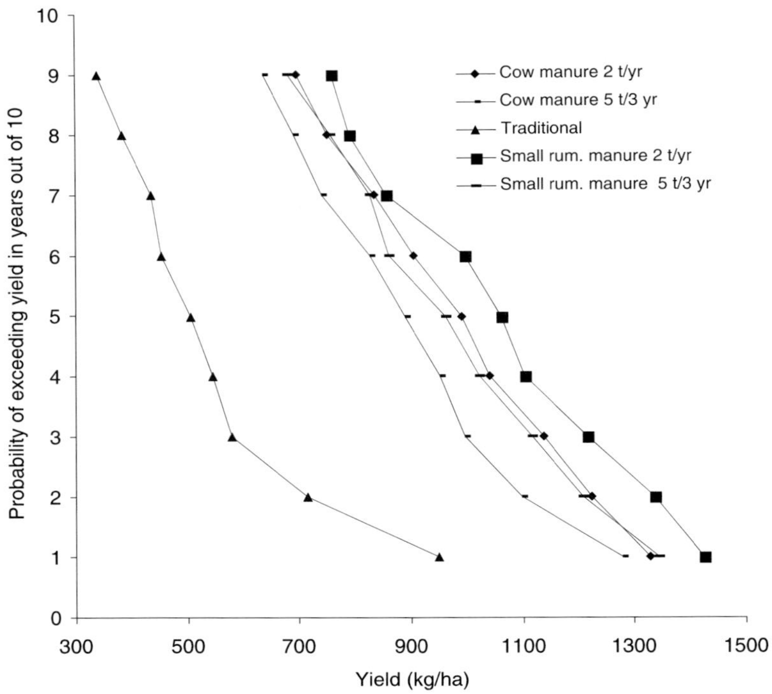

Fig. 11.4. Yield probability functions for organic fertilization practices.

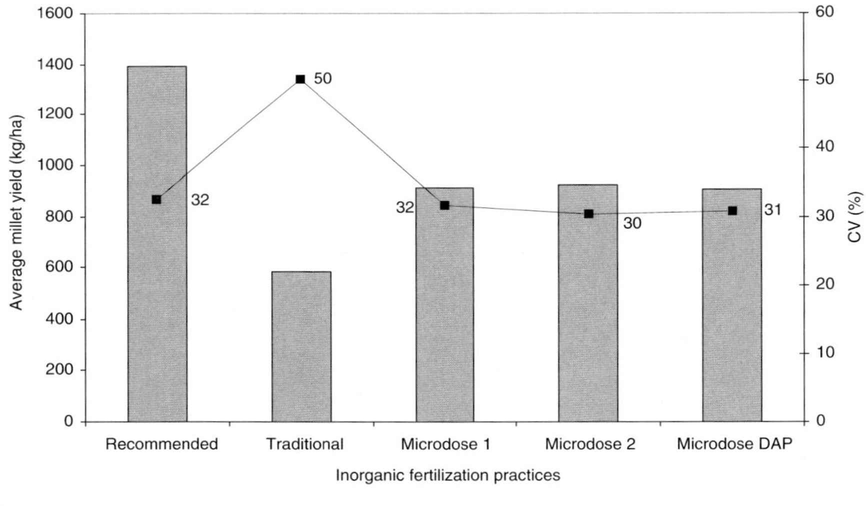

Fig. 11.5. Millet long-term average yield and variation (CV%) for inorganic fertilization practices.

Complex cereal NPK (15-15-15) at 10 days after emergence) has an average yield only slightly higher than Microdose complex 1 (60 kg/ha Complex cereal NPK at planting) and Microdose DAP (20 kg/ha of diammonium phosphate at planting, and 10 kg/ha of urea at 30 days after emergence). All three practices have a similar CV of about 31%

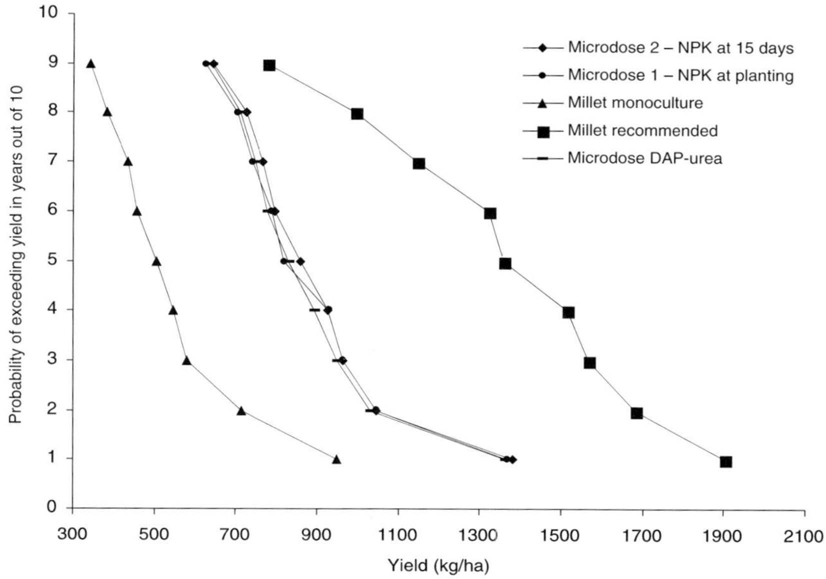

Fig. 11.6. Yield probability functions for inorganic fertilization practices.

and an efficiency level compared to attainable yield of about 65% (Table 11.2). This represents a yield improvement of about 64% over the traditional practice.

The yield probability distribution functions (Fig. 11.6) give a small edge to Microdose 2, which slightly dominated the other two Microdose practices at most probability levels. Although Microdose 1 and 2 have the same total amount of fertilizer input (60 kg/ha), the difference is that Microdose 2 was implemented at 10 days after emergence instead of at planting as done with Microdose 1. Despite a lower total input in fertilizer (30 kg/ha), which is only half of both Microdose 1 and 2, Microdose DAP has an average yield similar to Microdose 1 and only a difference of 16 kg/ha from Microdose 2. Furthermore, even with lower fertilizer input it has shown a slight SSD to Microdose 1. This result strengthens the importance of date of fertilizer placement and type, which was picked up by the model. Economic analyses as presented in Chapter 12 should establish the financial viability of this finding for decision makers.

Sustainability and biophysical viability assessment of soil fertility management practices

Sustainability analysis of all management practices

Erosion rates are determinants of soil fertility status and water holding capacity that are critical to plant production. The amount of soil eroded from a plot of cropped land is a primary determinant of sustainability of crop production. Sustainability is concerned with whether the production level can be maintained over the years at the site. Although some erosion during cropping cycles is inescapable, all sites have some tolerance to erosion. A strict tolerance with the soil loss rate equal to or less than the soil formation rate would allow production to continue indefinitely. Generally, estimates of soil formation rates at undisturbed sites are near 0.1–0.2 mm depth per year (i.e. about 1.3–2.6 t/ha/year). Rates increase with rainfall (Loomis and Connor, 1992), and many other soils in Madiama have soil loss above these figures. The soil type Ultic Haplustalfs (tropical ferruginous soils) used in the present analysis has exhibited one of the lowest soil loss rates with an overall average of about 1 t/ha under the parameters used in the modelling.

In the present analysis, the rate of soil loss is used as an indicator of sustainability. Cumulative (30-year) soil loss was computed for each management practice and comparisons were made to determine the least prone to soil loss. The most sustainable management is considered as the one that would have less total soil loss over the 30 years and would therefore maintain production level over years at the site.

Figure 11.7 shows that in the course of 30 years the soil fertility management of providing 2 t of small ruminant manure every year (PR 2t) has the least cumulative soil loss (9 t/ha on soil t2), compared to all other practices. It is followed by Bov 2t (2 tonnes cow manure every year), PR 5t, and in fourth place, the recommended fertilization. The least effective practices are the traditional continuous cropping and the rotations. Figure 11.8 shows the long-term trend of cumulative soil loss, and Fig. 11.9 shows the soil loss probability levels. Overall, during the course of 30 years manure application at 2 t/year has shown less soil loss at all probability levels. The least sustainable (i.e. with higher soil loss) are the rotations and monoculture.

The results show that simulation modelling can be a good tool to help show the long-term impact of management on soil loss and sustainability. The ranking of the practices as related to the level of soil loss shows that improvement of both soil physical and nutrient status is paramount to conserving the soil resource over the long-term. Practices that yield more crop biomass and residue, such as the recommended optimum fertilization practice, and improve soil binding and water holding capacity, such as organic fertilization through manuring of fields, are better in sustaining the resource production base, which is the soil. In general, manure, which allows binding of soil particles and an improvement of soil water holding capacity and fertility status, has shown superiority in maintaining the soil resource.

Overall ranking of the biophysical viability of soil fertility management practices

Combining yield efficiency, stability and sustainability levels of all compared management practices, the technologies have been ranked from most viable to least viable as individual practices on biophysical grounds (Table 11.3). Although optimal inorganic fertilization allows higher yields, organic fertilization appears to be the most

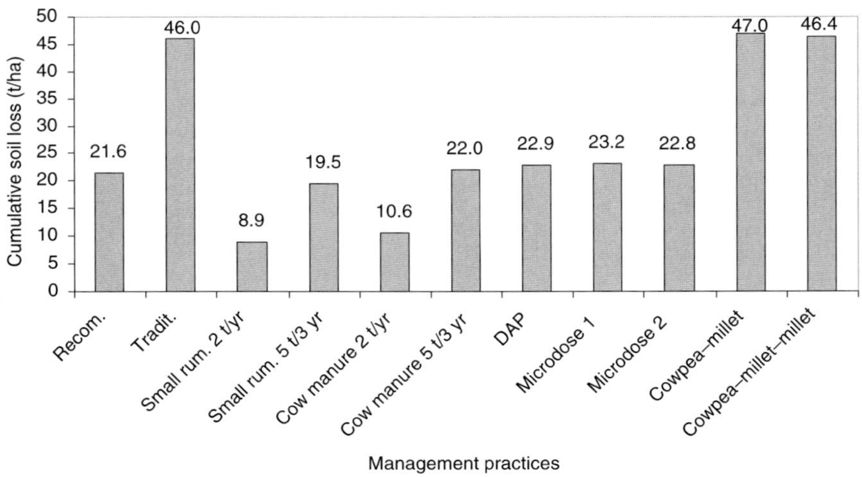

Fig. 11.7. 30-year cumulative soil loss for different management practices on Ultic Haplustalfs soil in Madiama commune.

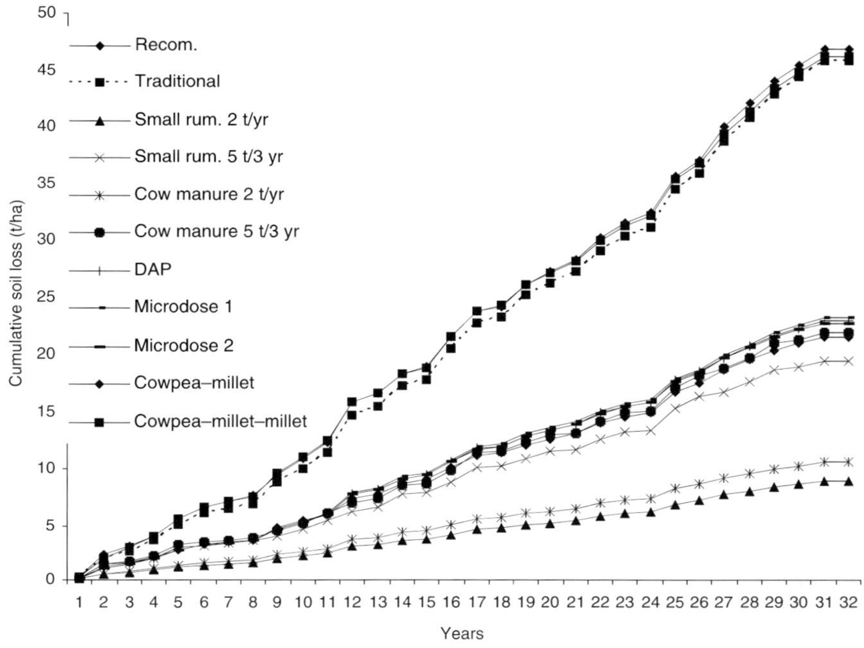

Fig. 11.8. Long-term cumulative soil loss for all management practices on Ultic Haplustalfs soil in Madiama commune.

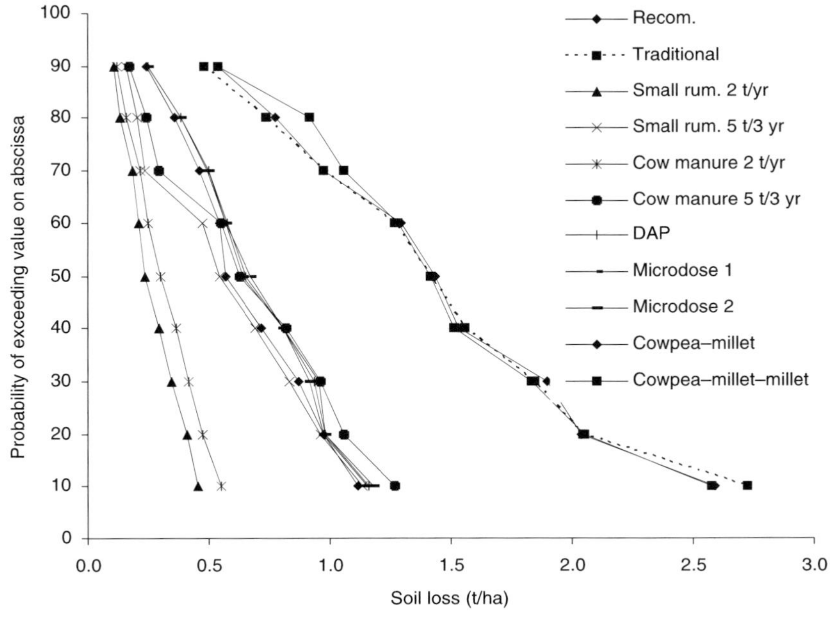

Fig. 11.9. Soil loss probability functions for all management practices on Ultic Haplustalfs soil in Madiama commune.

Table 11.3. Biophysical viability ranking of soil fertility management practices effect on millet yield in Madiama commune.

Fertilization practices	Efficiency rank (1)	Stability rank (2)	Sustainability rank (3)	Total (1+2+3)	Viability rank
Small R. manure (2 t/year)	2	2	1	5	1
Cow manure (2 t/year)	3	1	2	6	2
Small R. manure (5 t/3 years)	4	6	3	13	3
Recommended	1	8	4	13	4
Microdose 2	6	3	6	15	5
Cow manure (5 t/3 years)	5	7	5	17	6
Microdose DAP	8	4	7	19	7
Microdose 1	7	5	8	20	8
Traditional (monoculture)	11	9	9	29	9
Millet in cowpea–millet	9	10	10	29	10
Millet in cowpea–millet–millet	10	11	11	32	11

viable practice. The 2 t/year of small ruminants' manure, followed by the 2 t/year of large ruminants' manure applications and the 5 t every 3 years of small ruminants' manure have demonstrated higher overall long-term viability than the optimum recommended fertilization. In all criteria, rotation practices together with low input cereal monoculture appear to be the least viable technologies in terms of yield efficiency, stability and soil sustainability.

Conclusions

One of the greatest challenges in the semi-arid Sahel region is to develop cropping strategies that provide the most efficient level of nitrogen uptake for long-term yield stability and soil resource base sustainability. The SANREM experience in Mali shows that soil, weather, crop and management databases, in combination with cropping systems simulation modelling, offer better insights about what crop, technology or cropping system would best fit a specified landscape. Through long term screening and analysis plant breeders can be provided with a better representation of the actual situation faced by farmers for whom breeding programmes are undertaken. In addition, the extrapolation of findings to other similar environments can be facilitated. It is expected that such tools will assist individual researchers and community decision makers in Madiama and in the region in reaching decisions about how to best plan their land use and manage their resources. Also, information generated from these analyses provides improved insight into crop production responses to various agricultural practices and soil management technologies across the Sahel. The end result is identification of suitable practices based on performance evaluation through field monitoring and simulation modelling.

The short-term field results and long-term simulation outputs presented here have shown that all alternative management technologies contributed to substantial increase in yield. Long-term yield efficiency levels were between 42% for low input traditional practice and 77% for small ruminants' manure at an application rate of 2 t/ha every year. Overall, by considering both productivity (yield efficiency and stability) and sustainability (soil loss rates), the 2 t/year of small and large ruminants' manure applications, the 5 t every 3 years of small ruminants' manure, and the optimum recommended fertilization are the most viable technologies in the long-term. The least viable individual technologies that showed both lower yields and higher rates of soil loss were the cowpea–millet rotations and the low input traditional practices.

Under the variable conditions of rain-fed agriculture, nitrogen management is an essential component of the production process. This can be easily accomplished with external fertilizer inputs, because it is possible to control their availability in relation to crop requirements and weather. However, because of high costs or lack of availability of inorganic fertilization, a number of soil fertility management alternatives were compared to the unimproved low input monocropping practice. All alternative technologies including the rotations exhibited higher overall biophysical viability than the unimproved practice. Organic and inorganic fertilization were the technically superior alternatives, because legumes in rotations provide uncertain additions of nitrogen relative to crop requirements due to the variable impact of weather on their growth. However, manuring in general, and especially small ruminants' manure applied even at reduced rate of 2 t/year, offers a sustainable and viable alternative to inorganic fertilization. Although each of these technologies was tested individually higher returns would be anticipated by combining them in more integrated packages.

Organic fertilization is a traditional practice in the Sahel, but the difficulties in producing the required amount for optimum crop growth limits its availability and use. Research and development efforts must emphasize strategies and methods to produce more manure and other organic fertilizers. The successful application of techniques for the mass production of organic fertilizers and manure using 'composting peats' and 'animal stabulation' in other similar environments such as Burkina Faso (Badini, 1993) indicate that future efforts in soil fertility improvement should be directed towards the technologies of production of organic fertilizers. The long-term impact analysis performed here confirms that organic fertilization is good for crop productivity as well as soil sustainability.

References

Anderson, J.R., Dillon, J.L. and Hardaker, J.B. (1977) *Agricultural Decisions Analysis.* Iowa State University Press, Ames, Iowa.

Badini, O. (1993) Propositions de technologie en milieu reel: Conditonnement de boeufs de trait et production de fumier en saison seche au Campement de Dogoma (région Ouest du Burkina Faso). In: Lawrence, P.R., Lawrence, K., Djikman, J.T. and Starkey, P.H. (eds) *Proceedings of the Fourth Workshop of the West Africa Animal Traction Network, Kano, Nigeria.* The International Livestock Center for Africa (ILCA), Addis Ababa, Ethiopia.

Badini, O. and Dioni, L. (2001) *Etude Morphopédologique de la Commune de Madiama, Cercle de Djenné* (Technical report SANREM West Africa). Washington State University, Intenational Programs, Pullman, Washington.

Badini, O., Stöckle, C.O. and Franz, E.H. (1997) Application of Crop Simulation Modeling and GIS to agroclimatic assessment in Burkina Faso. *Agriculture, Ecosystems & Environment* 64, 233–244.

Johnson, A.K.L., Cramb, R.A. and McAlpine, J.R. (1994) Integrated land evaluation as an aid to land use planning in Northern Australia. *Journal of Environmental Management* 40, 139–154.

Loomis, R.S. and Connor, D.J. (1992) *Crop Ecology: Productivity and Management in Agricultural Systems,* 1st edn. Cambridge University Press, Cambridge, UK.

Marten, G.G. (1988) Productivity, stability, sustainability, equitability and autonomy as properties for agroecosystem assessment. *Agricultural Systems* 26, 291–316.

Ruthenberg, H. (1980) *Farming Systems in the Tropics.* Clarendon Press, Oxford, UK.

Stöckle, C.O. (1989) Simulation models in agriculture: From cellular level to field scale. In: Clema, J.K. (ed.) *Proceedings of the 1989 Summer Computer Simulation Conference,* Austin, Texas (24–27 July), pp. 639–644.

Stöckle, C.O and Nelson, R.L. (1996) *CropSyst (Cropping Systems Simulation Model) Users' manual (Version 2.0).* Biological Systems Engineering Dept, Washington State University, Pullman, Washington.

Traoré, B. (2002) *Techniques de conservation et d'amélioration de la fertilité des terres cultivées en zone sahélienne du Mali: Synthese bibliographique SANREM CRSP West Africa* (Working Paper No. 03–01). Virginia Tech, Office of International Research, Education and Development, Blacksburg, Virginia.

12 An Economic Evaluation of Alternative Soil Management Technologies

Peter Wyeth,[1] Boureima Traoré,[2] Oumarou Badini,[1] Mohamed Sidi Mohamed Touré[2] and Charlene Brewster[3]

[1]International Programs-Development Cooperation, Washington State University, Pullman, WA 99164–5121, USA; [2]Centre Régional de Recherche Agricole, Institut d'Economie Rurale, BP 205, Mopti, Mali; and [3]Office of International Research, Education and Development, Virginia Tech, Blacksburg, VA 24061–0378, USA

This chapter evaluates alternative soil management techniques that were tested in Madiama. Trials of these techniques were developed following the identification of declining soil fertility and crop yields as matters of special concern among farmers. As Chapter 10 makes clear, farmers already had a sound grasp, in general qualitative terms, of the relative benefits of different techniques for conserving soil fertility. What they wanted from the research was some level of quantification of their relative returns and costs, and providing this is the objective of this chapter.

The evaluation employs a three-pronged approach. First, it takes account of the outcomes of on-farm trials. Second, it relates them to the output of biosystems modelling reported in Chapter 11 so that the two sets of results can be compared. Third, it provides an assessment of the economic viability of the various techniques by deriving a notional profit for each technique. Most households consume much more of their crop production than they sell, but attributing values to outputs (crop yields) and inputs (labour and materials such as fertilizer or manure) provides a means of weighing the benefits farmers derive from the practices against the costs they incur in implementing them.

Although we compare profits among practices, we do not conclude that farmers should necessarily all adopt the practice that brings the greatest net returns. Different farmers have different means at their disposal, including cash for buying chemical fertilizer or access to animal manure. They also have different personal preferences for one technique or another. For instance, an interest in cowpeas, because their families can eat the peas and they can feed the leaves to their livestock, may lead farmers to grow them in rotation with millet, even though this is one of the least effective techniques. Although different farmers will continue to follow different practices, we hope that the information in this chapter will allow their decisions to be better informed.

This chapter is a truly multi-disciplinary effort. The farmers in Madiama not only worked with the research agronomist in determining what trials should be run; they also did all the field work and assisted in evaluating the results. The Institut de l'Economie Rurale (IER) research agronomist and technicians oversaw the implementation of the trials and collected the data. The biosystems engineer provided the modelling input, and the economists contributed the economic elements of the analysis. The Natural Resources Management Advisory Committee (NRMAC) in Madiama served as the mechanism of communication between the researchers and the farmers. It was NRMAC members who selected the participating farmers, in several instances volunteering themselves. Besides communicating individually with the farmers, researchers also reported their analytical findings to the NRMAC. Its members actively responded with their own assessments and considerable insights and this chapter includes their views in its analysis.

Analytical Method

The results from the on-farm trials and the modelling complement one another in important ways, while each on its own would have serious limitations. Regarding the on-farm data, they are available for, at best, only the 3 years from 2001 to 2003. For some practices there are fewer years than that. Given the kind of variations that occur in the Sahel due to rainfall, and in on-farm trials due to the difficulty of holding constant everything other than the treatments, 3 years worth of data are not enough. The rainfall in Madiama in 2001, 2002 and 2003 was 671 mm, 396 mm and 825 mm, respectively. This was a wide variation (unfortunately so for the farmers, who suffered crop losses due to the drought in the second year) but still a small sample of the possibilities.

Although everything is done to try to keep the variations in on-farm conditions to a minimum, *Striga hermonthica* (a parasitic weed) can unexpectedly affect one plot, an unaccustomed amount of rain can flood another, and animals can graze another before harvest. However, trials are conducted on-farm because conditions are more realistic than they are on research stations. Because of this, and because farmers can see the results for themselves and know first-hand what went into obtaining them, on-farm trials have a credibility that research station trials do not. However, it is also true that the greater realism comes at a cost. Because uncontrolled differences in conditions are far from equally dispersed among the trial plots, variables other than the treatments are not held constant and inter-farm comparisons are often difficult.

One advantage of biosystems modelling is that control over all variables is perfect, and contamination of results by random events such as floods and invasions by animals is unknown. Further, the model can provide as many observations as the researcher wants, and this chapter makes use of yields simulated over a period of 30 years, using actual historical data from Madiama on the volume and timing of rainfall. Such advantages, however, do not allow modelling to stand on its own, independent of field observations of some kind. As a matter of principle, the modellers need farm trial results to calibrate the model. As a matter of practice, farmers will be sceptical of modelling results that they cannot clearly relate to field experience.

Unfortunately, both farm trial and modelling results are not available for all treatments. One of the commonest treatments in the farm trials was the application of PNT, a natural phosphate rock from Tilemsi in Mali, as a fertilizer. However, the model could not accommodate this because not enough is known either about the rock's chemical make-up or how it breaks down and makes its nutrients available to plants. Therefore, trial results alone must be relied on for PNT. The same applies to the intercropping of cowpeas and millet, which the model saw only as side-by-side pure stands. A benefit of intercropping might be the reduction of *Striga*, a parasitic weed which devastates millet but cannot survive on cowpeas, but the model looks only at soil nutrients and water, not the impact of weeds.

On the other hand, only modelling results are available to assess full application rates of chemical fertilizer and a 2 year cowpea–millet rotation. The first was not included in farm trials because it would have involved an expensive cash outlay beyond the means of most farmers. The second was omitted because farmers and researchers preferred the 3 year rotation and there was a limit to the number of treatments that could be tested.

In Chapter 11, Badini and Traoré refer to efficiency, stability and sustainability as criteria for evaluating techniques. The information focused on in this chapter can easily be related to the same concepts. Badini and Traoré take yield as one measure of efficiency, and we look at that here also, but the more comprehensive measure of efficiency is profits because, in one concise figure, it takes account of not only benefits but also costs.

Stability, or keeping risk down, is crucial for any farmer, but especially for those with limited means. They cannot afford to follow practices that do very well in some years and very poorly in others, because storage facilities are inadequate and no one can know how many bad harvests there might be in a row. For the trials, we illustrate stability by reporting average yield data for each treatment for each of the 3 years that they were run. The considerable variation from one year to the next in these results will be evident. However, the long series of results from the modelling output allow a different view of stability. Because it is helpful for farmers to know how badly they are likely to do in years when rainfall is poor, average and good, the minimum yield and profit that farmers can expect 75%, 50% and 25% of the time, is reported assuming that the weather pattern of the last 30 years remains more or less typical is reported.

The modelling also permits an assessment of sustainability. Whereas Chapter 11 reported on this from the point of view of soil fertility, with modelling it is possible to graph profits over the 30 year period, which brings certain trends out clearly. At the end of the chapter, the average discounted profits for the period are reported to take account of the fact that poorer people, of whom there are many in Madiama, generally discount the future more heavily than those who are better off because the needs of the poor are, by definition, more immediate.

Techniques Tested

Researchers and farmers decided on the following four main types of intervention:

- Applications of animal manure to crops, either by collecting it and transporting it to the fields, or by agreeing with herders that they may graze their animals on crop residue in the farmers' fields if they leave the manure there.
- Applications of small amounts (microdoses) of chemical fertilizer. To minimize costs while increasing yields, farmers apply a few grammes of fertilizer right next to the seed or the plant.
- Application of PNT, a natural phosphate rock from Tilemsi available in local markets, as a low-cost fertilizer.
- Rotating and intercropping millet and cowpeas (*niébé*) to diversify farm diets, increase production of better quality forage that the cowpea residues (*les fanes de niébé*) supply, and rest the soil between millet crops, thereby taking advantage of the nitrogen fixation capacity of cowpeas.

The trials were conducted on small plots in farmers' fields (although the results are reported on a per hectare basis), and the farmers provided the labour and kept the harvest. The project provided the chemical inputs (fertilizers and insecticides) and plant material, and IER researchers and technicians were present both to advise on techniques of implementation and to record labour time.

There were ten 'unimproved' control plots that received no applications of manure and chemical fertilizer and used no intercropping and no rotations. This treatment, or lack of treatment, of the control plots closely resembles the farmers' current practices, except that the farmers generally spread some amount of manure on their fields. However, in practice the amounts are ill defined and highly variable. Using a control plot with no soil fertility practices at all provides a consistent benchmark that all farmers readily understand. Table 12.1 describes in detail the trials analysed in this report. All quantities are per hectare, and there are four replications in each of 3 years, except where otherwise noted.

One advantage the fourth treatment set, corralling animals on a field, has over transporting manure from elsewhere is that the

Table 12.1. Treatment sets.

Treatment set	Treatment
1. Millet with animal manure – *farm trials and modelling*	• 2 tonnes cattle manure every year • 5 tonnes cattle manure every 3 years • 2 tonnes small ruminant manure every year • 5 tonnes small ruminant manure every 3 years
2. Sorghum with animal manure – *farm trials only*	• 2 tonnes cattle manure every year (2 farmers) • 5 tonnes cattle manure every 3 years • 2 tonnes small ruminant manure every year
3. Millet with animal manure and PNT – *farm trials only*	• As for Set 1, but with 300 kg of PNT in the first year for each treatment.
4. Millet after corralling animals on the field before planting, first year only – *farm trials only*	• 1667 cattle/ha (25/150 m^2) for 2, 5 and 10 nights • 3333 small ruminants (50/150 m^2) for 2, 5, and 10 nights (2002 and 2003 only)
5. (a) Millet with microdoses of DAP (diammonium phosphate, 18-46-0) and/or urea (46-0-0) – *farm trials only* (b) Millet with microdoses of NPK (15-15-15) – *modelling only* (c) Millet with recommended full doses of DAP and urea – *modelling only*	• 2 grammes DAP/pocket each year at planting (2003 only) • 2 grammes DAP/pocket each year, planting time ***plus*** 1 gramme urea at tillering (2003 only) • 6 grammes NPK/pocket each year at planting • 6 grammes NPK/pocket each year at tillering • 100 kg of DAP at planting, 50 kg of urea at 10 days and 50 days after emergence
6. Sorghum with PNT or microdosing with NPK (15-15-15) or DAP – *farm trials only (3 trials with PNT not reported)*	• 6 grammes NPK/pocket each year at planting • 6 grammes NPK/pocket at planting (2001) ***or*** 2 grammes DAP/pocket at planting (2002, 2003) • 6 grammes NPK/pocket at tillering (2001) ***or*** 2 grammes DAP/pocket at tillering (2002, 2003)
7. Cowpeas–millet intercropping – *modelling only*	• No PNT • 300 kg PNT in the first year • 600 kg PNT in the first year
8. (a) Cowpeas–millet–millet 3 year rotation – *farm trials and modelling*	• No PNT • 300 kg PNT in the first year • 600 kg PNT in the first year
(b) Cowpeas–millet 2 year rotation – *modelling only*	• No PNT

field benefits not only from the animals' faeces but also from their nitrogen-rich urine. Another benefit is that the manure does not have to be collected and transported, thereby eliminating the labour costs and losses entailed. In the calculations, a charge for a herder is deducted as a labour cost, but in practice compensation to the animals' owners may take the form of crop residues that the animals are allowed to consume. There are two points to make regarding the large numbers of animals per hectare. First, because the densities that occur in practice are large, these may not be unrealistic. (For cattle the figure translates to one every 6 square metres and for small ruminants one every 3 square metres.) Second, the same result can be had by reducing the number of animals and increasing the number of nights they are on the field. (Three hundred cattle on a plot for 2 nights is the equivalent of 150 for 4 nights.)

Assumptions

Tables 12.2–12.5 show the assumptions regarding labour per hectare, prices of crops and inputs, and input quantities per hectare.

The field technicians who supervised

Table 12.2. Person-days per hectare.

	Millet	Intercrop cp-m	Cowpeas	Sorghum
Labour to plough, weed, guard				
Ploughing (every year)	2	2	2	2
Planting (every year)	1	1.5	1	1
Weeding 1 (every year)	3.75	4.2	4	4
Weeding 2 (every year)	2.25	3.4	1	2.5
Guard corral 2 nights (1st year)	2			
Guard corral 5 nights (1st year)	5			
Guard corral 10 nights (1st year)	10			
Labour to spread				
2 t animal manure (every year)	2			2
5 t animal manure (1st year)	3			3
300 PNT (1st year)	1.5	1.5	1.5	1.5
600 PNT (1st year)		2	2	
Microdoses NPK/DAP (every year)	1			1
Microdoses urea (every year)	1			1
Recommended rate (every year)	1			
Pesticide treatment (every year)		0.5	1	

the on-farm trials kept records of the amounts of labour used as the tasks were performed. These records did not directly yield days of labour per task because often a set number of individuals did not perform the tasks. More typically, the farmer might go to his field with two of his younger children in the morning. After a period of time an older child might join them and the four of them work together before the younger ones return home. Later an older member of the household might also come out to the field for a while. Every day could see different individuals appearing for different periods of time, and each farmer's situation differed from every other's. We looked carefully at this data and then, in consultation with the technicians and the farmers, derived reasonable assumptions of person-days per task, shown in Table 12.2. Harvest labour is typically paid a share of the crop, here assumed to be 10%.

There were similar variations in the records regarding the prices at which farmers bought and sold inputs and crops. Crop prices also varied through the year, being lowest just after harvest and higher later. Again, working with the technicians and farmers, we took the data and derived the representative figures in Table 12.3. However, when deriving profits over 30 years using the modelling results, we assumed that the prices for the year 2000 prevailed throughout the period.

There were no good market prices in Madiama for animal manure. In Table 12.4, the estimated values derived for it are based on the content in the manure of nitrogen (N), phosphorus (in the form P_2O_5) and potassium (in the form K_2O) and the prices of these chemical fertilizers. The nutritive content of the manure was determined in IER laboratories. (Although this is quite variable, the content assumed here is the same as that assumed by the model in Chapter 11.) While the prices of manure arrived at in this way might reflect the nutritive value of the manure, they almost certainly overstate the opportunity cost of the manure to farmers by a considerable margin, as its value in other uses (e.g. in plaster for house construction) would be much less. Therefore, in addition to the full value calculated in Table 12.4, we have also calculated the profits using half the value. This table shows both sets of results. Table 12.5 shows the quantities per hectare of seed and fertilizer.

Results

We developed budgets for all treatments and controls based on the assumptions above

Table 12.3. Prices for inputs and crops.

	Unit	2000–1	2001–2	2002–3	Assumptions
Millet (selling)	kg	110	150	90	
Millet (buy seed)	kg	165	165	225	
Sorghum (selling)	kg	110	150	90	
Sorghum (buy seed)	kg	165	165	225	
Cowpeas (selling)	kg	300	no harvest	250	
Cowpeas (buy seed)	kg	450	450	675	
Cowpea residue (sell)	kg	125	no harvest	120	
Cattle manure (full)	kg	15.99	15.99	15.99	
Cattle manure (half)	kg	7.99	7.99	7.99	
Small ruminant man. (full)	kg	22.13	22.13	22.13	
Small ruminant man. (half)	kg	11.06	11.06	11.06	
NPK	kg	225	not used	225	
DAP	kg	Not used	260	260	
Urea	kg		200	200	
PNT	kg	75	75	75	
Pesticide treatment	hectare	12,000	12,000	12,000	
Ploughing	½ hectare	4,000	4,000	4,000	(2 days for 1 ha)
Harvest mil/sorghum	% harvest	10%	10%	10%	
Harvest cowpeas	% harvest	10%	10%	10%	
Other labour	Pers. day	750	750	750	
Transport manure	kg	250	250	250	200 kg donkey cart
Transport production	kg	250	250	250	200 kg donkey cart
Sack	1	275	275	275	yield × 90% ÷ 80 kg @250–300 FCFA/sack

Table 12.4. Pricing manure.

Fertilizer	Element in the manure (%)	Coefficient (kg fertilizer/ kg elem)	Fertilizer in the manure (%)	Price fertilizer (FCFA/kg)	Value manure (FCFA/kg)
Cattle manure					
kg. Urea	0.980	2.18	0.021	220	4.70
kg. P_2O_5	0.580	2.27	0.013	220	2.90
kg. K_2O	1.680	2.27	0.038	220	8.39
					15.99
Small ruminant manure					
kg. N	1.250	2.18	0.027	220	6.00
kg. P_2O_5	0.830	2.27	0.018	220	4.15
kg. K_2O	2.400	2.27	0.054	220	11.99
					22.13

together with the results of the farm trials and the modelling. The tables that follow are taken from much larger spreadsheets that laid out prices and quantities of yields and both labour and material inputs in detail. They show the input costs and days of labour (excluding harvesting) in order to indicate the investment farmers need to make to obtain the profits shown. The crop yields include the 10% assumed to go to harvest labour, but the monetary profit or loss is the net of this payment.

Table 12.5. Input quantities per hectare.

Input	Quantity	Units
Cattle manure	2–5	mt/ha
Small ruminant manure	2–5	mt/ha
Seed, millet	5	kg/ha
Seed, cowpeas	20	kg/ha
NPK 15-15-15, microdoses	60	(6 g/pocket x 10,000 pockets/ha)
DAP, microdoses	20	(2 g/pocket x 10,000 pockets/ha)
NPK 15-15-15, recommended	100	kg/ha
DAP, recommended	100	kg/ha
Urea, microdoses	10	(1 g/pocket x 10,000 pockets/ha)
Urea, DAP	100	kg/ha
PNT	300–600	kg/ha

In the tables for the farm trials, the average yields for each year's replications only take account of the middle two results, leaving aside the highest and the lowest, because variations in yield among replications are commonly very great indeed. We wanted to reduce the possibility that the averages would be distorted by a single unrepresentative result, but did not want to substitute a different distortion by deciding on an ad hoc basis which results were outliers and which were not. As the tables show, even after trimming the means, plenty of variation among yields remains.

In the tables for the model results, we also report three results for each treatment. As the model took account of 30 years of real rainfall data, it was able to generate a frequency distribution of yields. Farmers can expect the low yield or more for 23 or 24 years out of 30, the middle yield or more for 15 years out of 30 and the high yield or more for only 7 or 8 years out of 30.

In the tables for the farm trials we report annual costs of inputs for each year, as they too vary significantly from one year to the next for some treatments. The costs of purchased inputs are those that cannot be provided on farm. Here this means chemical fertilizer, PNT and pesticide treatment for cowpeas. Some farmers might purchase seed and animal manure, but typically they obtain these out of their own resources. Cash costs are important because some farmers are cash poor, particularly at planting time. Costs, excluding harvest, include labour costs incurred before harvest as well as seed and manure costs. They provide a fairly complete accounting of the resources farmers must put into each treatment.

Treatment sets 1, 2 and 3: millet with animal manure, with and without PNT in the first year; sorghum with animal manure

The trials using animal manure taken out to the fields were both with and without 300 PNT applied in the first year. The figures in Table 12.6 suggest that for millet the results with PNT are considerably better than those without. Also, the results without PNT are sometimes worse than the no-fertilizer control, but it is not clear how much weight should be put on this outcome. There were two farmers involved, one who carried out all the trials with PNT (accounting for the cash costs in 2001) and another who implemented them without PNT. Differences in soils or farm practices could account for some of the yield differences. Furthermore, the negative results of using manure without PNT are not borne out either by the field trials for sorghum or the modelling, results for which are shown in Tables 12.7 and 12.8. Because of the cost of PNT and the value attributed to manure, the overall pre-harvest costs noted here are very high in the first year for several of the treatments, especially where manure is priced at the value arrived at in Table 12.4. In fact, in some instances total pre-harvest costs are higher than those

Table 12.6. Millet: animal manure – trials.

		Unim-proved	C man. 2 t	C man. 5 t	SR man. 2 t	SR man. 5 t	Unim-proved	C man. 2 t+ 300PNT	C man. 5 t+ 300PNT	SR man. 2 t+ 300PNT	SR man. 5 t+ 300PNT
Yield (kg/ha)	2001	217	859	1,437	1,281	734	258	1,006	1,562	1,562	1,441
	2002	0	0	0	0	0	120	360	560	390	458
	2003	672	625	594	734	500	672	1,552	1,552	1,781	1,333
	Avg.	***296***	***495***	***677***	***672***	***411***	***350***	***973***	***1,225***	***1,244***	***1,077***
Total cost excl. harvest (FCFA)	2001	14,075	50,048	102,508	62,326	133,203	14,075	73,673	126,133	85,951	156,828
	2002	14,075	50,048	14,075	62,326	14,075	14,075	50,048	14,075	62,326	14,075
	2003	14,375	50,348	14,375	62,626	14,375	14,375	50,348	14,375	62,626	14,375
Cost purchased inputs (FCFA)	2001	0	0	0	0	0	0	22,500	22,500	22,500	22,500
	2002	0	0	0	0	0	0	0	0	0	0
	2003	0	0	0	0	0	0	0	0	0	0
Avg. days labour	*p/day*	*9*	*11*	*10*	*11*	*10*	*9*	*12*	*11*	*12*	*11*
Profit/loss (FCFA)	2001	6,359	30,966	33,019	58,488	(63,946)	10,258	21,174	21,184	61,365	(20,955)
	2002	(14,075)	(50,048)	(14,075)	(62,326)	(14,075)	1,563	(3,136)	58,900	(11,504)	45,608
	2003	36,897	(2,653)	30,936	(6,584)	23,781	36,897	68,095	104,068	73,305	87,375
Full cost manure	***Avg.***	***9,727***	***(7,245)***	***16,627***	***(3,474)***	***(18,080)***	***16,239***	***28,711***	***61,384***	***41,055***	***37,343***
Half cost manure	***Avg.***	***9,727***	***8,742***	***29,949***	***18,651***	***358***	***16,239***	***44,698***	***74,706***	***63,181***	***55,781***

Without PNT, the most profitable treatment is on average 71% more profitable than the unimproved practice with full cost manure, 208% more profitable with half priced manure.
With PNT, the most profitable treatment is on average 278% more profitable than the unimproved practice with full cost manure, 359% more profitable with half priced manure.

Table 12.7. Sorghum: animal manure – trials.

		Unim-proved	2 t CM each yr	5 t CM 3 yrs	Unim-proved	2 t CM each yr	5 t SRM each yr
Yield (kg/ha)	2001	688	1,469	1,375			
	2002	56	220	280	280	1,040	1,140
	2003	594	828	819	594	1,313	1,221
	Average	***446***	***839***	***825***	***437***	***1,176***	***1,180***
Total cost excl. harvest (FCFA)	2001	14,450	50,423	102,883			
	2002	14,450	50,423	15,450	14,450	50,423	62,701
	2003	14,375	50,348	14,375	14,375	50,348	62,626
Cost purchased inputs (FCFA)	2001	0	0	0	0	0	0
	2002	0	0	0	0	0	0
	2003	0	0	0	0	0	0
Avg. days labour	*p/day*	*10*	*12*	*11*	*10*	*12*	*12*
Profit/loss (FCFA)	2001	50,390	88,098	26,797			
	2002	(7,153)	(21,754)	22,038	22,038	85,102	85,855
	2003	30,936	12,848	48,106	30,936	49,812	30,532
Full cost manure	***Average***	***24,724***	***26,397***	***32,313***	***26,487***	***67,457***	***58,194***
Half cost manure	***Average***	***24,724***	***42,384***	***45,636***	***26,487***	***83,444***	***80,319***

for the recommended rates of chemical fertilizer (see below).

Among the manure treatments for millet, and taking the overall average of the three years, the most favourable results in the trials were obtained with 5 t of cattle manure spread at the beginning of the three year period. (Yields were as good with 2 t of small ruminant manure per year, but profits were not as good because of poor rainfall in the second year, when the 2 t were spread and the 5 t were not.) Farmers expected the results from small ruminant manure to be better than those from cattle manure, because they are aware that small ruminant manure is richer in nutrients. This advantage does not show up consistently in trial yields, and although it did in the modelling (see below), the higher cost attributed to the small ruminant manure because of its nutrient superiority reduced or wiped out any superiority in profits.

Table 12.8 shows the modelling results. These are consistent with those from the trials in that results of all four treatments with manure are of the same magnitude and better than the no-manure control. As mentioned earlier, the way to interpret the percentiles is that farmers can expect that yield or level of profit, *or more*, for the stated percentage of seasons. The differences between using manure and not using it increase from the 75th to the 50th to the 25th percentile as the rainfall that these percentiles are based on increases.

Figure 12.1 illustrates the profits over 30 years for the unimproved control and for small ruminant manure. (The cattle manure results were omitted so as not to further clutter the figure, but the results are similar to those for small ruminant manure. The profits graphed are those calculated using manure at half its full nutrient value.) For the treatments requiring 5 t of manure every 3 years, the downward spike in profits at those times is very evident. However, the general trend for each of the treatments is similar. The trend for 'unimproved' millet, on the other hand, is clearly downward.

Treatment set 4: millet after animals have been corralled on the field before planting in the first year

The first conclusion in this treatment set is well in line with expectations and holds every year: the more nights the animals

Table 12.8. Millet: animal manure – model results.

		Unimpoved	C man. 2t	C man. 5t	SR man. 2t	SR man. 5t
Yield (kg/ha)	75%	418	756	720	818	765
	50%	***505***	***991***	***875***	***1,057***	***959***
	25%	642	1,196	1,050	1,303	1,123
Total cost excl. harvest		825	32,798	27,469	45,076	37,701
Cost purchased inputs		0	0	0	0	0
Avg. days of labour	*p/day*	*9*	*11*	*10*	*11*	*10*
Profit/loss (FCFA)	75%	25,303	21,258	24,314	14,790	18,333
Full cost manure	***50%***	***33,574***	***43,381***	***38,942***	***37,319***	***36,627***
	25%	46,476	62,758	55,441	60,535	52,091
Half cost manure	75%	25,303	37,244	37,636	36,915	36,771
	50%	***33,574***	***59,367***	***52,264***	***59,445***	***55,065***
	25%	46,476	78,744	68,763	82,660	70,529

The most profitable treatment with median rainfall is 29% more profitable than the unimproved practice with full cost manure, 77% more profitable with half cost manure.

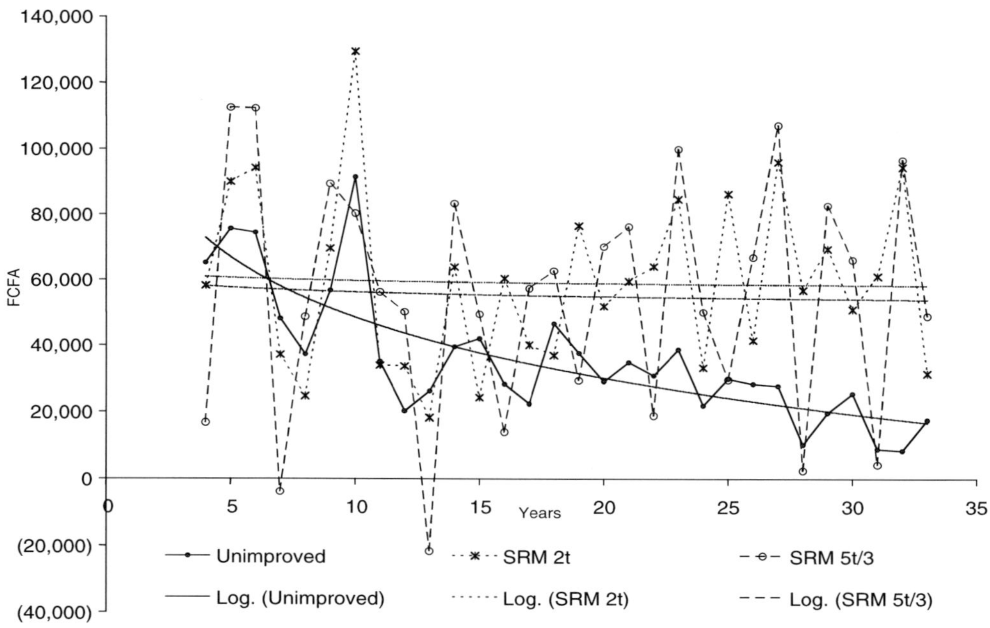

Fig. 12.1. Millet profits with organic fertilizer.

spend on the fields the higher the yields and the profits. The second more interesting conclusion is that this result holds even 3 years after the treatment was applied, which considerably reinforces the value of the practice. On the fields that had cattle, yields were even higher in the third year than the second, presumably because of the higher rainfall in the third year. Profits were lower that third year only because the price of millet was FCFA 90 rather than FCFA 150. Once again, there is not much to choose between the treatment with cattle and that with small ruminants. (There are only 2 years' worth of

Table 12.9. Millet: night corralling – trials.

		Unim-proved	(Cattle 1667 head)			Unim-proved	(Small rum. 3333 head)		
			2 nights	5 nights	10 nights		2 nights	5 nights	10 nights
Yield (kg/ha)	2001	219	469	875	1,375				
	2002	560	700	660	1,120	440	500	1,000	1,300
	2003	375	500	813	1,250	375	884	750	938
	Average	***385***	***556***	***782***	***1,248***	***408***	***692***	***875***	***1,119***
Total cost excl. harvest (FCFA)	2001	14,075	15,575	17,825	21,575				
	2002	14,075	14,075	14,075	14,075	14,000	15,500	17,750	21,500
	2003	14,375	14,375	14,375	14,375	14,375	14,375	14,375	14,375
Cost purchased inputs (FCFA)	2001	0	0	0	0	0	0	0	0
	2002	0	0	0	0	0	0	0	0
	2003	0	0	0	0	0	0	0	0
Avg. days labour	*p/day*	9	*10*	*11*	*12*	*9*	*10*	*12*	*14*
Profit/loss (FCFA)	2001	6,548	28,626	64,667	108,073				
	2002	58,900	77,144	71,931	131,875	43,338	49,656	112,563	147,906
	2003	14,242	23,781	47,629	81,016	14,242	53,113	42,859	57,168
	Average	***26,563***	***43,184***	***61,409***	***106,988***	***28,790***	***51,385***	***77,711***	***102,537***

For cattle, the most profitable treatment is on average 303% more profitable than the unimproved practice.
For small ruminants, the most profitable treatment is on average 256% more profitable than the unimproved practice.

Table 12.10. Millet: microdoses of chemical fertilizer – trials.

		Unimproved	DAP	DAP + Urea
Yield (kg/ha)	2003	463	1,190	1,257
Total cost excl. harvest	2003	14,375	20,325	23,075
Cost purchased inputs	2003	0	5,200	7,200
Avg. days of labour	*p/day*	*9*	*10*	*11*
Profit/loss (FCFA)	2003	20,983	70,487	72,824

The most profitable treatment is on average 247% more profitable than the unimproved practice.

Table 12.11. Sorghum: microdoses of chemical fertilizer – trials.

		Unimproved	NPK	Unimproved	NPK/ DAP 1	NPK/ DAP 2
Yield (kg/ha)	2001	688	1,256	442	1,031	906
	2002	56	520	56	510	20
	2003	594	875	594	1,063	641
	Avg.	***446***	***884***	***364***	***868***	***522***
Total cost excl. harvest (FCFA)	2001	14,450	28,700	14,450	28,700	28,700
	2002	14,450	30,800	14,450	20,400	20,400
	2003	14,375	30,725	14,375	20,325	20,325
Cost purchased inputs (FCFA)	2001	0	13,500	0	13,500	13,500
	2002	0	15,600	0	5,200	5,200
	2003	0	15,600	0	5,200	5,200
Avg. days labour	*p/day*	*10*	*11*	*10*	*11*	*11*
Profit/loss (FCFA)	2001	50,390	89,780	27,236	68,536	56,747
	2002	(7,153)	36,963	(7,153)	46,059	(17,794)
	2003	30,936	36,048	30,936	60,757	28,563
	Avg.	***24,724***	***54,264***	***17,006***	***58,451***	***22,505***

For the trials on the right, DAP was substituted for NPK in the second and third years because it is cheaper. The NPK/DAP 1 treatment was at applied at planting, the NPK/DAP 2 treatment was applied at tillering.

data on the latter because this treatment was first applied in the second year.)

Treatment sets 5 and 6: millet and sorghum with microdoses of chemical fertilizer

There was only one year of trial results for millet using microdoses of chemical fertilizer. (It was tried in 2002, but there was a complete crop failure due to low rainfall.) The result was that microdosing clearly did well in terms of both physical yield and profit, and better than the control. The yield using urea (46-0-0) as well as DAP (18-46-0) was sufficiently higher than that using DAP alone to pay for the urea and the labour to spread it, but not much more. Table 12.11 shows that microdosing also paid off for sorghum, though not as consistently. In fact, the application at tillage only did better than the unimproved control in 1 year.

The modelling results include not only DAP and urea but also two treatments with microdoses of NPK (15-15-15), the first being applied at planting and the second at tillage. According to the results in Table 12.12 and Fig. 12.2 there is little to choose between the three. NPK is cheaper than DAP and slightly more expensive than urea (see Table 12.3), but much more is used (6 g of NPK per pocket instead of 2 g of DAP and 1 g of urea), so the DAP–urea combination is slightly more profitable. The model calculated yields for the recommended quantities of DAP and urea (100 kg of each,

Table. 12.12. Millet: chemical fertilizer – model results.

		Unimproved	NPK1	Microdoses NPK2	DAP + urea	Recom. rate DAP + urea
Yield (kg/ha)	75%	418	707	745	733	1,000
	50%	***505***	***817***	***840***	***809***	***1,358***
	25%	642	975	978	966	1,603
Total cost excl. harvest		14,075	28,325	28,325	22,775	57,325
Cost purchased inputs		0	13,500	13,500	7,200	42,500
Avg. days of labour	*p/day*	*9*	*10*	*10*	*11*	*10*
Profit/loss (FCFA)	75%	25,303	38,401	41,923	46,353	37,032
	50%	***33,574***	***48,718***	***50,852***	***53,549***	***70,749***
	25%	46,476	63,613	63,867	68,317	93,837

DAP + urea microdosing (with median rainfall) is 59% more profitable than the unimproved practice.

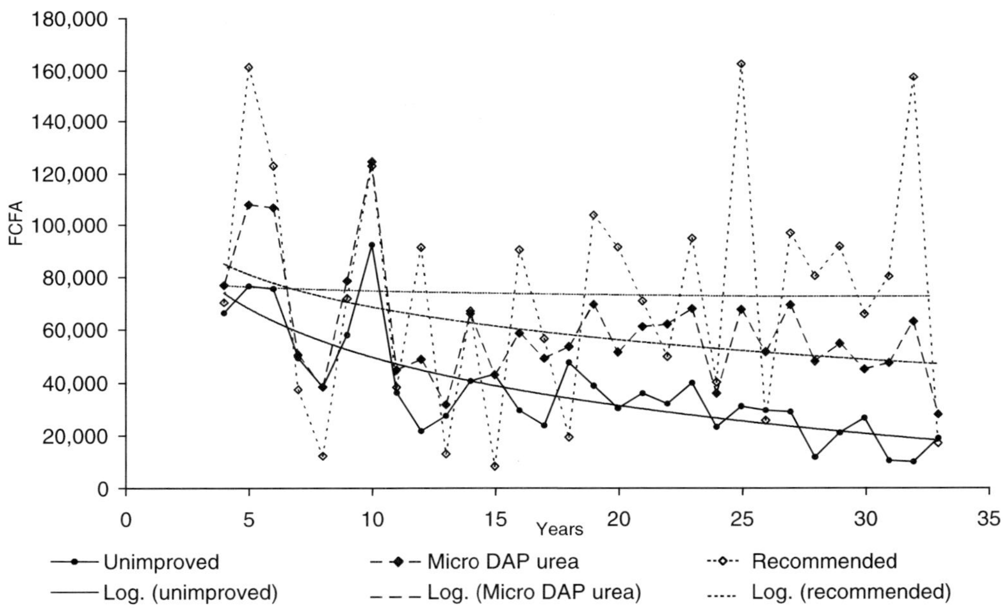

Fig. 12.2. Millet profits with chemical fertilizer.

the first at planting and the second at tillage). This was not tried in the fields, on the grounds that few farmers could afford the cash outlay but, if any could, the model shows it would be clearly the most profitable.

Similar conclusions are implicit in Fig. 12.2. The curves and the trends for microdosing and for recommended rates are clearly above that for the unimproved control. What is interesting and significant is that the microdosing trend is clearly still downward. Microdosing is effective at raising short term profits, but not to sustain yields and profits over a long period.

Treatment set 7: cowpeas and millet intercropped

Table 12.13 shows the results from the intercropping trials. The control was millet on its own and the treatments intercropped

Table 12.13. Millet and cowpeas: intercropping – trials.

			Without PNT				300 kg PNT				600 kg PNT			
		Unim-proved	Millet	C'peas	C'pea residue	Total grain	Millet	Cow-peas	C'pea residue	Total grain	Millet	Cow-peas	C'pea residue	Total grain
Yield (kg/ha)	2001	734	594	120	594		734	248	906		484	205	937	
	2002	460	380	0	0		660	0	0		660	0	0	
	2003	750	203	22	53		216	47	244		313	57	175	
	Average	***648***	***392***	***47***	***216***		***537***	***98***	***383***		***486***	***87***	***371***	
Total cost excl. harvest (FCFA)	2001	14,075	7,825	14,888		22,713	19,638	30,100		49,738	31,075	41,538		72,613
	2002	14,000	7,788	18,288		26,075	7,825	18,288		26,113	7,825	18,288		26,113
	2003	14,375	7,975	14,688		22,663	7,975	20,538		28,513	7,975	20,538		28,513
Cost purchased inputs (FCFA)	2001	0	0	6,000		6,000	11,250	17,250		28,500	22,500	28,500		51,000
	2002	0	0	6,000		6,000	0	6,000		6,000	0	6,000		6,000
	2003	0	0	6,000		6,000	0	6,000		6,000	0	6,000		6,000
Avg. days labour	*p/day*	*9*	*6*	*6*		*12*	*6*	*6*		*12*	*6*	*6*		*12*
Profit/loss (FCFA)	2001	55,182	48,165	16,862	64,005	65,027	49,588	35,609	97,660	85,197	14,572	12,763	101,056	27,335
	2002	45,944	41,731	(18,288)	0	23,444	78,181	(18,288)	0	59,894	78,181	(18,288)	0	59,894
	2003	42,859	7,526	(9,868)	5,488	(2,342)	8,480	(10,210)	25,182	(1,730)	15,873	(8,090)	18,080	7,783
	Average	***47,995***	***32,474***	***(3,765)***	***23,164***	***28,709***	***45,416***	***2,370***	***40,947***	***47,787***	***36,209***	***(4,538)***	***39,712***	***31,671***

Even the most profitable treatment is not, on average, more profitable than the unimproved practice, unless account is taken of the value of the cowpea residues for fodder.

millet with cowpeas in alternating rows. One treatment was without PNT, one with 300 kg spread the first year, and one with 600 kg the first year. These results are the least satisfactory among the treatment sets. The millet half of the treatment performed reasonably well in the sense that the millet yield was always greater on a per hectare basis than the pure millet yield in the control. (In the table, the millet yields in the treatments are for half a hectare while the control is a full hectare.) However, the cowpeas failed completely in the second year, presumably because of low rainfall, and were very poor in the third, in spite of excellent rain. If profits are reckoned on the basis of millet grain and cowpeas alone, the profits are less for every treatment than they are for the control. The treatment with 300 kg of PNT performed better than that for 600 kg, though it is not clear why.

If account is also taken of the value of the cowpea leaves or residues, the profit situation for intercropping is rescued, as these are quite valuable as fodder, around FCFA 125/kg, in monetary terms. Not a great deal of weight should be put on the precise figures for the residues, as the ratio of residue to grain varies greatly among replications and treatments, but the general point remains that the value of the residues is an important element in determining the value of the crop.

Treatment set 8: cowpeas and millet in rotation, with and without PNT in the first year

As with intercropping, rotation treatments included one without PNT, one with 300 kg the first year, and one with 600 kg the first year. The trials shown in Table 12.14 used a 3 year, cowpea–millet–millet rotation. As the intercropping results just showed, the first year might have been the best year of the three to grow them. This was fortunate for experimental purposes, but the treatment might be risky, because if yields are poor when the cowpeas are grown the technique will be a loss maker. In the second year, the yields from the treatments were of roughly the same magnitude as those from the control, but in the third year the treatment results were much better than the control. The delayed reaction might have arisen because the PNT takes time to break down into a form useable by the crop, but whatever the reason, the consequence is that the three year rotation with 300 kg of PNT performed best. It was better than either the unimproved millet on millet or the rotation without PNT and, as was the case with the intercropping techniques, it also did better than the treatment with 600 kg. Again, adding the value of the cowpea residues adds greatly to revenues and profits.

Although the model did not consider the application of PNT, it was able to look at not only the 3 year cowpeas–millet– millet rotation but also the 2 year cowpeas–millet–cowpeas–millet variation. In Table 12.15, the left hand block of results shows millet under no rotation, 3 year rotation and 2 year rotation. The second block of results shows the same for cowpeas: cowpeas every year, every 3 years and every 2 years. The last two columns on the right show the average annual profits for the two rotations.

Millet yields are best with the 2 year rotation and cowpea yields with the 3 year rotation. Consumption preferences for millet may lead farmers to prefer the 3 year rotation, but the 2 year rotation is the more profitable, not only because millet yields are better but also because cowpeas are more profitable as a crop than millet.

In fact, according to the model, growing cowpeas all the time is as good as any rotation from a cash return point of view. This is the one set of treatments where results from the farm trials and modelling might be in conflict. In the intercropping trials cowpeas failed outright in the second year, and in the third did poorly in spite of good rain. By contrast, the model suggests that year by year yields of cowpeas are actually less variable than yields of millet. It gives coefficients of variation of 19–22% for cowpeas and of 40–48% for millet. If the opportunity presents itself, this discrepancy would be worth investigating to find out whether it is the farmers who are too easily discouraged by cowpeas or the model is overlooking an

Table 12.14. Cowpeas and millet: rotation – trials.

			C-M-M rotation without PNT				C-M-M rotation +300 PNT				Rotation +600 PNT			
		Unim-proved	Millet	Cowpeas	Cowpea residue	Total grain	Millet	C'peas	C'pea residue	Total grain	Millet	C'peas	C'pea residue	Total grain
Yield (kg/ha)	2001	578		350	1,400			450	1,742			362	1,775	
	2002	244	214				244				280			
	2003	141	266				1,375				563			
	Average	***321***	***240***	***350***	***1,400***		***540***	***450***	***1,742***		***281***	***362***	***1,775***	
Total cost excl. harvest (FCFA)	2001	14,075		34,250		34,250	0	34,500	0	34,500	0	57,000	0	57,000
	2002	14,075	14,075			14,075	0	0	0	0	0	0	0	0
	2003	14,375	14,375			14,375	0	0	0	0	0	0	0	0
Cost purchased inputs (FCFA)	2001	0	12,000	0		12,000	0	0	0	0	0	0	0	0
	2002	0	0	0		0	0	0	0	0	0	0	0	0
	2003	0	0	0		0	0	0	0	0	0	0	0	0
Avg. days labour	*p/day*	*9*	*9*	*9*		*9*	*9*	*11*		*11*	*9*	*11*		*11*
Profit/loss (FCFA)	2001	40,438		58,521	150,938	58,521		61,427	187,755	61,427		15,382	191,367	15,382
	2002	17,721	13,812			13,812	17,721			17,721	22,413			22,413
	2003	(3,644)	5,896			5,896	90,555			90,555	28,551			28,551
	Average	***18,172***	***9,854***	***58,521***	***150,938***	***26,076***	***54,138***	***61,427***	***187,755***	***56,568***	***25,482***	***15,382***	***191,367***	***22,115***

The most profitable treatment is on average 211% more profitable than the unimproved practice.

Table 12.15. Cowpeas and millet: rotation – model results.

		Millet	(Cowpea –mil–mil) Millet	(Cowpea –mil) Millet	C'peas	(Cowpea –mil–mil) C'peas	(Cowpea –mil) C'peas	(Cowpea –mil–mil) All grain	(Cowpea –mil) All grain
Yield (kg/ha)	75%	418	419	460	331	366	331		
	50%	***505***	***550***	***628***	***379***	***430***	***375***		
	25%	642	721	851	440	481	452		
Total cost excl. harvest		825	825	825	21,000	21,000	21,000	7,483	10,913
Cost purchased inputs		0	0	0	12,000	12,000	12,000	3,960	6,000
Avg. days of labour	*p/day*	*9*	*9*	*9*	*10*	*10*	*10*	9	10
Profit/loss (FCFA)	75%	25,303	25,442	29,317	52,790	62,114	52,790	37,544	45,715
	50%	***33,574***	***37,798***	***45,179***	***65,441***	***79,090***	***64,388***	***51,425***	***62,135***
	25%	46,476	53,928	66,196	81,624	92,699	84,813	66,722	79,448

Difference between most profitable treatment and the profit of the unimproved practice (with median rainfall): 53% with the 3 year rotation, 85% with the 2 year rotation.

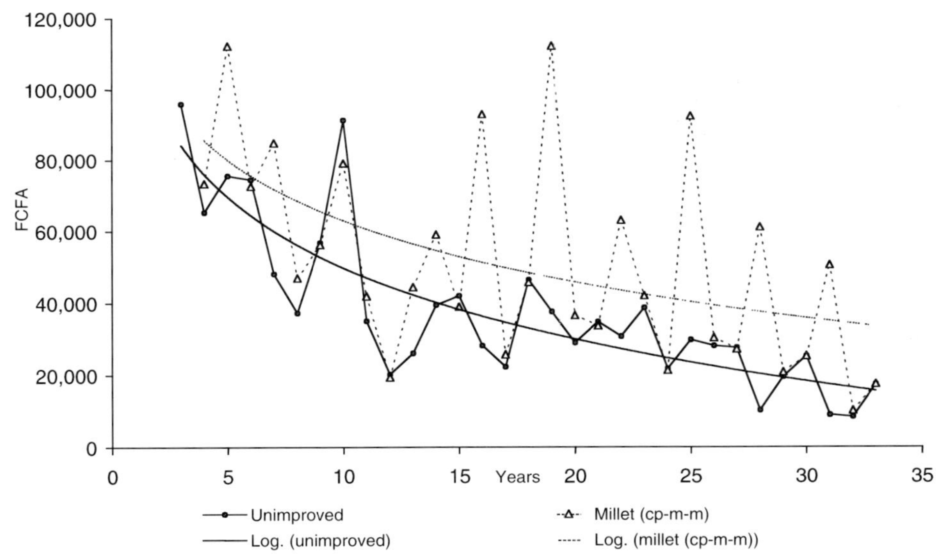

Fig. 12.3. Millet profits with rotations.

aspect of the Madiama environment that is not favourable to this crop.

The most significant point about the rotations is evident in Fig. 12.3, namely that, like microdosing, rotating cowpeas is not a practice that sustains profits over time. To keep it simple, the figure shows only the curves for millet-on-millet and the cowpeas–millet–millet 3 year rotation, but the 2 year rotation produces similar results after 30 years. The upward spikes in the curve for the rotation are generally, though not always, in those years when the higher priced cowpeas are produced. Any positive impact on millet might be understated in that the model does not take account of any beneficial reductions in *striga* that the rotation might produce, but such an impact would only eliminate occasional crop failures due to *striga* infestations. It would not prevent millet yields from steadily declining.

Comparisons Among Treatment Sets

In this section, the results of the various treatment sets are compared to see if one or another clearly stands out from the others, either as especially profitable or as particularly inferior. Figure 12.4 shows the 3-year averages for millet obtained from the trials. Yields are measured in bars against the left hand *y* axis and profits are represented by line graphs measured against the *y* axis on the right. For the manure treatments, the solid line shows profits calculated using the lower manure cost while the dotted line shows profits figured at the higher cost per kilo.

The only treatments that are clearly worse than the average unimproved control are those using animal manure without PNT. However, as already noted, the results for these treatments might not be trustworthy. In the trial results for sorghum shown above in Table 12.7, as well as the modelling results presented in Table 12.8, animal manure in the absence of PNT gives clearly better returns than the control.

The earlier analysis in this chapter and Figs 12.4 and 12.5 also suggest the following conclusions.

Profits generally follow yields

This means that higher yields generally pay for the increased costs of obtaining them. (However, since this analysis has not shown the marginal returns to input use, there is no indication of whether the amounts of inputs used in the treatments here are economically optimal.)

Comparisons between treatments hold up across low, medium and high rainfall years for most practices

That is, if a treatment is better than another in a low rainfall year, it is also likely to be so

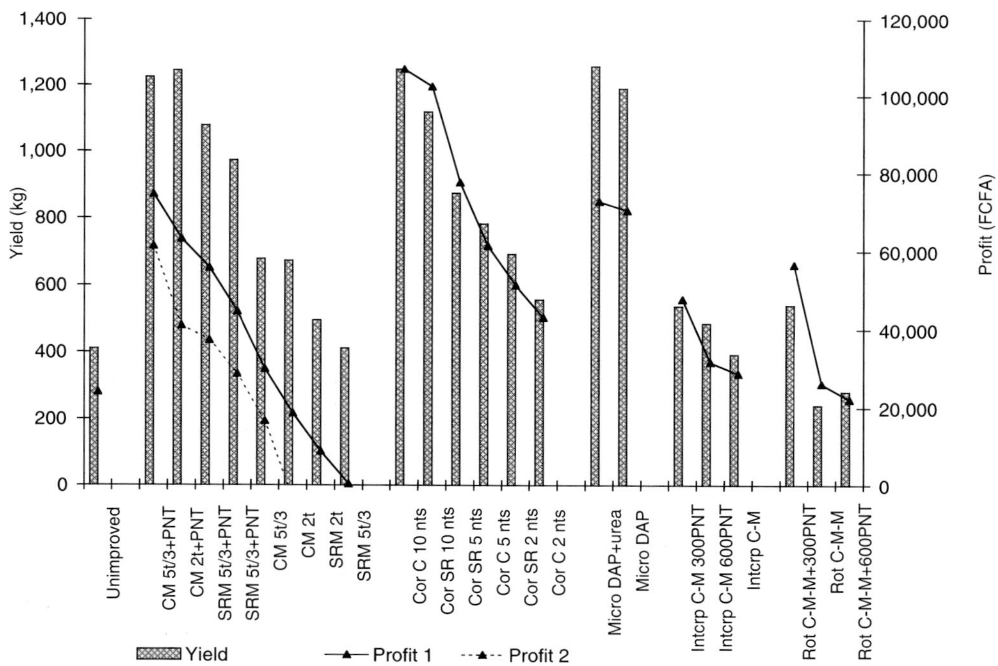

Fig. 12.4. Millet average profits and yields – trials.

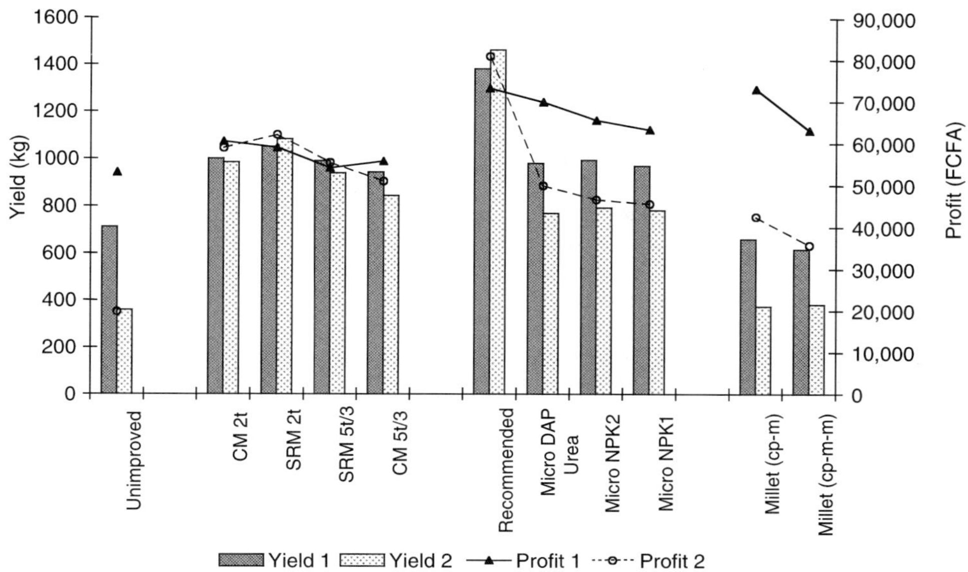

Fig. 12.5. Millet yield and profit – first 10 and last 10 years.

in medium and high rainfall years as well. (This conclusion derives from the tables shown earlier in the chapter, not from the figures in this section.)

The best results in terms of profit were obtained from three treatments

From the trials, the best results in terms of profit obtained from the three treatments (in descending order of profitability) were:

1. Corralling animals on the fields at night prior to planting.
2. Microdoses of chemical fertilizer.
3. Spreading animal manure on fields with PNT.

Overall costs are low for the first two. For the corralling treatment there need be no cash costs at all if the farmer can use his own animals or come to some arrangement with a livestock owner to pay in some way other than a cash fee, for instance by allowing the animals to feed on crop residues. There are some unavoidable cash costs for the microdosing treatments, but the cost per hectare is only FCFA 7200 for DAP and urea because of the small amounts involved. Costs are much higher for animal manure, at FCFA 22,126 for 2 tonnes of small ruminant manure (using half the full values attributed to fertilizer in Table 12.4) and FCFA 22,500 for 300 kg of PNT. However, the true opportunity cost for the manure could be lower and the farmer would have to buy the PNT only every 3 years, usually for much less than a full hectare. It is also important to note that although microdosing is more profitable than applying manure, modelling results suggest that the difference is not sustainable.

From the modelling, spreading animal manure (without PNT) or the recommended rates of fertilizer produce the most sustainable results

Figure 12.5 illustrates these conclusions. As in Fig. 12.4, yield is measured against the *y* axis on the left and profit against the one on the right. The left-hand, darker shaded bar for each treatment shows the average yield for the first 10 years of the 30 year period, and the right hand, lighter shaded bar shows yield for the last 10 years. The solid line shows average profits for the first 10 years and the dashed line that for the second 10.

If farmers grow millet year after year without doing anything to conserve soil fertility, the model says that yields and profits in the last 10 years will be half those in the first ten. In contrast, if farmers use any of the manure treatments or the recommended rates of fertilizer, their yields and profits will fall very little or not at all. There is not a lot to choose between 2 t every year or 5 tonnes every 3 years, nor between the manure from cattle and small ruminants.

The model is clear that the full, recommended rate of fertilizer performs the best. There is a common belief that animal manure is better than chemical manure because of its high organic matter content and, at higher rates of application than assumed here, no doubt that is true. However, the model assumes that 30% of the vegetable matter in the crop is returned to the soil every year. Therefore, although chemical fertilizer contains no organic matter of its own, it still results in more organic matter being incorporated into the soil. As emphasized earlier, management of crop residues is an important part of farming.

It remains true that applying recommended rates of chemical fertilizer is expensive. Although the model takes account of the high risk due to variable rainfall, applying chemical fertilizer at recommended or microdosing rates is a higher risk practice than applying manure. In a drought year the nitrogen in chemical fertilizer is largely lost, whereas more of it remains tied down in manure, to the benefit of crops in subsequent years when rainfall is better.

The lack of sustainability of microdosing chemical fertilizer is clear. Yields start out about the same as for spreading manure and profits start out higher, but over the last 10 years both yields and profits are lower than those for the manure treatments.

The cowpea–millet crop rotations show a pattern similar to that for microdosing, except that yields do not start out as high and both yields and profits fall to lower levels. Millet yields are no higher with rotations

Table 12.16. Millet: differences in average discounted profits between improved and unimproved practices.

	Discount rate							
	0%		15%		25%		50%	
	FCFA	Ratio	FCFA	Ratio	FCFA	Ratio	FCFA	Ratio
Unimproved	35,888	1.0	11,394	1.0	7,686	1.0	4,249	1.0
Cattle manure 2 Mt a year	58,468	1.6	13,666	1.2	8,709	1.1	4,619	1.1
Small ruminant manure 2 Mt a year	58,988	1.6	13,406	1.2	8,510	1.1	4,503	1.1
Small ruminant manure 5 t every 3 years	55,173	1.5	12,556	1.1	7,827	1.0	3,925	0.9
Cattle manure 5 Mt every 3 years	53,854	1.5	12,776	1.1	8,056	1.0	4,117	1.0
Recommended rates chemical fertilizer	72,653	2.0	16,805	1.5	10,943	1.4	6,054	1.4
Micro DAP urea	58,936	1.6	15,454	1.4	10,082	1.3	5,452	1.3
Micro NPK 2	54,973	1.5	14,529	1.3	9,523	1.2	5,180	1.2
Micro NPK 1	53,579	1.5	14,229	1.2	9,369	1.2	5,125	1.2
Millet (cp-m)	59,393	1.7	16,510	1.4	11,036	1.4	6,175	1.5
Millet (cp-m-m)	51,299	1.4	14,470	1.3	9,563	1.2	5,257	1.2

than with unimproved monocropping, and profits are only higher because the cowpeas are worth more. In other words, cowpeas contribute to conserving soil fertility chiefly through slowing down its decline by keeping millet off the fields every so often.

If poorer farmers discount the future at higher rates, they have less incentive to adopt improved practices

Every producer discounts future profits to some degree, meaning they value profits earned sooner more highly than the same profits earned later. Poorer producers are generally supposed to discount the future at higher rates, because their immediate needs are more urgent. A more sustainable practice, by definition, is one that shows good returns in the future rather than the present and therefore, by its nature, not as attractive to a poor farmer as to a rich one.

Table 12.16 illustrates this phenomenon for the practices discussed in this chapter. The leftmost pair of columns refers to the average undiscounted profits over 30 years for each practice. Since there is always discussion about what the appropriate discount rate is, three, 15%, 25% and 50%, are applied in the next three pairs of columns. At least the 25% and 50% rates would be very high for a western farmer, but they are not unrealistic for poor farmers in the Sahel. As the discount rate increases, the discounted returns decline rapidly, and there is also some reduction in the relative differences between the returns to the treatments as shown by the columns of ratios. This table illustrates the common observation that, while the actual benefits of sustainable practices clearly benefit the land whether it belongs to a rich or a poor farmer, the perceived benefits are distinctly lower for the latter. This makes it especially important to educate poor farmers in the returns they can get from those improved practices that require the lowest outlays, not only in cash but also in kind and in labour.

Conclusion

The combination of results from field trials and computer modelling has yielded useful insights into what farmers can do to improve their soil fertility and raise incomes. Without the trial results there would be doubt about the applicability of the modelling and without the modelling there could be some question about how robust the trial results are.

It is notable that the modelling results are less erratic than the field trial results. At least part of the explanation is that the model does not include certain elements of risk, such as *Striga* infestations, floods, shortages of labour, and invasions by animals that farmers face every year. We do not wish to underestimate the harshness of farming conditions in Madiama, or the riskiness due to the unpredictable variations in rainfall. Nevertheless, the results of both the on-farm trials and the modelling indicate that techniques such as corralling animals on fields, spreading animal manure, or microdosing with chemical fertilizer pay off. These practices are within the means of most farmers, and we hope that the documentation of relative costs and returns will help farmers to be better able to choose among the alternatives.

13 Economic Linkages Among Occupational Groups in the Madiama Commune

Charlene Brewster,[1] Daniel Kaboré,[2] Michael Bertelsen[3] and Peter Wyeth[4]

[1]Office of International Research, Education and Development, Virginia Tech, Blacksburg, VA 24061, USA; [2]Institut de l'environnement et des recherches agricoles, Burkina Faso, 04 BP 7192, Ouagadougou, Burkina Faso; [3]College of Agricultural and Life Sciences, Virginia Tech, Blacksburg, VA 24061–0334, USA; and [4]Washington State University, Pullman, WA 99164–5121, USA

In the Niger Delta Region of Mali there is a long history of shared resource use among farmers, fishermen and pastoralists. Like most areas of sub-Saharan Africa, the region is characterized by a combination of growing population pressure, low and declining agricultural productivity and low soil fertility. With increasing natural resource depletion, the area has witnessed heightened conflict between affected groups over access to cropping and grazing lands and water. In the Madiama Commune, the major river (the Bani) supplies less water and cropping activities suffer each year because of decreasing rainfall in the area. Crop yields have declined significantly over the years, and farmers have expanded cultivated lands to compensate for yield losses. The increased pressure on the natural resource base has led to greater potential for conflict as families try to achieve better health, food security and education for their members in the face of grinding poverty. This poverty itself is one of the greatest contributors to resource degradation. The poor do not invest in soil- and water-conserving technologies not only because they cannot afford to, but also because they discount the future more heavily than those with monetary resources. Therefore, resource management and poverty alleviation and economic growth must be addressed hand in hand. Economic growth within the commune is one way of relieving the pressure on natural resources, but this can only happen if such growth does not preclude the alternatives of future generations. If it is to be sustainable, economic growth must lead to sustainable use of natural resources.

It is difficult to design and implement pro-growth policies that will positively impact natural resource users unless linkages within the local economies are properly understood. The objective of this study is to model these linkages and analyse the potential impact of targeted growth strategies on different groups of natural resource users. Using information from the Participatory Landscape/Lifescape Appraisal (PLLA) and an extensive data set collected in the study area, a Social Accounting Matrix (SAM) model for the Madiama commune was developed. The SAM analysis was used to: (i) analyse the linkages among stakeholders and institutions and (ii) trace the path through which growth strategies may affect different sectors and institutions within the commune.

Economic Actors

The commune, the newly created and lowest decentralized administrative unit in Mali, was chosen as the relevant geographical survey unit. Within this study area, farmers (Marka, Dogon and Bambara), herders (Peul) and fishermen (Bozos) are now in permanent or semi-permanent residence (Sada Sy, 1994). In the past, most of the land in the commune was flooded by the end of the rainy season, greatly benefiting agriculture (rice, sorghum) due to the enhancement of soil fertility and provision of adequate moisture for plants. The government built a system of canals to capture floodwaters as part of the effort to promote wetland recessional rice cropping. Fishermen were able to fish for nearly the whole year, because water in the river never reached a level that prevented it (Daget, 1994). Peul herdsmen followed the rains south during the rainy season and from harvest time to early in the cropping season moved north to take advantage of crop residues (rice, sorghum in farmers' fields), the *bourgou* (*Echinocloa stignina* – a wetland grass) and the pasture along the river when the water level was low. Farmers cultivated their crops during the rainy season and raised animals. Their cattle were mostly entrusted to Peul herders who removed them from the area during the rainy season. During the dry season, farmers benefited from manure deposited during the herds' stay on their fields. A barter system usually existed between the different classes of resource users: farmers trading cereals for milk or manure from the Peul who gained water and pasture access while the Bozos traded fish for the products of the others.

This pattern of natural resource use began to change a few decades ago as a consequence of climate variability and human activities that combined to progressively deplete the resource base. The Niger River, as well as one of its main tributaries, the Bani River, supplies less and less water each year because of decreasing rainfall and up-stream dam activities. Rice farmers often experience severe water shortages that prevent rice cropping. Crop yields have decreased and farmers have had to expand cultivated lands not only to compensate for the drastic yield decline, but also to increase total production in an attempt to keep pace with the increasing population. This extensive agriculture (made possible by the availability of animal traction) resulted in the virtual disappearance of fallow, the traditional means of resting land and restoring soil fertility.

Over time, unsuccessful fishermen have become farmers (Lae and Weigel, 1994), increasing the pressure on agricultural resources, especially cropland. Herders also have progressively had more difficulty finding grazing land for animals during the cropping season along the river and/or on the farmers' fields. In addition, they have had less chance to prevent crop damage caused by their animals because of larger fields that have developed along their migration routes. The depleted natural resource base has resulted in high rates of livestock mortality during drought with little opportunity for herd reconstitution. For these reasons, many herders have also become farmers, keeping all of their livestock manure for themselves. Similarly, bad crop yields over time have led farmers to a diversification strategy, adding livestock that remains on farm year round to their enterprise mix. Based on the history and social structure of the region, four groups of natural resource users were characterized:

1. Farmers: agricultural production systems are the major activities of this group of individuals.
2. Agropastoralists: these individuals rear animals in conjunction with farming activities, as part of a risk reduction strategy.
3. Sedentary pastoralists: their animals graze both within and outside of the commune, but the household is sedentary. In addition, these individuals usually carry out some type of farming activities.
4. Transhumant pastoralists: these individuals travel inside and outside of the commune seeking water and pasture for their animals. Generally, they move around the Niger River Delta region's water sources during the dry season and then move south or east during the rainy season.

A 2-week PLLA in the commune provided preliminary data that were used to select villages for more formal production and income-expenditure structural surveys. In-depth structural surveys were carried out in February–March and in September 1999.

Information from the PLLA, as well as secondary data (Mali 1996 Census data), was used to select five study villages. These five of the ten villages in the area for study represent the demographic distribution of stakeholders: Madiama, the headquarters of the commune; Nerokoro, a pastoralist village; Promani, a village of sedentary and transhumant pastoralists and farmers; Tombonkan, a farmers' village; and Tatia-Nouna, a village of farmers and agropastoralists. The list of the households in the commune was obtained for each village from the 1996 census data available from government offices (Dept of Sofara, 1996). The sample size was 120 households randomly selected and representing about 10% of the commune households in 1996. The distribution of the four groups of stakeholders in the sample is presented in Table 13.1.

Survey participants were randomly selected from the village list. A meeting was held in each village with the farmers/herders in order to explain the objectives of the study and to facilitate their patience and openness to the questions.

Two types of data were collected. The first group of data was related to household characteristics, production and consumption, factors exchanged (labour, equipment, land, money). The origin and source of factors exchanged were also recorded. Four enumerators and two supervisors administered the questionnaire. Both the enumerators and the IER researcher attended a 2-day training session in Madiama in order to learn the objectives of the study, the SAM approach and the requirements of this kind of survey. The survey was conducted over a 2-month period, took 3–4 hours for each respondent, and was scheduled with a break after 2 hours for Muslim prayer or lunchtime. Each enumerator surveyed a maximum of two households per day in order to avoid weariness and preserve data quality.

The second group of data was collected on microenterprise activities in the commune that are income generating activities, such as food processing, handcrafts, retail trade, livestock trading, cereal trading and so forth. This data was collected in September 1999 on a sub-sample of 60 households drawn from the larger sample of 120 households. A pre-test was implemented with a few respondents in order to: (i) correct/improve the questionnaire and (ii) give enumerators survey experience. The training, the pre-test and the questionnaire improvement took 1 week (Kaboré *et al.*, 2000).

Table 13.1. Sample distribution by activity and village, Madiama Commune, Mopti Region, 1999.

	Group by main activity				
			Pastoralists		
Village	Farmers	Agro-pastoralists	Sedentary	Transhumant	Total
Madiama	12	20	0	0	32
Promani	08	05	10	0	23
Tombonkan	09	01	0	0	10
Tatia-Nouna	14	06	04	0	24
Nerekoro	0	01	09	21	31
Total	43	33	23	21	120
	(36)	(27)	(19)	(18)	(100)

Note: sample size as a percentage of total population is expressed in parentheses.

The SAM Framework and Multiplier Decomposition

A SAM model was developed to meet the objectives of the study. The SAM model is a modified input/output model that shows the flow of income and expenditure among production activities and economic actors. It is a very flexible and powerful tool that can be used to analyse economies in diverse social and cultural settings (Taylor and Adelman, 1996). It has been used in analysis of village economies (Subramanian, 1988) as well as for nationwide studies (Dorosh *et al.* 1991; Arndt *et al.* 1998; Pradham *et al.*, 1999). Given its flexibility, the SAM can also be developed to address specific topics such as environmental issues (see Miller and Blair, 1985) or migration (Adelman *et al.*, 1988).

The SAM is organized as an accounting matrix of modeller-selected endogenous and exogenous sectors' inflows and outflows. It is based on the assumption that production activities are endogenous and demand-driven. Endogenous accounts are those for which changes in the level of expenditure directly follow any changes in income. The endogenous accounts typically include: (i) production activities (the input–output sub-matrix), (ii) factors (labour) and (iii) institutions (households and firms). Exogenous accounts are typically those for which it is assumed that expenditures are set independently of income. These consist of: (i) government, (ii) capital, and (iii) the rest of the world. By convention, receipts (incoming value added) to a sector are listed in the rows while expenditures (outgoing) appear in the columns (see Table 13.2). Table 13.2 contains transformation matrices involving the endogenous accounts: $\mathbf{A}_{11}$ gives the intermediate input requirements, i.e. the input–output transaction matrix; $\mathbf{A}_{13}$ reflects the expenditure pattern of the institutions (households) for different commodities they consume from the production sector. $\mathbf{A}_{21}$ dispatches the value added generated by production activities by factor, while $\mathbf{A}_{32}$ captures the income distribution to the various institutions. $\mathbf{A}_{33}$ maps the income transfers within and among households (institutions). The row totals received by endogenous accounts are given by $\mathbf{y}_{n}$, which consist of: (i) expenditures by endogenous accounts ($\mathbf{A}_{nn}$) summed up as $\mathbf{n}$ and (ii) expenditures by the exogenous accounts ($\mathbf{A}_{nx}$) summed up as $\mathbf{x}$ and referred to as injections. Finally $\mathbf{y}_{n} = \mathbf{n} + \mathbf{x}$ and similarly, income received by exogenous accounts $\mathbf{y}_{x}$ is $\mathbf{y}_{x} = \mathbf{l} + \mathbf{t}$. Since Table 3.2 is a SAM, rows and columns should balance; that is: $\mathbf{y}_{n} = \mathbf{y}'_{n}$ and $\mathbf{y}_{x} = \mathbf{y}'_{x}$ where $\mathbf{y}'_{n}$ and $\mathbf{y}'_{x}$ are column analogues of row sums $\mathbf{y}_{n}$ and $\mathbf{y}_{x}$, respectively.

For the purposes of the SAM analysis, the commune economy is assumed to consist

Table 13.2. Simplified SAM structure.

	Expenditures (outflows)				
	Endogenous			Exogenous	
Receipts (inflows)	Production	Factors	Institutions	Sum of other Accounts	Totals
Endogenous					
Production	$\mathbf{A}_{11}$	–	$\mathbf{A}_{13}$	$\mathbf{x}_1$	$\mathbf{y}_1$
Factors	$\mathbf{A}_{21}$	–	–	$\mathbf{x}_2$	$\mathbf{y}_2$
Institutions	–	$\mathbf{A}_{32}$	$\mathbf{A}_{33}$	$\mathbf{x}_3$	$\mathbf{y}_3$
Exogenous					
Sum of other Accounts	$\mathbf{l}'_1$	$\mathbf{l}'_2$	$\mathbf{l}'_3$	$\mathbf{t}$	$\mathbf{y}_x$
Totals	$\mathbf{y}'_1$	$\mathbf{y}'_2$	$\mathbf{y}'_3$	$\mathbf{y}_x$	

of five account categories: production activities, commodities, factors, institutions, capital account and the rest of the world (see Table 13.7). The community's economy has been organized around ten main activities/commodities. These are cereals other than rice (millet, sorghum, maize and fonio), rice; vegetables, legumes, livestock, fish, natural resources, retail trade, durables and government services (schools, dispensaries and so forth). Activities/commodities result from an aggregation of agricultural activities and their associated microenterprises. For example, the livestock activity includes not only livestock production, but also traded livestock and produced cereals and traded cereals are combined. Factors include land, capital (equipment), hired labour, and family labour. Institutions consist of farmers, agropastoralists, sedentary pastoralists, transhumants and government services. The capital account includes investment/saving in physical assets (agricultural and non-agricultural equipment) and livestock investment. Table 13.7 outlines the basic SAM structure for the commune.

Exogenous shocks or changes introduced into the commune economy will alter the circular flows of resources among sectors within the commune. Examples of exogenous variables include government input subsidies or taxes. The magnitudes of these impacts (measured by multipliers) will depend upon the strength of the linkages between the sectors. Similarly, the impact (and multipliers) of a decision made within the Madiama commune will be related to endogenous variables such as changing land distribution among sectors or changing transhumance scheduling through commune taxation of grazing lands. The multipliers provide important information to commune level decision-makers on the prospects for, and differential impacts of, economic growth on the various sectors. This has direct implications for who benefits and loses from policy changes and, by extension, how conflicting sectors benefit and how they might accordingly react to changes.

Multiplier analysis allows decision makers to fully trace the effects of interventions in one sector or industry. It can provide decision-makers (at the commune level) with information about the differential impacts of economic growth on various sectors within the commune. Three types of multipliers are calculated using the method developed by Pyatt and Round (1979) and illustrated by Holland and Wyeth (1993):

1. Direct effects multipliers (transfer effects): these capture the interactions within each category of accounts, e.g. intersectoral input–output elements. For example in the case of an exogenous shock such as an increase in the demand for a commodity (e.g. small ruminants), the direct effect multiplier demonstrates how the shock affects other production activities in the commune.

2. Indirect or open loop multipliers: when an exogenous shock affects one sector, indirect multipliers illustrate how the impacts are transmitted to the other types of accounts; for example, they may show how a shock in production affects factor incomes. These are one-way effects, and any impacts on the originating sector are excluded. For example, indirect multipliers may be used to illustrate the impact of changes in production activities on factors such as labour, or changes in incomes of the different occupational groups within the commune.

3. Closed loop multipliers: this group of multipliers shows the full circular effect of a shock. They demonstrate how a shock to a sector travels outward to other sectors and then back to the original point of origin.

Economic Interdependencies within the Commune

When describing the economic interactions within the commune, two sectors can be distinguished. The tradable sector, as the name suggests, consists of goods and services that are traded with regions external to the commune. The price of tradable goods is determined outside of the commune where there is a larger effective demand for these commodities. The commune can sell all of its production at this price. However, as production rises, the cost of producing additional units of production also rises (upward sloping supply curve). If these costs rise above the market price, then the commune will no longer be able to sell this commodity to the external world without sustaining economic losses. Growth strategies aimed at expanding the tradable sub-sector must, therefore, focus on improving technologies and management systems so that production costs are lowered. The non-tradable sector consists of goods and services that are not traded outside the wider area, although there may be some trade within the commune itself. In Madiama, the non-tradable goods include meat, fish, millet, milk, sorghum and legumes. Labour and manure are non-tradable inputs, because their use is confined to the commune. Expansion of the non-tradable sector is constrained by low levels of demand, rather than problems of supply. More of these goods and services can be provided (their supply is elastic), but the level of demand does not justify this.

An examination of the sales data for primary agricultural products reveals the importance of exports in the trade of primary livestock products (Fig. 13.1). The livestock trade is particularly important to sedentary pastoralists, and for this group over 50% of sales is to areas outside of the commune. The major tradable crop products include rice and vegetable crops, primarily gumbo (okra).

Figure 13.2 describes the pattern of microenterprise activity in the commune. Sales in the tradable sector account for 20%

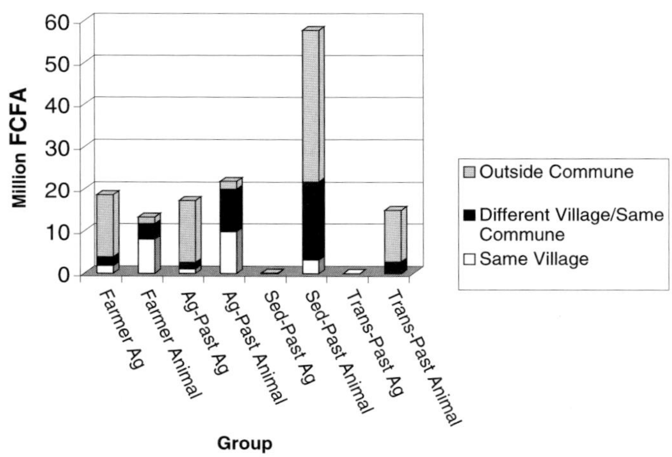

Fig. 13.1. Sales of primary crop and animal products.

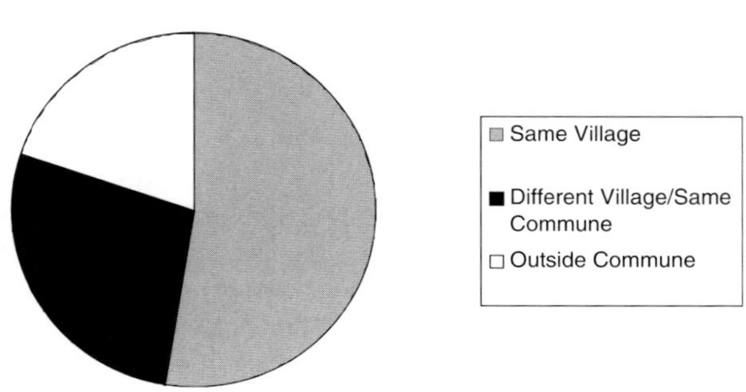

Fig. 13.2. Distribution of microenterprise sales volume in Madiama commune.

Table 13.3. Microenterprise activity for Madiama commune ('000 CFAF).

Type of activity	Farmers	Agro-pastoralists	Sedentary pastoralists	Transhumants
Same village sales	3025.6	21098.7	24525.9	
Sales within commune – different village	2570.4	4541.0	18132.0	
Sales outside commune	10710.0	3641.4	2934.5	1299.9
Totals	16306.0	29281.1	455924.4	1299.9

Table 13.4. Selected direct effect multipliers for Madiama commune.

Introduction of shock	Sector of impact	Own effect multiplier
Rice	Rice	1.17
Small ruminants	Small ruminants	0.16
Microenterprise	Small ruminants Microenterprise	1.17
Large ruminants	Large ruminants	0.19
Microenterprise	Large ruminants Microenterprise	1.17
Poultry microenterprise	Poultry microenterprise	1.37

of all microenterprise activity, with sedentary pastoralists accounting for the largest share of activity and transhumants the least (Table 13.3).

This pattern of trade suggests a direction for growth strategy. Over the years, development agencies have increasingly looked towards the microenterprise sector as a vehicle for growth. The existence of a tradable microenterprise sector (though relatively small) suggests a starting point. One way of increasing incomes is to reduce the supply constraints associated with the tradable sector so that more commodities and services can be produced and traded. Thus, research investments that focus on lowering livestock production costs through, for example, improved pasture management will be instrumental in raising incomes. This highlights the importance of the SANREM pasture productivity research (see Chapters 14 and 15). However, since much of the commune economy still revolves around the non-tradable sector (millet, sorghum and legumes such as cowpea) research is needed to increase productivity in these areas also. Much of the SANREM soil fertility work is focused on the non-tradable sector (see Chapters 11 and 12).

Results and Discussion

The direct effect multipliers describe the impacts of a shock to one sector or industry, on the other sectors. Table 13.4 illustrates the effect of exogenous increases in demand for selected commodities. Only impacts on sectors that are affected are shown and the impacts on the other sectors are small. The largest multipliers are those associated with rice and livestock. Interpretation of these multipliers is straightforward. For example, if there was an increase in rice demand of 1 million FCAF, then interindustry transactions alone would induce increases in output of 1.17 million FCAF. Increases in demand (or investment) in the livestock and livestock microenterprise sectors would also induce large increases in output.

Open loop effects are shown in Table 13.5. This group of multipliers describes the impact of a shock to one sector on other sectors outside of that block. The open loop multipliers in Table 13.5 demonstrate the impact of exogenous changes on incomes of stakeholders within the commune. For example, a 1 million CFAF increase in demand for millet will result in increases in income of 216,000 CFAF for farmers,

Table 13.5. Selected open loop effects for Madiama commune.

	Millet	Rice	Small ruminants	Large ruminants	Natural resources and food products
Farmers	0.216	0.199	0.230	0.221	0.203
Agropastoralists	0.355	0.340	0.395	0.380	0.269
Sedentary pastoralists	0.224	0.207	0.238	0.230	0.132
Transhumants	0.125	0.108	0.122	0.117	0.068

Table 13.6. Closed loop multipliers for institutions within Madiama commune.

	Farmers	Agropastoralists	Sedentary pastoralists	Transhumants
Farmers	1.34	0.34	0.34	0.38
Agropastoralists	0.58	1.58	0.58	0.66
Sedentary pastoralists	0.35	0.35	1.35	0.40
Transhumants	0.18	0.18	0.18	1.20

355,000 FCFA for agropastoralists, 224,000 FCFA for sedentary pastoralists and 125,000 FCFA for transhumant pastoralists. It can be seen that in general, the multipliers for transhumant pastoralists are significantly less than those for the other groups of stakeholders. In addition, agropastoralists would benefit the most from increases in commodity demand, followed by sedentary pastoralists, farmers and then transhumants. It is also noteworthy that the open loop multipliers for the natural resources and food products category is large, illustrating the potential for economic growth through sustainable exploitation of natural resources.

Selected closed loop effects are shown in Table 13.6. These multipliers show how a shock in one sector is transmitted through all the other blocks and back to where it started. The multipliers in Table 13.6 show how the effects of changes in income of one group will work through the system and impact incomes of all groups in the commune. For example, an exogenous increase of 1 million FCAF in the income of farmers, will stimulate an additional increase of 34,000 FCAF to farmers, 54,000 FCAF for agro-pastoralists, 35,000 FCAF for sedentary pastoralists and only 18,000 FCAF for transhumants.

This lower level of impact for the transhumants group holds in every case (Table 13.6) and these results show that in general transhumants benefit the least from injections of income into the system.

Conclusions

The study illustrated the linkages between socioeconomic groups in the commune and the potential to use SAM analysis for community level economic studies. An analysis of production and sales activity demonstrated the relatively closed nature of the commune; i.e. the tradable sector is relatively small compared to the non-tradable sector. Impacts from exogenous changes in demand for the commune's agricultural and livestock production, as measured by multipliers, are shared differentially among socioeconomic groups with the transhumant group always benefiting least. This reinforces earlier conclusions that interventions need to specifically target this group.

These results were discussed at face-to-face meetings with community representatives of the occupational groups. Leaders were intuitively aware of these results, since they already knew that the transhumant group is the least integrated into the commune economy. Given this knowledge,

Table 13.7. Social accounting matrix outline, Madiama commune, Mopti, Mali, 1999 – million FCFA.

	ACTIVITIES									
	Cereals	Rice	Veget	Legumes	Livestock	Fish	Retail	NatRes	Gvt Services	Durables
Activities										
Cereals	0	0	0	0	0	0	0	0	0	0
Rice	0	0	0	0	0	0	0	0	0	0
Veget	0	0	0	0	0	0	0	0	0	0
Legumes	0	0	0	0	0	0	0	0	0	0
Livestock	0	0	0	0	0	0	0	0	0	0
MeatFish	0	0	0	0	0	0	0	0	0	0
Retail	0	0	0	0	0	0	0	0	0	0
NatRes	0	0	0	0	0	0	0	0	0	0
Gvt Services	0	0	0	0	0	0	0	0	0	0
Durables	0	0	0	0	0	0	0	0	0	0
Commodities										
Cereals	8.21	0	0	0	11.38	0	0	0	1.107	0
Rice	0	28.05	0	0	0.907	0	0	0	2.214	0
Veget	0	0	1.1493	0	0.074	0	0	0	0	0
Legumes	0	0	0	0.3923	2.4454	0	0	0	0.5536	0
Livestock	0	0	0	0	3.5513	0	0	0	1.107	0
Fish	0	0	0	0	0	0.6771	55.6915	0	0.5535	0
Retail	3.8535	1.5	4.19	0	44.6285	0	77.2578	0	1.845	0
NatRes	0	0	0	0	10.0975	0.31598	0	2.37549	0	0
Gvt Services	0	0	0	0	0	0	0	0	0	0
Durables	0	0	0	0	0	0	0	0	0	1.059
Factors										
Family Labour	324.2573	472.4437	65.53844	11.0417	546.459586	24.31131	110.6296	3.3009	0	0
Hired Labour	13.16603	13.4241	0.05156	0.00125	26.468702	0	0	0	0	0
Capital(Eqpt)	2.331826	0.27765	0	0.33446	2.576114	0.179311	0.135302	0.3	0	0
Land	0.63998	2.278	0	0	0	0	0	0	0	0

Table 13.7. (continued)

	ACTIVITIES									
	Cereals	Rice	Veget	Legumes	Livestock	Fish	Retail	NatRes	Gvt Services	Durables
Institutions										
Farmers	0	0	0	0	0	0	0	0	0	0
AgroPast	0	0	0	0	0	0	0	0	0	0
SedPast	0	0	0	0	0	0	0	0	0	0
Transhum	0	0	0	0	0	0	0	0	0	0
Central Govt	0	0	0	0	0	0	0	0	0	0
Total Institutions										
Capital										
Savings	0	0	0	0	0	0	0	0	0	0
ROW										
Imports	0	0	0	0	0	0	0	0	0	0
Totals	352.458636	517.97345	70.9293	11.76971	648.588102	25.483701	243.714202	5.97639	7.3801	1.059

	COMMODITIES									
	Cereals	Rice	Veget	Legumes	Livestock	Fish	Retail	NatRes	Gvt Services	Durables
Activities										
Cereals	195.9733	0	0	0	0	0	0	0	0	0
Rice	0	205.0028	0	0	0	0	0	0	0	0
Veget	0	0	70.9293	0	0	0	0	0	0	0
Legumes	0	0	0	11.7706	0	0	0	0	0	0
Livestock	0	0	0	0	410.2522	0	0	0	0	0
Fish	0	0	0	0	0	25.4837	0	0	0	0
Retail	0	0	0	0	0	0	68.4864	0	0	0
NatRes	0	0	0	0	0	0	0	5.6764	0	0
Gvt Services	0	0	0	0	0	0	0	0	7.38	0
Durables	0	0	0	0	0	0	0	0	0	1.059

Commodities										
Cereals	0	0	0	0	0	0	0	0	0	0
Rice	0	0	0	0	0	0	0	0	0	0
Veget	0	0	0	0	0	0	0	0	0	0
Legumes	0	0	0	0	0	0	0	0	0	0
Livestock	0	0	0	0	0	0	0	0	0	0
Fish	0	0	0	0	0	0	0	0	0	0
Retail	0	0	0	0	0	0	0	0	0	0
NatRes	0	0	0	0	0	0	0	0	0	0
Gvt Services	0	0	0	0	0	0	0	0	0	0
Durables	0	0	0	0	0	0	0	0	0	0
Factors										
Family Labour	0	0	0	0	0	0	0	0	0	0
Hired Labour	0	0	0	0	0	0	0	0	0	0
Capital(Eqpt)	0	0	0	0	0	0	0	0	0	0
Land	0	0	0	0	0	0	0	0	0	0
Institutions										
Farmers	0	0	0	0	0	0	0	0	0	0
AgroPast	0	0	0	0	0	0	0	0	0	0
SedPast	0	0	0	0	0	0	0	0	0	0
Transhum	0	0	0	0	0	0	0	0	0	0
Central Govt	0	0	0	0	0	0	0	0	0	0
Total Institutions										
Capital										
Savings	0	0	0	0	0	0	0	0	0	0
ROW										
Imports	313.1809	20.76969	3.795578	59.96981	276.26701	102.675597	315.431425	48.3300977	0	59.45546
Totals	509.1542	225.77249	74.724878	71.74041	686.51921	128.159297	383.917825	54.0064977	7.38	60.51446

Table 13.7. (continued)

	FACTORS				Institutions					Capital	ROW	Total
	Family Labour	Hired Labour	Capital (Eqpt)	Land	Farmers	AgroPast	SedPast	Transhum	Central Gvt		Exports	
Activities												
Cereals	0	0	0	0	0	0	0	0	0	0	156.4853	352.4586
Rice	0	0	0	0	0	0	0	0	0	0	312.9707	517.9735
Veget	0	0	0	0	0	0	0	0	0	0	0	70.9293
Legumes	0	0	0	0	0	0	0	0	0	0	0	11.7706
Livestock	0	0	0	0	0	0	0	0	0	0	238.3359	648.5881
Fish	0	0	0	0	0	0	0	0	0	0	0	25.4837
Retail	0	0	0	0	0	0	0	0	0	0	175.2278	243.7142
NatRes	0	0	0	0	0	0	0	0	0	0	0.3	5.9764
Gvt Services	0	0	0	0	0	0	0	0	0	0	0	7.38
Durables	0	0	0	0	0	0	0	0	0	0	0	1.059
Commodities												1885.3334
Cereals	0	0	0	0	226.5395	204.3053	47.2603	10.35212	0	0	0	509.15422
Rice	0	0	0	0	98.40711	75.19646	15.75832	5.2396	0	0	0	225.77249
Veget	0	0	0	0	23.1158	42.0647	1.07559	7.245488	0	0	0	74.724878
Legumes	0	0	0	0	58.83713	5.3408	3.75848	0.4127	0	0	0	71.74041
Livestock	0	0	0	0	49.6067	134.2907	148.525	71.15771	0	278.2808	0	686.51921
MeatFish	0	0	0	0	33.76504	25.20564	8.810493	3.456024	0	0	0	128.1593
Retail	0	0	0	0	96.7902	76.248	62.2226	15.382225	0	0	0	383.91783
NatRes	0	0	0	0	16.644515	17.6468	3.9249	3.0013127	0	0	0	54.006498
Gvt Services	0	0	0	0	0	0	0	0	7.38	0	0	7.38
Durables	0	0	0	0	0	0	0	0	0	59.4555	0	60.5145
Factors												2201.8893
Family Labour	0	0	0	0	0	0	0	0	0	0	0	1557.9825
Hired Labour	0	0	0	0	0	0	0	0	0	0	0	53.111642
Capital(Eqpt)	0	0	0	0	0	0	0	0	0	0	0	6.134663
Land	0	0	0	0	0	0	0	0	0	0	0	2.91798

Institutions												1620.1468
Farmers	516.117	16.1903	2.6165	1.1	31.4773	2.89231	0.25796	0.22309	0	0	143.7609	714.63536
AgroPast	570.503	17.775	2.7143	1.3	2.7675	32.56083	1.336477	0.22309	0	0	129.3527	758.5329
SedPast	313.3308	14.6849	0.3995	0.51798	0.97222	2.319956	19.68656	0.59809	0	0	74.7116	427.22161
Transhum	158.032308	4.4614	0.4044	0	2.52381	3.566856	3.91458	5.78819	0	0	50.06093	228.75247
Central Govt	0	0	0	0	3.31	4.201837	1.914634	0.664	0	0	0	10.090471
Total Institutions												2139.2328
Capital												
Savings	0	0	0	0	60.9544	131.90265	102.71844	104.4368	2.7105	0	0	402.72279
ROW												
Imports	0	0	0	0	8.9242	0.790075	6.05726	0.572105	0	64.9865	0	1281.2057
Totals	1557.983108	53.1116	6.1347	2.91798	714.635425	758.532914	427.221594	228.752545	10.0905	402.7228	1281.2058	9530.5309

however, they argued that perhaps the more important concept at hand was the notion of inter-relationships between the group, and the knowledge that 'if one group benefits, then we all will benefit' as stated by the transhumants representative.

References

Adelman, I., Taylor, J.E. and Vogel, S. (1988) Life in a Mexican Village: a SAM perspective. *Journal of Development Studies* 25, 5–24.

Arndt, C., Jensen, T.H and Tarp, F. (1998) *Structural Characteristics of the Economy of Mozambique: a SAM-based analysis.* DERG, University of Copenhagen, Copenhagen, Denmark.

Daget, J. (1994) *La pêche dans le Delta Central du Niger-Approche pluridisciplinaire d'un système de production halieutique,* Avant-propos, Jacques Quensière ed. IER-ORSTOM, Karthala, Paris, France.

Department of Sofara. 1996 Census Unpublished, Djenné Cercle, Mali.

Dorosh, P.A., Bernier, R., Randrianarivony, A.R. and Rasolomanana, C. (1991) *A Social Accounting Matrix for Madagascar: Methodology and Results* (Working Paper #6). Cornell Food and Nutrition Policy Program, Ithaca, New York.

Holland, D and Wyeth, P. (1993) *SAM Multipliers: Their Decomposition, Interpretation and Relationship to Input–Output Multipliers* (Research Bulletin XB1027). Washington State University, College of Agriculture and Home Economics Research Center, Pullman, Washington.

Kabore, P.D., Traoré, B., Taylor, B.D., Bertelsen, K.D. and Holland, D. (2000) Modeling District-level Socio-economic Linkages and Growth: Towards Sustainable Natural Resource Management under Environmental Stress and Conflict in the Niger Delta Region of Mali. Selected paper presented at AAEA Annual Meeting, 30 July–2 August 2000 at Tampa, Florida.

Lae, R. and Weigel, J.-Y. (1994) Adaptabilité des pêcheurs aux changements environnementaux et économiques. In: Quensière, J. (ed.) *La pêche dans le Delta Central du Niger-Approche Pluridisciplinaire d'un Système de Production Halieutique.* I.E.R.-ORSTOM, Karthala, Paris, France, pp. 295–310.

Miller, R.E. and Blair, P.D. (1985) *Input–Output Analysis: Foundations and Extensions.* Prentice Hall, Englewood Cliffs, New Jersey.

Pradham, B.K., Sahoo, A. and Saluja, M.R. (1999) A Social Accounting Matrix for India, 1994–95. *Economic and Political Weekly* 34(48), 3378–3394.

Pyatt, G. and Round, J.I. (1985) Accounting and fixed-price multipliers in a Social Accounting Matrix framework. In: Pyatt, G. and Round, J.I. (eds) *Social Accounting Matrices: a Basis for Planning.* World Bank, Washington, DC, pp. 186–206.

Sada Sy, B. (1994) Préface. In: Quensière, J. (ed.) *La pêche dans le Delta Central du Niger-Approche Pluridisciplinaire d'un Système de Production Halieutique.* IER – ORSTOM, Karthala, Paris, France.

Subramanian, S. (1988) Production and Distribution in a Dry-land Village Economy in the West Indian Deccan. Ph.D. dissertation. University of California, Berkeley, California.

Taylor, J.E. and Adelman, I. (1996) *Village Economies: the Design, Estimation, and Use of Village Wide Economic Models.* Cambridge University Press, Cambridge, UK.

14 Dry Season Feed Supplements: the Potential Role of *Cassia tora*

Meriem El Hadj,[1] Ozzie Abaye,[1] Amadou Kodio[2] and Moussa Keïta[2]

[1]Department of Crop and Soil Environmental Science, Virginia Tech, Blacksburg, VA 24061–0404, USA; [2]Centre Régional de la Recherche Agronomique, Institut d'Economie Rurale, B.P. 205, Mopti, Mali

During the summer of 2002, we sat down with women from three villages (Madiama, Siragourou and Nérékoro) within the Madiama Commune, Mali. The objectives of this meeting were to pursue questions on pasture resources, a priority concern of the community, and to understand women's food production constraints. Almost all the women expressed interest in resolving livestock feed problems, particularly for small ruminants during the dry season. This was most important to the women in Nérékoro, who are from a different ethnic group (Peul) and, unlike the women from Madiama and Siragourou who help plant as well as harvest field crops, do not participate in fieldwork. As the Nérékoro women explained, their responsibilities include raising children, and buying and fattening small ruminant animals for market and milk production. In terms of fattening animals as well as increasing milk production, they said the critical period of feed shortage is during the dry season (November–May). With this began an exploration of ways to preserve some of the forage plants for later use in the dry season. When asked about available plant species that remain green late after the rainy season has passed that might be improved through ensiling, the women pointed to the *Cassia tora* that filled their degraded pastures.

Cassia tora (*C. tora*, *Cassia obtusifolia* (*Senna obtusifolia*)) is a leguminous plant that fixes atmospheric nitrogen. Hoveland *et al.* (1976) found that among other tropical legumes *C. occidentalis* L. and *C. obtusifolia* L. (sicklepod) were the most tolerant to low available soil P. For this reason *C. tora* has an important role to play in soil fertility enhancement. However, due to its aggressive nature that excludes grasses from being grown with it or in the surrounding areas, and its unpalatability for animals, it imposes more problems than potential benefits. In most cases, grazing animals are used to control undesired species in pasture by varying the stocking rate. In this case, however, this option is not available since animals do not graze on fresh *C. tora* as it is unpalatable to all livestock. Since the use of herbicide to control this plant is unlikely, some alternatives are necessary to reduce its propagation or improve its quality for animal feed.

Keeping in mind that the research goal was to increase forage resources and provide a source of feed for the dry season, we decided to investigate the hypothesis that ensiling *C. tora* could provide a nutritious source of feed. The objectives of the experiment were:

1. To evaluate the chemical characteristics (crude protein, *in vitro* dry matter digestibility, fibre, etc.) of ensiled vs. fresh *C. tora*.

2. To examine the effect of additives (water or honey) and other forages (grasses) on the quality of the ensiled material.

Identifying Characteristics and Distribution of *C. tora*

Cassia (*Senna*) *tora* L., *foetid cassia*, *C.* (*S.*) *obtusifolia* L., sicklepod, coffee weed, etc., are very closely related annual weedy shrubs belonging to the family Caesalpiniaceae in the order Leguminosae (Cock and Evans, 1984). *Cassia* is a weed of 26 crops in 67 countries and is more prevalent in soybeans, groundnuts, pastures, cotton and sugarcane than other crops. *Cassia obtusifolia*, commonly known as coffeeweed, sicklepod, or coffeepod, is the most prevalent species of *Cassia*, can reach heights of 1.8–2.1 m, and produces sickle-shaped seedpods. The teardrop-shaped leaves grow in clusters of four to six leaflets. It is often confused with *C. tora*. In fact, it is difficult to distinguish between the two (Brenan, 1958). Some investigators claim to have seen plants intermediate between the two species. Bhandari (1978) treated *C. tora* as the synonym of *C. obtusifolia*. Others distinguished the two on the basis of epidermal structure and seed analysis (Mall, 1957; Pandey, 1970; Mathur, 1985). Randell (1995) discussed the possibility that *C. tora* evolved in Asia from *C. obtusifolia*. *Cassia obtusifolia* is well documented as a weed of groundnuts and soybeans in the southern USA, and is widespread throughout the tropical world. *C. tora* is becoming increasingly prominent as a pasture weed in the southwestern Pacific, Asia and Africa. De Wit (1955) characterized *C. tora* as a weed species throughout the tropics, probably of South American origin. Singh *et al.* (1970) characterized it as a minor weed of various crops but of greatest importance as a pasture weed and suggested that it became established in response to overgrazing.

Invasive Species and Biological Significance

Most plant species have evolved with predators that eat them. This raises the question of what happens if someone takes a plant and moves it to a place where its predator does not exist. To keep its population under control, the plant needs its predator. When a plant is reproducing without control and pushing out native plant species, it has become a weed or an invasive exotic plant. Ehrenfeld (2003) sees the invasive species not as a threat to biodiversity and ecosystem stability, but rather as a potential for nutrient cycling processes in the soil. Ehrenfeld reviewed data on invasive species and soil nutrient cycling and suggested that invasive plant species frequently increase N availability, alter N fixation, and produce litter with higher decomposition rates than coexisting native species. Most of the data was based on observational studies, and there is very little information on the time since invasion, the causes of invasion and plant densities. Davis (2003) states that extinctions often result from the combined effects of multiple processes, such as over-harvesting and habitat loss, and therefore, although competition from introduced species might not be the primary cause of native species extinction, it is a contributing factor. Davis (2003) sustains the idea of *dispersal* by deriving from Hubbell (2001) the theory of 'ecological drift' as a positive factor of global biodiversity and that competition from introduced species is not likely to be a common cause of extinctions of long-term resident species at global, metacommunity, and even community levels. Chandrasekaran and Swamy (2002) studied the influence of an herbaceous weed community on natural secondary forest and man-modified ecosystems. Man-modified ecosystems showed significantly greater weed density than the natural ecosystem; the researchers attribute this difference to a higher incidence of exotic weed species in man-modified ecosystems due to a higher degree of disturbance and open environment. In addition, the sites invaded by the exotic weeds showed higher herbaceous litter production. Uhl and Jordan (1984) attribute the production of successional forests to the nutrients released from the decomposition of herbaceous litter. These studies suggest that exotic plant invasions influence the ecosystem structure, species composition and aboveground biomass.

Cassia tora, if managed, can improve pasture productivity by providing more litter and increasing organic matter in the soil as well as recycling essential nutrients. It can also provide a source of feed during the dry season when used as silage.

Cassia as Forage

Although *Cassia tora* is considered a poisonous plant, the toxic element in *Cassia* has not been clearly defined. The seeds appear to exert their toxicity upon the skeletal muscles, kidney and liver. The leaves and stem also contain toxin, whether green or dry. The plant can poison animals if they consume it in the field, in green chop, in hay or in grain containing the seed. Perkins and Payne (1985) suggest that anthraquinones found in the leaves are the toxic compounds of this plant. Others report that the seeds of the plant contain emodin, other anthraquinones and xanthones (Manjunath and Subbajois, 1930; Gupta and Sharma, 1965; Rangaswami, 1963). Tiwari and Behari (1972) have found anthraquinones in the roots of *C. tora*. Pal *et al.* (1977) have also found emodin from the leaves of *C. tora*. Toxicity has been observed in cattle and broilers, and other animals are also susceptible to the effects of this plant. In cattle, diarrhoea is usually the observable first symptom of poisoning. Later, the animals go off feed, appear lethargic, and tremors appear in the hind legs, indicating muscle degeneration. As the symptoms progress, the urine darkens to the colour of coffee, and the animal becomes recumbent and is unable to rise. Death often occurs within 12 h after the animal goes down.

In their study, Sood *et al.* (1990) used *C. tora* seeds as a potential poultry feed. The seeds were treated with sodium hydroxide followed by rectified spirit (alcohol) to lower the toxicity. *Cassia tora* could be fed up to 10% level. Other researchers suggested the presence of tannins and saponins in the seed of *C. tora*. Lower feed intake, poor growth response coupled with enlarged and fatty livers, reduced spleens and haemolysed erythrocytes (destruction of red blood cells) suggested the presence of these toxins in *Cassia* seeds (Katoch *et al.*, 1978; Katoch and Bhowmik, 1983). Chand and Katoch (1983) reported that 0.1 M sodium hydroxide treatments to *C. tora* meal resulted in 77.27% tannin extraction and 25% saponin extraction, and treatment with rectified spirit eliminated 35.61% and 58.33%, respectively. However, more research needs to be done in order to characterize the exact cause of *Cassia* toxicity, particularly because, as some herders in Madiama pointed out, the small ruminants eat *C. tora* in the pasture fresh or dry if nothing better is available.

Cassia tora, known as Chakwar in India and Jue-ming-zi (seeds of the legume) in China, has many virtues, attributed to the anthraquinones present within the plant, and is used as a medicinal plant in these countries as well as in Africa. The seed of *C. tora* has physiological functions as an antiseptic, diuretic, diarrhoeal, antioxidant, and antimutagen (Wu *et al.*, 2001). The water extract of *C. tora* has been used as a health beverage in China. In India, *C. tora* is also utilized to feed livestock, because Chakwar (*C. tora*) seeds have been found to be very rich in protein. Patel *et al.* (1971) evaluated *Cassia* seeds in partial replacement of concentrate mixture from the ration of milking cows. *Cassia* seeds were boiled in water, dried and ground. The results showed that 15% boiled *Cassia* seeds could be included in the ration of cows without any adverse effect, and a net profit per animal and per year was observed when using this ration.

Singh *et al.* (2001) conducted an experiment on broilers looking at the impact of *C. tora* seed on their growth and meat production. The results indicated that *C. tora* seed could be safely incorporated up to the 10% level in broiler mash for maximum return on least expenditure. This study confirmed results found previously by Pandit *et al.* (1979). These reports indicate that *Cassia* seeds could be incorporated in animal feed, but it would be even better if the entire plant could be utilized as a source of feed.

Previous studies by Gupta *et al.* (1970, as cited in Ranjhan *et al.*, 1971) have shown that a neglected summer legume that is not accepted in the green stage, when conserved into silage with or without molasses was readily accepted by the livestock and was quite nutritious. Ranjhan *et al.* reviewed the chemical composition and nutritive value of *C. tora* fed as hay and compared it to the *C. tora* fed as a green plant and/or as silage. The green plant was collected at flowering stage, air dried for a day, and stored in the shed for a day until the dry matter content of the fodder reached approximately 75%. The hay was evaluated after a month and fed to adult sheep. Palatability was determined using four yearling rams and four 3-year-old male buffalo calves. Gupta *et al.* (as cited in Ranjhan *et al.*, 1971) observed that the fodder was palatable and could be compared to the dry matter consumption observed on feeding *C. tora* silage. The chemical composition and nutritive value of *C. tora* hay was found to contain 12.70, 26.80, 1.76, 46.95, 2.41 and 0.64% of crude protein, crude fibre, ether extract, nitrogen free extract, calcium and phosphorous, respectively. The digestibility coefficients for organic nutrients in *C. tora* hay were significantly more than the silage Gupta *et al.* (as cited in Ranjhan *et al.*, 1971) reported earlier. The data for the chemical composition and digestibility of *C. tora* hay as compared to silage are given in Tables 14.1 and 14.2.

The silage concept is more prevalent in temperate regions, with their distinct seasons, than in the tropics. Nevertheless, silage production in the tropics has become more relevant to fulfil the forage needs of smallholder farmers (FAO, 2000). Silage making is less dependent on weather conditions than haymaking. More research needs to be done to determine the potential of *C. tora* conservation as silage and to evaluate *Cassia* nutrition and toxicity in silage.

Ensiling *Cassia tora*

The goal of silage preservation is conversion of moist, chopped forage with a short storage

Table 14.1. Per cent chemical composition of Chakunda hay as compared to Chakunda silage (from Ranjhan *et al.*, 1971).

Treatment	Dry matter	Crude protein	Ether extract	Crude fibre	Total	Nitrogen free extract	Calcium	Phosphorus
Chakunda hay	84.50	12.70	1.76	26.80	11.79	46.95	2.41	0.64
Chakunda green	20.90	12.92	4.12	24.35	12.17	46.44	2.19	0.39
Chakunda silage (without molasses)	23.75	10.51	3.43	30.96	11.06	44.05	2.22	0.43
Chakunda silage (with molasses)	28.82	12.72	2.41	23.55	10.10	51.22	2.33	0.40

Table 14.2. Digestibility coefficients of various nutrients in Chakunda (from Ranjhan *et al.*, 1971).

Treatment	Dry matter	Crude protein	Crude fibre	Ether extract	Nitrogen free extract
Chakunda hay	60.66 ± 1.96	57.09 ± 3.33	68.52 ± 2.37	34.83 ± 1.46	57.00 ± 3.20
Chakunda silage (without molasses)	59.20 ± 1.34	39.74 ± 1.11	49.66 ± 1.21	15.36 ± 3.15	71.13 ± 2.89
Chakunda silage (with molasses)	55.80 ± 1.01	40.69 ± 0.86	59.41 ± 1.96	7.51 ± 3.50	68.62 ± 1.82

life to preserved forage that can be fed to livestock as needed. Most legume and grass crops can be ensiled successfully. The key to good silage management is air exclusion to obtain the best preservation of forage as silage. The production process of silage can be divided in four stages: forage harvesting, wilting, chopping, compacting and sealing (air-tightness). The proper execution of these stages has a big impact on the success or failure of the fermentation and the quality of the silage.

The most important treatment after harvesting the forage is chopping, followed by wilting and conditioning. Chopping is necessary to obtain good compaction to exclude air in order to promote a rapid initiation of the microbiological processes of fermentation. Chopping to between 2–4 cm lengths has the additional benefit of ease of ingestion, regurgitation and posterior rumination.

Wilting and conditioning forage before ensiling is also highly beneficial to avoid production of effluents[1] or development of undesirable microorganisms. Dry matter (DM) levels between 30% and 35% will promote a better fermentation and an increased intake by animals. Anti-nutritional metabolites (e.g. tannins and alkaloids) in certain forages (e.g. herbaceous and woody legumes and cassava leaves) will be eliminated or reduced. The same should be expected with *C. tora*. The field-drying time required to reach an optimum DM content depends on the species and on the weather conditions. The time may vary from 4 to 24 hours depending on the thickness of the stems.

The method of compaction depends on the silo (cylindrical structure, pit or bag) dimensions. In vertical silos of 2 t or less, a person walking over the successive layers of forage can achieve compaction. In pit (horizontal) silos, less than 4 metres wide, animals or people walking over the material can achieve compaction. Mud or water accumulation around the silos must be avoided to prevent contamination of the forage, but water also provides weight for more effective packing. The main problem with a small-scale container is to know when adequate compaction has been obtained. In well-chopped forages, placed in shallow layers, adequate compaction is obtained when the loose green material under the person's foot or animal's hoof does not exceed 2 cm. With wilted forages, adequate expelling of air can be expected only if the particles are less than 2 cm long. The recommendation for longer pieces is to add a final layer (10–15 cm) of fresh green forage to weigh down the material before closing the container.

Although harvesting forages at the right stage of maturity and moisture content maximizes the nutritive value of the forage, several types of additives for enhancing the quality of hay crop silage are available. The addition of honey, which was used instead of molasses, because it was readily available in the village, can improve silage quality by providing fermentable sugars. Fermentable sugars are an important factor in fermentation, because forages, such as most legumes, including *C. tora*, are typically low in fermentable sugars and higher in buffering capacity. As a result, they have slower rates of pH decline during ensiling, which translates into poor fermentation of the fresh material.

Materials and Methods

Summer 2002

Cassia tora was collected from three locations (Siragourou, Nérékoro and Madiama villages) within Madiama Commune. Most of the *C. tora* plants were at an advanced stage of maturity (more stem and fruits than leaves). Prior to ensiling the plant, samples were collected and separated into stem, leaf and seedpod, and dried to determine the percentage dry weight of each component. Per cent stem, leaf and seedpod were 80%, 19.6% and 0.4%, respectively (Table 14.3). In a 'normal' growing season (in terms of rainfall and temperature), *C. tora* would be at an early to late flowering stage at the end of October. However, during the 2002 growing season only 430 mm of rainfall was recorded in Madiama, which was 350 mm less than the previous year. Due to the extreme drought conditions, the plants were

Table 14.3. Percentage stem, leaf, and seedpod of *Cassia tora* prior to ensiling, Madiama, October 2002.

Plant parts	Fresh weight (g)	Dry weight (g)	Dry matter (%)	Plant parts (%)
Stem	86.4	70.1	81	80
Leaves	24.8	18.4	74	19.6
Seedpod (fruit)	5.0	0.3	60	0.4

forced to enter the reproductive stage earlier in the season.

Wilting was not necessary, because the plants had an adequate dry matter level. Prior to placing the chopped material in the bags that were used as silos, a sub-sample of fresh *C. tora* was taken from each location for dry matter determination and chemical analysis. The moisture content of the ensiled material varied slightly among locations. In general, the moisture content of the fresh chopped plant (no treatment added) was much lower for the Siragourou and Nérékoro locations than for those of Madiama. To facilitate chopping for proper cutting length, which was done with a long sharp machete, the plant was placed on a wooden block. In Madiama, the women chopped the plant to a length of 0.5–1.5 cm; in the other locations, the villagers participated in the chopping and the length varied between 1 and 3 cm. To enhance the quality of the silage and improve its nutritional value, two treatments were incorporated prior to ensiling. The treatments consisted of fresh material only (control), and fresh material with water or honey added. Since the fresh material was drier than desired for ensiling (less than 70% moisture), the added water might help facilitate the ensiling process (fermentation) and improve silage quality.

Chopped samples from each location were placed in a 31 × 51 cm, 4-mil polyethylene bag. Air was excluded from samples in the bag by exerting weight on the plastic bags, and the top of the bags were folded, gathered and wrapped with masking tape.

Fig. 14.1. Women chopping *Cassia tora* for ensiling.

All samples were grouped by location and placed in a larger plastic bag to start the fermentation process for either 60 or 90 days. After 60 or 90 days, the ensiled samples were opened and immediately given visual, colour and mould ratings, as well as sensory scores for sharpness (acidity) or off odours.

Summer 2003

As in 2002, *C. tora* was harvested from three locations (Madiama, Siragourou and Nérékoro). In Nérékoro, the *C. tora* sample was brought to a central location for processing. Most of the *C. tora* was at vegetative/early flower bud stage. Two men from the village did the initial chopping while the women reduced the size of the chopped material to a more desired size for ensiling. The length of the chopped material for the Nérékoro location ranged from 1 to 3 cm. Prior to placing the chopped material in the bags, sub-samples of fresh *C. tora* were taken from each location for dry matter determination and chemical analysis. *Cassia* was ensiled as is, with honey, or with 20% chopped grasses. The grass components for this location were 40% *Dactyloctenium aegyptium*, 10% *Digitaria longitudinalis*, 20% *Elucine indica* and 30% *Panicum laetum*. For the Siragourou location, the grass components were 50% *Bracaria* sp., 30% *Elucine indica*, and 20% *Dactyloctenium aegyptium*, while at the Madiama village location, the components were 10% *Elucine*, 10% *Panicum laetum*, 40% *Dactyloctenium aegyptium* and 40% *Digitaria horizontalis*.

The same treatments, ensiling and sampling procedures used at Nérékoro were used for both the Madiama and Siragourou locations. At the Siragourou location, ensiling and preparation of *C. tora* was done efficiently and effectively, because the women came prepared to do the job. They not only chopped the *C. tora* at a 0.5 cm length (which is the most desired size for ensiling) but also asked about what the experiment was expected to accomplish. After they finished chopping, they volunteered to pack the chopped materials into the small plastic bags. The chopped materials were treated and ensiled as they were at the Nérékoro location. Like the Siragourou women, the Madiama women came prepared with their sharp machetes and asked a number of questions, such as why were they ensiling *Cassia tora*, why was it necessary to chop it at this length, and why add additives. The women chopped the *Cassia* per instruction and socialized for a while before they left.

Since no difference in silage quality between the 60 and 90 day ensiling period was observed in 2002, in 2003 only a 60 day period was used to ensile the *C. tora*. After 60 days, bags were opened and visually assessed. Sub-samples were obtained from each bag for quality analysis.

Laboratory analysis

Samples in 2002 were dried in a forced-air oven at 60°C, ground (1-mm screen) in a stainless steel Wiley Mill,[2] and analysed for acid detergent fibre (ADF) (Van Soest, 1963), neutral detergent fibre (NDF) (Van Soest and Wine, 1967; Goering and Van Soest, 1970), *in vitro* dry matter digestibility (IVDMD) (Tilley and Terry, 1963), and crude protein (CP).[3] In 2003, samples were analysed for CP, ADF and total digestible nutrients (TDN).

Statistical analysis

Three treatments (control, honey and water or grass) were applied at three levels. Level 0 represents the fresh samples (not ensiled), level 60 represents the samples ensiled for 60 days, and level 90 represents the samples ensiled for 90 days, done only in the first year (2002). A randomized complete block design (RCBD) was used to perform the analysis.[4]

Results and Discussion

Crude protein (CP)

In 2002, *C. tora* CP varied from 9% to 10%. The addition of honey prior to ensiling slightly increased CP compared to the control or the water treatments (Fig. 14.2). There was no significant difference between fresh and ensiled samples. The percentage of CP in the August 2003 fresh and ensiled *C. tora* was up to 50% higher than the October 2002 samples (Fig. 14.3). The average CP in the 2003 *Cassia* samples ranged from 15–22%. This significantly higher CP content in the 2003 *C. tora* was due to the fact that the fresh material ensiled in August was mostly leafy compared to the October 2002 samples that were mainly stem (80%, 19.6%, 0.4% stem, leaf, seed pod, respectively) (see Table 14.3). The 2003 *C. tora* prior to ensiling was approximately 50% stem, 40% leaf and 10% immature seedpod. Leaves of grasses or legumes have more CP and less fibre compared with the lignified part of the plants. As the plant matures, the CP content declines, while the fibre components of the plant increase proportionally. In addition, the total amount of rainfall for the 2003 growing season exceeded the 1999 to 2002 average seasonal rainfall. The total annual rainfall for 1999, 2000, 2001, 2002 and 2003 were 602.1, 507.4, 678.1, 374.5 and 1130 mm, respectively. There was no difference in CP among treatments in 2003, and fresh *Cassia* had significantly higher CP than ensiled *Cassia*. The CP values reported for 2002 are consistent with the slightly higher values Ranjhan *et al.* (1971) previously reported. Differences in CP between 2002 and 2003 (significantly higher values) may be attributed to the time the *Cassia* was harvested and the stage of maturity of the plant prior to ensiling.

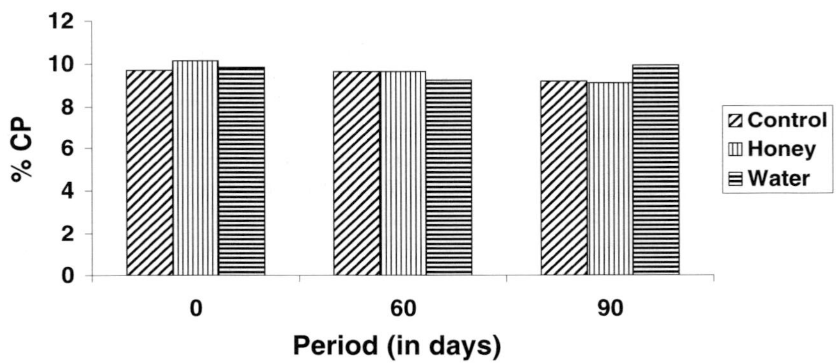

Fig. 14.2. Percentage CP in fresh vs. ensiled *Cassia tora*, October 2002.

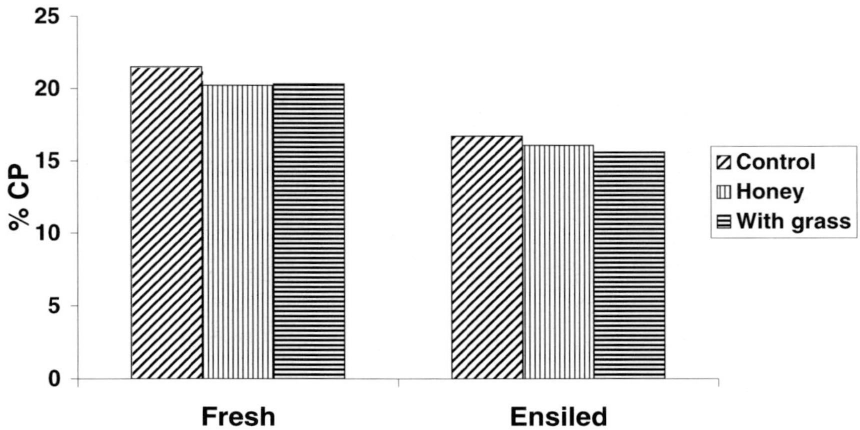

Fig. 14.3. Percentage CP in fresh vs. ensiled *Cassia tora*, August 2003.

Neutral detergent fibre and acid detergent fibre

Neutral detergent fibre (NDF) and acid detergent fibre (ADF) are two laboratory methods that chemically distinguish the readily available and soluble cell contents from the less digestible portion of the cell walls. NDF consists of cell walls while ADF represents lignified cellulose (indigestible) fibres. NDF is negatively related to forage intake by the animal, while ADF is negatively correlated to digestibility. NDF and ADF values of less than 40% and 30%, respectively are considered prime values. As the plant matures from vegetative stages to flowers and seed head production, the ADF and NDF values will increase proportionally.

In 2002, the ADF values ranged from 34% to 42% (Fig. 14.4). The combined effect of ensiling and honey treatment reduced ADF values (Fig. 14.4), a trend also observed by Ranjhan *et al.* (1971). The addition of honey to *C. tora* prior to ensiling decreased ADF values by eight percentage points (42% for the control vs. 34% for honey treated silage). The ADF values observed for the ensiled *C. tora* are comparable to mature lucerne hay. Generally, the ADF values of the fresh *C. tora* were much lower in 2003 than in 2002 (Fig. 14.5). This was not surprising, because the *Cassia* sample in August of 2003 was much higher in leaf (low in ADF) than stem and seedpod content. The percentage ADF of the ensiled *C. tora* was much higher than the fresh *C. tora* (Fig. 14.5). The addition of honey to the ensiled *C. tora* decreased ADF by three percentage points (38% for the grass treated vs. 34.6% for the honey treated silage). Overall, for both 2002 and 2003, the ADF values of the ensiled *C. tora* were acceptable values to maintain a reasonable production level.

Ensiling *C. tora* decreased the amount of NDF compared with the fresh *C. tora* (Fig. 14.6). The ensiled *Cassia* with the addition of honey was significantly lower in NDF than the control and the water-treated *C. tora*. The percentage NDF was 57%, 50% and 52% for the untreated control, honey and water treated *Cassia*, respectively. NDF values of grass or grass legume hay, depending on stage of maturity, range from 40 to 65%. As the plant matures from vegetative stages to reproductive stages, NDF values increase significantly.

When compared to the fresh *C. tora*, ensiling increased IVDMD (Fig. 14.7). Averaged over all treatments, the IVDMD of the ensiled *Cassia* was six percentage points higher than the fresh *C. tora*. The addition of honey as an additive significantly improved the IVDMD of the ensiled *Cassia*. The IVDMD of the honey treated *C. tora* was 63% compared with 54% for *Cassia tora* ensiled fresh (with no additives). These values are comparable to typical warm and cool-season forage crops grown in temperate regions.

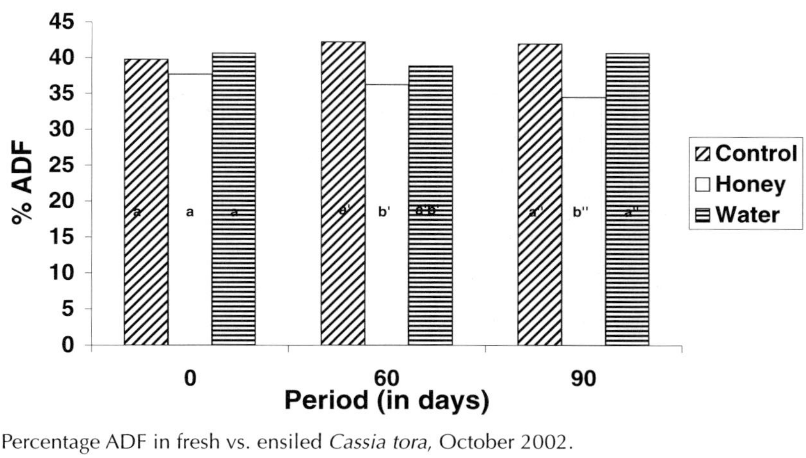

Fig. 14.4. Percentage ADF in fresh vs. ensiled *Cassia tora*, October 2002.

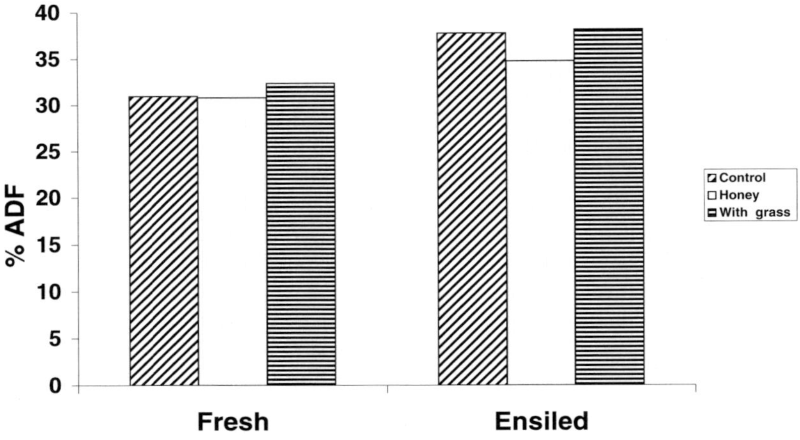

Fig. 14.5. Percentage ADF in fresh vs. ensiled *Cassia tora*, August 2003.

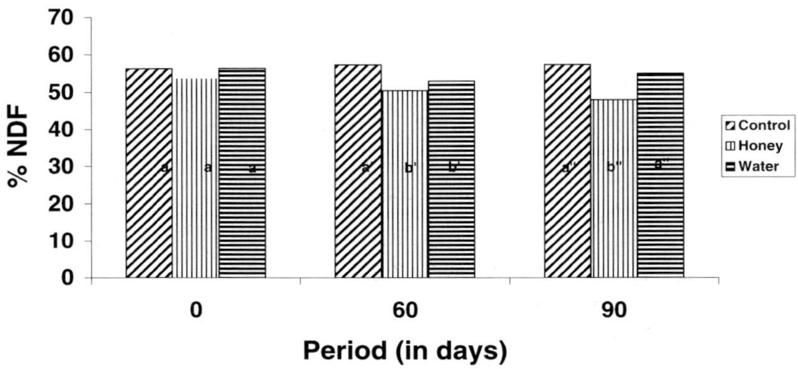

Fig. 14.6. Percentage NDF in fresh vs. ensiled *Cassia tora*, October 2002.

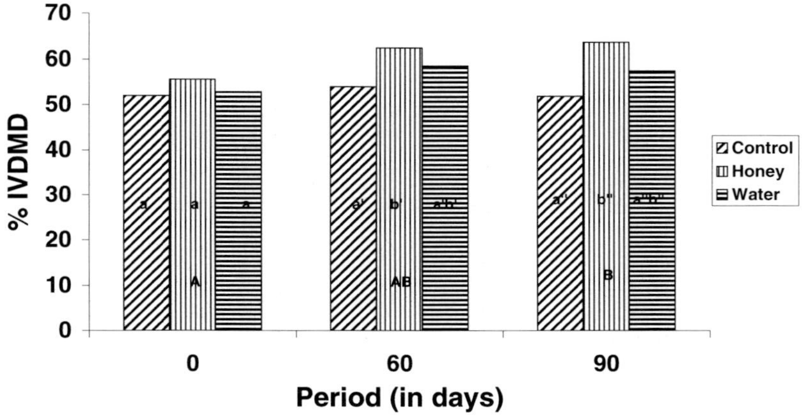

Fig. 14.7. Percentage IVDMD in fresh vs. ensiled *Cassia tora*, October 2002.

Total digestible nutrient (TDN) is an approximate measure of feed value based on feed energy content. Energy is the most yield- (milk, meat, wool) limiting factor in livestock with mainly forage-based diet. The fresh *Cassia* was higher in TDN than the ensiled *C. tora* (Fig. 14.8). This might have been due to the fact that the fresh material was of a higher quality as indicated by its high leaf to stem ratio and some energy was lost during the fermentation process. Within the ensiled *Cassia*, the honey treated *Cassia* was higher in TDN than the control and grass treated *Cassia*. Although the addition of grass slightly reduced the TDN values of the ensiled *C. tora*, this could be explained by the advanced stage of maturity of the grass added.

Summary

In terms of the nutrient values, as indicated by acceptable crude protein, fibre, relatively high *in vitro* dry matter digestibility and total digestible energy values, ensiled *C. tora* has a great deal of potential as a livestock feed. Although the October 2002 ensiled *C. tora* had relatively low nutrient values due to a low leaf to stem ratio of the fresh *C. tora*, it would still make a reasonably good feed for livestock. However, the addition of honey, as the results in this study have shown, molasses or sugar improves the quality of the ensiled material, as the results of similar studies in Australia and other tropical regions have demonstrated (Muhlback, 2000). In general, the August 2003 ensiled *C. tora* had a substantially higher percentage CP and lower ADF. The August 2003 fresh *Cassia* was high in leaf content compared to the October 2002 fresh *C. tora*, but unless the raw material (the fresh sample) is of a reasonable quality, the ensiling process *per se* would not enhance the quality of the end product. However, if the fresh material is excessively dry and low in sugar, the addition of water or a carbohydrate source can enhance the quality of the materials by improving the ensiling processes (such as fermentation).

The goal for this project was to help Madiama Commune find a feed source they can tap into during the dry season when livestock feed is in short supply. In a much drier year like 2002, the entire feed source for the livestock was gone at least 2 months before the end of the 'normal' production season, and *C. tora* was the only live and visible plant in the entire Commune. *Cassia tora* even at its late stage of maturity could provide a good feed source with reasonably high nutrient content.

As for the question of toxicity, it is not clear how animals respond to a *Cassia* diet as evidenced by the various studies. Although anthraquinones are the major toxic compounds found in *C. tora*, other toxins have been identified as well. However, both Gupta *et al.* (as cited in Ranjhan *et al.*, 1971) and Ranjhan *et al.*

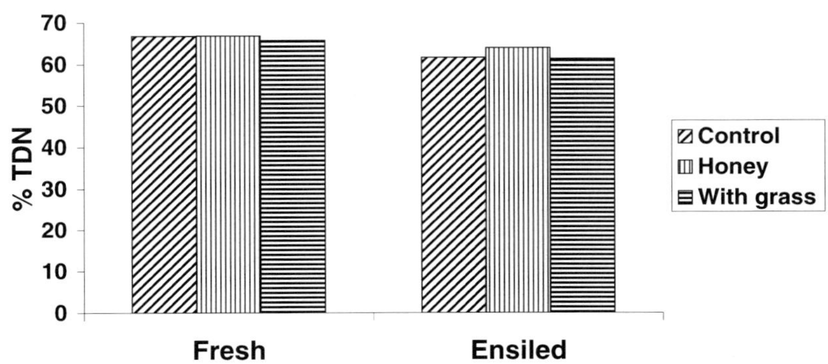

Fig. 14.8. Percentage TDN in fresh vs. ensiled *Cassia tora*, August 2003.

(1971) suggested that *C. tora* conserved as hay or silage was readily accepted by livestock. Nevertheless, toxicity analyses of *C. tora* ensiled are needed, and a Brazilian scientist is exploring this question.

Biological impact

If making *C. tora* into silage was to be adopted, it could affect the ecosystem within this region. As discussed earlier, exotic species can bring both positive and negative effects into an existing ecosystem. *Cassia tora* may have positive effects in the recycling of nutrients; however it is also aggressive in its growth and excludes grasses and other species from growing. In order to control *C. tora* without eliminating the plant, a cutting regime would be adequate. Ensiling *C. tora* could be a solution to two problems. If cut at the beginning of the growing season when still at the vegetative and leafy stage, the quality of the plant would be optimum for conservation and feeding to small ruminants during the dry season. In addition, it would allow other species to emerge and persist early in the season. Because *Cassia* has a prolific seed production, not allowing all the plants to go to the reproductive stage would control its growth and would make it ecologically more acceptable. In turn, a shifting in plant population with an increase and coexistence of various species might occur. Some might raise the question of the risk of exterminating *C. tora* over time if the plant is not allowed to enter the reproductive stage and regenerate. In order to prevent any loss in biodiversity, stakeholders need to understand that cutting or grazing any plant has to be done in a controlled and sustainable way, and like any other plant, *C. tora* must be managed in a sustainable way.

Stage of maturity is critical to the management of *C. tora* both for reproduction and ensilage. The results of the silage experiment show that even at a late stage of maturity, *Cassia* silage was of acceptable quality. However, in order to allow other plants to germinate and establish within the pasture plant community, cutting *Cassia* at a vegetative stage when it is not tall enough to over-shade other species would reduce competition. If the plants were cut at different stages of maturity, it would allow some to seed for regeneration. However, Hien (1997) states that *C. tora* produces heterogeneous seeds that can extend the germination process over several weeks. Thus, if the plants germinate and grow at different times, it would be hard to exterminate *Cassia*, even if it is cut at optimal times for making silage

Socio-economical impact

Livestock are central to the livelihoods and existing production systems in Madiama. Field crops and livestock are the dominant production systems, with each generating 46.2% and 34.4% of the community income, respectively (Kabore, 2001). Livestock provides the commune as well as the country with meat, milk, leather and wool products. Livestock exports to neighbouring coastal countries where livestock production costs are much higher are an important source of national income. In Madiama, goats and sheep are mostly kept in the villages and fed by women. In Nérékoro, women are responsible for the fattening of animals and for milk production. Increasing the feed supply, especially during the dry season, would increase meat, milk, skins and fibre production.

Brewster *et al.* (2001) have stated that any development strategy for Madiama economy should focus on the livestock sector. Such a focus points to the importance of pasture management. Improving pasture management improves pasture productivity and provides higher quantity and quality feed for livestock, thereby lowering livestock production costs and raising the women's income. The ensiling of *C. tora* requires few inputs and complicated technologies, because labour is the most important input in this process. The use of locally available equipment and the reliability and repeatability of the ensiling process makes *Cassia* silage a sustainable source of feed. A study shows that crossbred cattle can at the very least be maintained and possibly gain weight during the dry period in Guinea savanna zone by being fed ensiled cereal crop residues (Olayiwole and Olorunju, unknown date).

An economic analysis is necessary to characterize the potential economic benefits of *Cassia* silage. This feeding option would be a supplementary feed that could give rapid and significant returns on investments. In addition it is a reliable and repeatable process that could be done with any forage of relatively good quality. In conclusion, *Cassia* silage could be a sustainable and economically feasible solution to feed shortages during the dry season in Madiama. The social and economic impacts of using a simple tool could far exceed the inputs for *Cassia* silage. If the net benefits of using a production system with *Cassia* silage exceed the net benefits of using a production system without *Cassia* silage, then *Cassia* silage is economically beneficial. This experiment is the first step. The next step would be to conduct animal feeding trials to evaluate animal performance on *Cassia* silage during the dry season and perform an economic analysis.

Recommendations on future research

More research needs to be done to examine the toxicity of *C. tora* and to look at the effect of the ensiling process on these toxins. Research should not be limited to one species, because other weedy species could provide with a good source of feed as well. Animal trials and economic analysis are necessary to complete this research and to apply it to a larger scale.

Notes

1 The most common problem results when high moisture-silage forms a liquid effluent. Liquid effluents from silage are typically highly acidic and rich in organic matter, nutrients and salts and can be harmful to the environment.

2 Thomas-Willey Mill, Model ED-5, Arthur H. Thomas Co., Philadelphia, PA, USA.

3 Total N was determined colorimetrically (McKenzie and Wallace, 1954) with a Technicon Autoanalyzer (Technicon Industrial Systems, Tarrytown, NY; 1976).

4 Analysis of variance (ANOVA) procedures were performed on all data to test the treatment effects on various measured parameters using the GLM procedure of SAS (SAS Inst., 2001). The treatment means were compared using least significant difference (LSD) at $P = 0.05$.

References

Bhandari, M.M. (1978) *Flora of the Indian Desert Scientific* (United Book Traders, 4, 471p) Scientific Publishers, Jodhpur, India.

Brenan, J.P.M. (1958) New and noteworthy cassias from tropical Africa. *Kew Bulletin* 13, 231–252.

Brewster, C., Kabore, D., Bertelsen, M., Wyeth, P. (2001) *Economic interdependence among socio-economic groups in Madiama commune of the Niger delta of Mali.* SANREM CRSP Research Scientific Synthesis Conference.

Chand, S. and Katoch, B.S. (1983) Studies on the removal of tannins and saponins of some forest byproducts by chemical treatments. *Indian Journal of Poultry Science* 18, 9–12.

Chandrasekaran, S. and Swamy, P.S. (2002) Biomass, litterfall and aboveground net primary productivity of herbaceous communities in varied ecosystems at Kodayar in the western ghats of Tamil Nadu. *Agriculture, Ecosystems and Environment* 88(1), 61–71.

Charles, O.W. and Muller, H.D. (1975) Coffeeweed (*Cassia obtusifolia*) toxicity response of laying hens. *Poultry Science* 54, 1745 (Abstr.)

Cheeke P.R. (1998) *Natural Toxicants in Feeds, Forages, and Poisonous Plants.* Interstate Publishers, Danville, Illinois.

Cock, M.J.W. and Evans, H.C. (1984) Possibilities for biological control of *C. tora* and *C. obtusifolia. Tropical Pest Management* 30(4), 339–350.

Davis, M.A. (2003) Biotic globalization: does competition from introduced species threaten biodiversity? *BioScience* 53(5), 481–489.

De Wit, C.T. (1981) The energy spectrum: food and energy. Food security and a hungry world. pp. 66–70. *Proceedings of Wageningen Agricultural University Conference, San Francisco, CA* (4–6 March 1981). Wageningen Agricultural University, Wageningen, The Netherlands.

De Wit, H.C.D. (1955) A revision of the genus 'Cassia' (Caesalp.) as occurring in Malaysia. *Webbia* 11, 197–292.

Ehrenfeld, J.G. (2003) Effects of exotic plant invasions on soil nutrient cycling processes. *Ecosystems* 6(6), 503–523.

FAO (1971) *Improving soil fertility in Africa* (FAO Soils Bull. 14). Food and Agriculture Organization of the United Nations, Rome.

FAO (2000) Silage making in the tropics with particular emphasis on smallholders (FAO Plant Production and Protection Paper 161). *Proceedings of the FAO Electronic Conference on Tropical Silage,* 1 September–15 December 1999.

Goering, H.K. and Van Soest, P.J. (1970) Forage fibre analysis [microform]: Apparatus, reagents, procedures, and some applications. *Agriculture Handbook* 379 (20), ill.

Gupta, P.C. and Sharma, R.K. (1965) *Proceedings of the National Academy of Science India*, Sec. A, 26: 398.

Hien, F.G. (1997) Soil variability and effectiveness of soil and water conservation in the Sahel. *Arid Soil Research and Rehabilitation* 11, 1–8.

Hoveland, C.S., Buchanan, G.A. and Harris, M.C. (1976) Response of weeds to soil phosphorous and potassium. *Weed Science* 24(2), 194–201.

Hubbell, S.P. (2001) *The Unified Neutral Theory of Biodiversity and Biogeography*. Princeton University Press, Princeton, New Jersey.

Kabore, D. (2001) *Etudes socio-economiques menees dans la commune de Madiama.* Projet IER/SANREM-CRSP.

Katoch, B.S., Katoch, R.C., Mahajan, N.C. and Bhowmik, K.B.D. (1978) Utilization of Indian horse-chesnut (*Aesculus indica*) kernals and wild isbgol (*Plantago lanceolata*) plant-meal as poultry feed ingredients. *Indian Veterinary Journal* 55, 781–787.

Katoch, B.S. and Bhowmik. K.B.D. (1983) Biological evaluation of *Cassia* seeds (*C. tora* and *Cassia floribunda*) as poultry feed ingredients. *Indian Veterinary Journal* 60, 390–395.

Mall, L.P. and Raina, D.K. (1957) Autoecological studies on *Tridax procumbens* Linn. *Bulletin of the Botanic Society University of Sagar* 9, 35–54.

Manjunath, B.L. and Subbajois, H. (1930) Chemical examination of the seeds of *Cassia tora* Linn. *Journal of the Indian Chemical Society* 7, 521.

Mathur, M. (1985) Comparative study of epidermal structures and seed analysis of two *Cassia* species. *Geobios New Reports* 4(1), 96–97.

Mckenzie, H.A. and Wallace, H.S. (1954) The kjeldahl determination of nitrogen: a critical study of digestion conditions – temperature, catalyst, and oxidizing agents. *Australian Journal of Chemistry* 7, 55–70.

Mokwunye, A.U., de Jager, A. and Smaling, E.M.A (eds) (1996) *Restoring and Maintaining the Productivity of West African Soils: Key to Sustainable Development* (Miscellaneous Fertility Studies 14). International Fertility Development Center Africa, Lome, Togo.

Muhlback, P.R.F. (2000) Additives to improve the silage making process with tropical forages. In: L.'t Mannetje (ed.) *Silage making in the tropics with particular emphasis on smallholders* (FAO Plant Production and Protection, Paper 161, 151–164).

Olayiwole, M.B. and Olorunju, S.A.S. (no date) *Effects of ensilage, added nitrogen and energy on the utilization of grain sorghum stover by beef cattle.* National Animal Production Research Institute, Zaria, Nigeria.

Pal, M., Roy, D.K. and Pal, P.R. (1977) Emodin from the leaves of Cassia tora Linn. *Indian Journal of Pharmacology* 39(5), 116–117.

Pandey, Y.N. (1970) *Journal of the Indian Botanical Society* 49, 151–157.

Pandit, N.N., Singh, J. and Bhattacharjee, D.K. (1979) Impact of feeding chakwar (*Cassia tora*) seed on the growth of broilers. *Indian Journal of Poultry Science* 14, 176–178.

Patel, B.M., Patel, C.A., and Shukla, P.C. (1971) Evaluation of *C. tora* seeds in partial replacement of concentrate mixture from the ration of milch cows. *Indian Journal of Animal Sciences* 41(11), 1031–1034.

Penning de Vries, F.W.T. and Djiteye, M.A. (eds) (1982) *La productivite des paturages saheliens: Une etude des sols, des vegetations et de l'exploitation de cette resource naturelle.* PUDOC, Wageningen, The Netherlands.

Perkins, K.D. and Payne, W.W. (1985) *Guide to the Poisonous and Irritant Plants of Florida.* University of Florida, Florida Crop Extension Service, Institute of Food Agricultural Science, Gainesville, Florida.

Rangaswami, S. (1963) Crystalline chemical components of the seeds of *Cassia tora* Linn. Identity of tora substance C with Rubrofusarin and tora substance B with Nor-Rubrofusarin. *Proceedings of Indian Academy of Science* 57, 88–93.

Randell, B.R. (1995) Taxonomy and evolution of *Senna obtusifolia* and *S. tora. Adelaide Botanic Gardens* 16, 55–58.

Ranjhan, S.K., Gupta, B.S. and Chabbra, S.S. (1971) Chemical composition and nutritive value of a summer legume, chakunda (*C. tora* Linn) hay with special reference to metabolisable energy for sheep. *Indian Journal of Animal Health* 10(2), 217–221.

Sanchez, P.A. (2002) Soil fertility and hunger in Africa. *Science* 295, 2019–2020.

SAS Institute (2001) *SAS User's Guide: Statistics* (Version 8.2). SAS Institute, Cary, North Carolina.

Singh, J., Kumar, R. and Yadav, R.K. (2001) Impact of feeding chakwar (C. tora) seed on growth and meat production in broilers. *Indian Journal of Poultry Science* 36(3), 308–310.

Singh, M., Panday, R.K. and Shankarnarayan, K.A. (1970) Problems of grassland weeds and their control in India. In: *International Grassland Congress: Proceedings of the 11th International Grassland Conference.* Surfer's Paradise, Australia, pp. 71–74.

Sood, A.K., Sharma, M. and Katoch, B.S. (1990) Improvement in the nutritive value of Indian horse-chesnut kernel and cassia seeds by chemical treatments for feeding of chicks. *Indian Journal of Animal Sciences* 60(11), 1364–1369.

Steele, K.W. and Vallis, I. (1988) The nitrogen cycle in pastures. In: Wilson, J. (ed.) *Advances in Nutrient Cycling in Agricultural Ecosystems.* CAB International, Wallingford, UK, pp. 274–291.

Technicon Industrial Systems (1976) *Individual/simultaneous determination of nitrogen and/or phosphorus in BD acid digests* (Industrial Method No. 334–74A/A). Tarrytown, New York 10591.

Tilley, J.M.A. and Terry, R.A. (1963) A two stage technique for the *in vitro* digestion of forage crops. *Journal of the British Grassland Society* 18, 104–111.

Tiwari, R.D. and Behari, J.R. (1972) Chemical examination of the roots of *Cassia tora. Planta Medica* 21(4), 393–397.

Uhl, C. and Jordan, C.F. (1984) Succession and nutrient dynamics and burning in Amazonia. *Ecology* 65, 1476–1490.

Van der Pol, F. (1992) *Soil Mining, an Unseen Contributor to Farm Income in Southern Mali* (Bulletin 325). Royal Tropical Institute (KIT), Amsterdam, p. 47.

Van Soest, P.J. (1963) Use of detergents in the analysis of fibrous feeds. II. A rapid method for the determination of fibre and lignin. *Journal of the Association of Official Analytical Chemists* 46, 829.

Van Soest, P.J. and Wine, R.H. (1967) Use of detergents in the analysis of fibrous feeds. IV. Determination of plant cell wall constituents. *Journal of the Association of Official Analytical Chemists* 50, 50.

Van Soest, P.J. and Wine, R.H. (1968) Determination lignin and cellulose in acid detergent fibre with permanganate. *Journal of the Association of Official Analytical Chemists* 51, 780–785.

Wu, C.-H., Hsieh, C.-L., Song, T.-Y. and Yen, G.-C. (2001) Inhibitory effects of *C. tora* L. on Benzo[a]pyrene-mediated DNA damage towards HepG2 cells. *Journal of Agricultural and Food Chemistry* 49 (5), 2579–2585.

15 Controlled Grazing: Botanical Response and Animal Performance

Ozzie Abaye,[1] Meriem El Hadj,[1] Amadou Kodio[2] and Moussa Keïta[2]

[1]Department of Crop and Soil Environmental Science, Virginia Tech, Blacksburg, VA 24061–0404, USA; and [2]Centre Regional de la Recherche Agronomique, Institut d'Economie Rurale, BP 205, Mopti, Mali

The lack of pasture and forage resources has been a growing problem in the Sahel region of sub-Saharan Africa. Population growth, extensive row cropping, increasing herd size, high levels of exploitation of woody resources, and climatic fluctuations have all contributed to serious losses of plant biodiversity within this region. In addition, utilization of greatly reduced pastureland resources has become a major source of conflict between farmers and herders. This has certainly been the case in the Commune of Madiama. As a consequence, the Natural Resource Management Advisory Committee (NRMAC) requested SANREM researchers to focus their efforts on developing technologies to improve pasture management.

The unimodal annual rainfall pattern limits grazing potentials in Madiama, with respect to plant community composition, yield and quality. The perennial species, herbaceous as well as woody, have become scarce over the years. The less dependable annual grasses and other low forage quality species severely limit potential animal production. As perennial forages in these pastures have become nearly non-existent, annual grasses and some legumes have become the dominant species in the Madiama Commune. Seasonal movement of animals (transhumant practices) is also increasingly limited, resulting in overpopulation of animals and overgrazing. The existing low level of plant succession under the current management system has resulted in a severe botanical shift from dominantly perennial grasses to annual grasses and from a grass-legume mixture to almost pure grass sward.

To persist by reseeding or/and regrowth from existing plants, forage grasses and legumes must be given sufficient time between defoliations, because the time between defoliation helps the plant to produce and store enough energy for regrowth (Smith, 1962). Plant response to defoliation depends on plant species, the extent of removal of leaf area and growing points, and the amount of carbohydrate available for regrowth. Therefore, good management (such as rotational grazing) of these grasses should allow sufficient time between defoliations to restore carbohydrate reserves or leave enough leaf area for carbohydrate production. Because plants differ in their morphological and physiological responses to different management strategies, in a mixed species pasture it is possible to see the dominance of one or two species at any given time. In a grazing situation, livestock preference for one species or plant part over another depends on relative palatability, acceptability and overall availability of pasture species or parts.

Any plant that can adapt to defoliation in a pasture setting needs meristematic tissue close to the ground, out of danger of removal by livestock (Harris, 1978), because after defoliation leaves regrow from meristematic tissue at the base of each leaf and new leaves arise from auxiliary buds also located near the base of leaves (Smith, 1973). These tissues must be close to the soil surface to escape damage during defoliation in order to allow subsequent regrowth. Plants with upright growth patterns are more sensitive to defoliation, and recover more slowly than plants with a horizontal growth pattern (Butler *et al.*, 1959). Livestock grazing can influence grass/legume composition of a pasture (Schwinning and Parsons, 1996), because livestock will selectively graze more palatable legumes in preference to less palatable grasses. This selectivity can lead to grass domination in pastures, especially when continuously grazed (Watkin and Clements, 1974). Thornley *et al.* (1995) state that persistence of unfertilized mixtures of rye grass and white clover is not achievable, except when grazing eliminates the dominance of white clover. If a legume is highly competitive with grasses, the grass can often be maintained in the mixture only by the presence of grazing animals (Schwinning and Parsons, 1996). Depending on the severity of defoliation, root mass and depth can be reduced by grazing.

Root morphology is an important factor in successful competition for water and nutrients in a pasture (Evans, 1977). A more developed and deeper root system generally gives one pasture species an advantage over another (Davidson, 1978). In general, grasses have longer, thinner, more finely branched roots, and more root hairs, than legumes (Evans, 1978). These differences in root structure and root depth could give grasses a competitive advantage over legumes in nutrient and water uptake (Haynes, 1980), and could partially explain why grasses are better able to absorb limiting nutrients.

The importance of managing for natural succession in order to restore pastureland resources and determine indicators for improving pasture management practices is the purpose of this research. The specific objectives are: (i) to determine the influence of the tethered grazing method on plant diversity; and (ii) to evaluate animal performance in terms of weight gains under tethered grazing.

Materials and Methods

The study was conducted at Siragourou and Nérékoro, two villages within Madiama Commune. At both locations, an area was fenced to keep animals and people out of the experimental plots. Siragourou and Nérékoro villagers donated young sheep weighing 18–24 kg and 18–21 kg, respectively. Since the purpose of the experiment was to determine indicators for optimal pasture management, the experiment involved two treatments: 3 cm vs. 6 cm forage residual heights, each replicated four times (see Fig. 15.1). At each village, animals were weighed and tagged according to treatments and treated for internal parasites. Animals were randomized according to weight and treatment. Each animal in its grazing area represented an experimental unit; a total of eight animals per site were used. Residual plant height was determined based on measurements taken prior to the initiation of the experiment. The initial height of the forages at each site varied from 5 cm to 20 cm and 8 cm to 14 cm for the Siragourou and Nérékoro locations, respectively (see Table 15.1). Each animal was tethered separately to its own peg using a 6-m length of rope that allowed the animal to consume all forage within its range for the period of time determined by available forage and canopy residual height. Once each animal grazed its area to the 3 cm or 6 cm residual plant heights, they were then rotated to the next peg (see Fig. 15.1). Water was provided to the animals daily at the grazing location. At night, the animals were kept in a kraal. No additional feed was provided for the night.

Although the plan was to weigh the animals following each grazing cycle (after each animal had grazed at all eight pegs in their paddock), due to excessive forage production during the 2003 growing season no more than seven pegs within each paddock were

Fig. 15.1. Sheep pegged within paddocks.

needed during the season. Therefore, the animals were weighed only at the beginning and the end of the grazing season, instead of at the end of each grazing cycle. At both locations, guards were trained to watch the animals, provide water and take the animals home at night. The guards' knowledge of animals and their understanding of the process was impressive. After their training, they placed all the animals in their correct experimental plots without any further guidance. Animals remained on pasture for a total of 93 days.

Forage mass and botanical composition assessments

The botanical composition of pastures was determined by a non-destructive visual evaluation method using the Double DAFOR Scale (Brodie, 1985; Abaye *et al.*, 1995), where D = dominant, A = abundant, F = frequent, O = occasional, and R = rare are used to evaluate the relative abundance of species. Brodie (1985) describes a species as dominant if it covers most or all of the area. A ranking of abundant is given to species that cover approximately half to three-quarters of an area. Frequent ranking refers to species well scattered throughout a site but covering less than half the area. A species ranked as occasional occurs a few times, and a rare species is one that is present only once or twice. Brodie (1985) suggests that problems with this method include under-assessment of smaller species, over-assessment of conspicuous species and misidentification of species. Initially (August) the entire pasture was visually evaluated for both ground cover and species diversity. At the end of the grazing season (November) the grazed area (all paddocks and all treatments) and the ungrazed area (outside of the 6 m circled area grazed by animals) were evaluated using the DAFOR Scale.

Soil fertility

Soil samples to a depth of 10 cm were obtained prior to the start of grazing by taking six to ten soil cores from each treatment replication (areas with tethered grazing and ungrazed areas). Samples were analysed for N, P, K and pH in the IER Soils Laboratory at the CRRA/Sotuba.

Results and Discussion

Weather data for the Madiama Commune

The total amount of rainfall for the 2003 growing season exceeded the 1999 to 2002

Table 15.1. Plant height taken prior to initiation of the grazing experiment for the Siragourou and Nérékoro villages, August, 2003.

Siragourou			Plant height (cm)		Nérékoro			Plant height (cm)	
Replication	Treatment	Plot	←measurements→	Average	Replication	Treatment	Plot	←measurements→	Average
1	1	1	10.5, 8.2, 13, 6, 14	10.34	1	1	1	10.5, 8, 5, 16, 10.5	10.0
1	2	1	8.5, 5.5, 15.5, 10.5, 37	15.4	1	2	1	27, 8, 7, 10, 28	16.0
2	2	1	18, 31.5, 8, 6, 9	14.5	2	2	1	10.5, 8, 5, 10, 7	8.1
2	1	1	11, 9.5, 10, 5.5, 7	8.6	2	1	1	8, 10.5, 8, 12, 4	8.5
3	1	1	11.5, 10, 8, 39, 7.5	15.2	3	1	1	8, 12, 6, 7, 6	7.8
3	2	1	18.5, 31.5, 14, 19, 19	20.4	3	2	1	6.5, 17, 9.5, 17, 17	13.4
4	2	1	8, 8, 4, 12, 6.5	7.7	4	2	1	8, 15.5, 8.2, 27.5, 14.5	14.7
4	1	1	3.5, 6, 4, 8.5, 4	5.2	4	1	1	16, 16.5, 7, 8, 11, 38	16.0

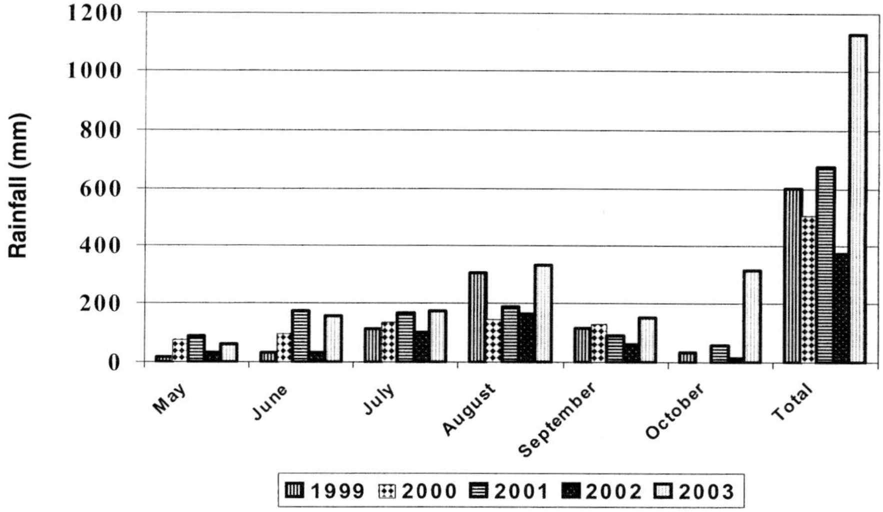

Fig. 15.2. Annual rainfall recorded at the Madiama Commune, Mali: 1999–2003.

average annual rainfall. The total annual rainfall for 1999, 2000, 2001, 2002 and 2003 was 602.1 mm, 507.4 mm, 678.1 mm, 374.5 mm and 1130 mm, respectively. The monthly rainfalls for 2003 exceeded all but the May and June rainfalls for 1999–2003 (Fig. 15.2). The rainfall for July, August, September and October of 2003 was higher than the previous 4 years. Rainfall usually begins declining in September and stops completely by the end of October. In 2003, 31.5 mm rain was recorded for November.

Forage diversity

Forage ground cover, species diversity and regeneration assessments

NÉRÉKORO LOCATION Prior to grazing, ground cover ranged from 45% to 100% at the Nérékoro location, where the predominant plant species were annual grasses. In the grazing area, the dominant grass species were *Dactyloctenium aegyptium* (Da), *Panicum laetum* (Pl) and *Bracaria* sp. (Bsp), which are all are highly palatable to livestock. However, the relative abundance of these species was less frequent in parts of the same grazing areas. Among the few legumes found at this site, *Cassia tora* (Ct) (sickle pod) was predominant. This annual legume, widely known in the USA as a noxious weed mostly found in soybean fields, is extremely aggressive. Most grasses do not compete well with *Cassia tora* (Ct) because of its aggressive and invasive nature. To a lesser extent, also present were other legumes including *Zornia* sp. (Zsp) and *Alysicarpus* sp. (Asp) (see Table 15.2).

The final visual assessment of ground cover and species diversity is shown in Table 15.3a and 15.3b. In most cases, ground cover was at an acceptable level (40–100%) regardless of treatments. In general, grazing appeared to increase species diversity when compared to initial assessments (Table 15.2 vs. Table 15.3a and 15.3b). *Dactyloctenium aegyptium* (Da) and *Panicum laetum* (Pl) were the dominant grass species. The dominance of *Dactyloctenium aegyptium* (Da) after severe grazing can be attributed to its morphological characteristic of having meristematic tissue (areas of rapid cell division) located close to the ground where it is protected from being removed by the

Table 15.2. Initial visual evaluation of ground cover and botanical composition, Double DAFOR scale, for the Nérékoro site, August 2003.

	Composition (%)						Forages				
Rep[a]	Trt[b]	Pr[c]	Gr[d]	Grass	Legume	Weed	D[e]	A[f]	F[g]	O[h]	R[i]
1	1	1	96	100	0	0	Da	Pl			
1	1	1	70	100	0	0	Pl	Da			
1	2	1	85	95	0	5	Da, Pl	Ct			
1	2	1	75	100	0	0	Da		Pl		
2	2	1	98	100	0	0	Da, Pl		Csp		Pl
2	2	1	50	100	0	0	Da, Bsp				
2	1	1	98	100	0	0	Da	Pl			Esp
2	1	1	98	100	0	0	Da		Pl		
3	1	1	99	100	0	0	Da		Pl		Zsp
3	1	1	45	100	0	0	Da			Pl	Esp
3	2	1	99	100	0	0	Da	Bsp			Pl
3	2	1	98	99	1	0	Da	Bsp			Pl
4	2	1	65	99	1	0	Pl	Da			Asp
4	2	1	50	100	0	0	Bsp	Pl, Da			
4	1	1	70	100	0	0	Csp	Pl	Bsp	Da	
4	1	1	100	100	0	0	Da	Pl			

Rep[a] = replication; Trt[b] = treatment (1 = grazed to 3 cm stubble height; 2 = grazed to 6 cm stubble height); Pr[c] = pasture; Gr[d] = ground cover; D[e] = dominant; A[f] = abundant; F[g] = frequent; O[h] = occasional; R[i] = rare.
Grasses: *Dactyloctenium aegyptium*: Da; *Panicum laetum*: Pl; *Eragrostis* sp.: Esp; *Bracaria* sp.: Bsp.
Legumes: *Zornia*: Zsp.; *Alysicarpus* sp.: Asp.; *Cassia tora*: Ct.

grazing animal. The *Bracaria* sp. (Bsp), dominant in part of the pasture prior to grazing, was less abundant at the end of the grazing season (Table 15.3a and 15.3b). The disappearance of this species was more evident where paddocks were grazed to 3 cm, but where they were grazed to 6 cm residual height some was left, an indication that grazing preference for this plant could be higher than for other species. In general, more grass species were observed where grazing occurred to 6 cm residual height than where it occurred to 3 cm residual height. Additionally, weedy types of plants such as *Cassia tora* (Ct) were more frequently observed where pastures were grazed to 3 cm residual height than to the 6 cm residual height. This indicates that grazing to 3 cm residual height is less damaging to invasive plants than to native species that are more desirable to animals.

Table 15.4a and 15.4b shows the end of season ground cover and species diversity assessment of the ungrazed area of each paddock. The ground cover and the type of species were similar to those shown in Tables 15.2, 15.3a and 15.3b. However, as shown particularly in replications 3 and 4, the diversity of plant species was greater than the grazed areas (Table 15.3a and 15.3b). The plants observed in these paddocks included both new seedlings from reseeding and first growth after dry season dormancy. In most cases, the presence of legume species was more evident in the ungrazed areas than in the grazed areas, an indication of animal preference for legumes rather than grasses.

SIRAGOUROU LOCATION The initial ground cover of the grazing area at this location, which was much less than at the Nérékoro location, ranged from bare to 80% (see Table 15.5). In most cases, the legume *Zornia* sp. (Zsp) was the most dominant species, followed by the grass *Bracaria* sp. (Bsp), which appeared in mixture with the legumes or pure stands. Among the weedy types of plant

Table 15.3a. Final visual evaluation of ground cover and botanical composition of the grazed area: Nérékoro site, November 2003, Replications 1–3.

			Composition (%)				Forages				
Rep[a]	Trt[b]	Pl[c]	Ground cover	Grass	Legume	Weed	D[d]	A[e]	F[f]	O[g]	R[h]
1	1	1	90	80	20	0	Pl	Da	Cor		Isp
1	1	2	99	90	10	0	Da	Dl			
1	1	3	85	99	1	0	Da	Dl			
1	1	4	100	100	0	0	Da	Dl, pp			
1	1	5	55	25	0	75	Ct	Dl	Da	Cor	Sid
1	2	1	98	98	1	1	Pl, Da				Cor
1	2	2	85	97	3	0	Pl	Da	Bsp		Cor, Isp
1	2	3	85	98	2	0	Da	Dl		Pl	
1	2	4	98	100	0	0	Da	Dl			
1	2	5	60	90	8	2		Bsp	Da	Ct	
2	1	1	80	99	1	0	Pl, Da			Bsp	Isp
2	1	2	65	90	10	0	Pl	Da	Cor	Isp	
2	1	3	90	100	0	0	Da	Dl			
2	1	4	85	100	0	0	Da	Pl	Bsp		
2	1	5	98	100	0	0	Da		Dl		Ct
2	2	1	90	95	5	0	Pl, Da				Isp
2	2	2	65	98	2	0	Da	Pl			Isp, Cor
2	2	3	98	99	0	1	Pl, Da			Esp	
2	2	4	99	100	0	0	Da				
2	2	5	100	100	0	0	Da	Dl	Pl		
3	1	1	80	99	1	0	Da	Pl	Bsp	Sg	Ct
3	1	2	99	98	2	0	Da	Pl			Isp
3	1	3	100	90	10	0	Da	Pl		Bro	Ct
3	1	4	70	100	0	0	Da	Bsp			
3	2	1	99	90	5	5	Da	Bsp		Isp	Cor
3	2	2	100	90	10	0	Da	Bsp	Isp	Dl	
3	2	3	40	94	1	5	Da	Esp, Pl			Ct
3	2	4	90	100	0	0	Da	Pl			

Table 15.3b. Final visual evaluation of ground cover and botanical composition of the grazed area; Nérékoro site, November 2003, Replication 4.

			Composition (%)				Forages				
Rep[a]	Trt[b]	Pl[c]	Ground cover	Grass	Legume	Weed	D[d]	A[e]	F[f]	O[g]	R[h]
4	1	1	85	9	5	2	Pl	Da		Dal	Cor, Isp
4	1	2	75	100	0	0	Da	Pl	Bsp		
4	1	3	85	98	1	1	Da	Pl			
4	1	4	98	95	5	0	Esp	Da	Isp		
4	2	1	40	98	2	0	Da	Bsp		Pl	Cor
4	2	2	50	98	1	1	Da	Pl		Ct, Isp	
4	2	3	99	88	2	10	Da	Pl		Bsp	Cor, Isp
4	2	4	15	0	100	0	Isp				

Notes for Tables 15.3a and 15.3b. Rep[a] = replication; Trt[b] = treatment; Pl[c] = paddock; D[d] = dominant; A[e] = abundant; F[f] = frequent; O[g] = occasional; R[h] = rare. *Grasses*: *Dactyloctenium aegyptium*: Da; *Panicum laetum*: Pl; *Eragrostis* sp.: Esp.; *Bracaria* sp.: Bsp.; *Pennisetum pedicellatum*: pp; *Brachiaria ramose*: Bro; *Schoenfeldia gracilis*: Sg; *Digitaria longitudinalis*: Dl. *Forbs*: *Ipomea* sp.: Isp.; *Cassia tora*: Ct; *Corchorus tridens*: cor; *Sida alba*: Sida.

Table 15.4a. Final visual evaluation of ground cover and botanical composition of the ungrazed area of the Nérékoro site, November 2003, Replications 1 and 2.

			Composition (%)				Forages				
Rep[a]	Trt[b]	Pl[c]	Gr.	Grass	Legume	Weed	D[d]	A[e]	F[f]	O[g]	R[h]
1	1	1	100	100	0	0	Pl	Da			
1	1	2	95	98	2	0	Pl	Da	Cor		Isp
1	1	3	95	10	0	0	Da	Dl		Esp	
1	1	4	100	100	0	0	Da	Pl, Dl			
1	1	5	99	90	10	0	Bsp	Dl		Da	Cor, Et, St
1	1	6	95	50	49	1	Dl		Ct	Isp	Cor, St
1	1	7	98	88	1	0	Bsp	Da		Esp	Cor
1	1	8	90	90	10	0	Da	Dl		Bsp	Cor, Isp, St
1	2	1	99	100	0	0	Dl	Da	Esp, Dl		
1	2	2	80	100	0	0	Da	Dl			
1	2	3	99	98	1	1	Pl	Da	Dl		
1	2	4	55	100	0	0	Da				
1	2	5	100	80	20	0	Da	Dl	Tp		
1	2	6	75	85	15	0	Da		Esp	Bsp	
1	2	7	95	80	20	0	Bsp	Dl	Da, Tp	Cor	
1	2	8	99	90	10	0	Da	Dl	Esp	Cor	Cyp
2	1	1	80	100	0	0	Da, Pl				
2	1	2	99	100	0	0	Da		Dl	Esp	Cl
2	1	3	100	99	1	0	Da		Bsp	Pl, Dl	Isp
2	1	4	100	100	0	0	Da	Dl	Esp		
2	1	5	100	100	0	0	Da				
2	1	6	98	75	20	5	Dl	Bsp	Da	St	
2	1	7	100	100	0	0	Da	Dl, Et			
2	1	8	100	99	0	1	Dl, Da		Bsp	Esp	
2	2	1	90	97	2	1	Pl, Da				Zsp, Cor
2	2	2	55	100	0	0	Esp	Da, Pl			
2	2	3	99	99	1	0	Pl, Da	Dl			Isp
2	2	4	100	94	5	1	Da	Pl		Isp	
2	2	5	98	100	0	0	Da	Bsp		Esp, Cyp	
2	2	6	75	60	37	3	Dl		Pl	Tp	Isp, Cor
2	2	7	100	98	2	0	Da	Esp		Sg	Ct, Isp
2	2	8	100	98	2	0	Da	Esp		Sg	Ct, Isp

species, *Cassia tora* (Ct) was most abundant (Table 15.5). As mentioned above, *Cassia tora* (Ct) is only marginally palatable to livestock. Animals consume *Cassia tora* (Ct) when it is at vegetative stage and in the absence of more desirable forage. Other grasses found in abundance to rare in this area were *Dactyloctenium aegyptium* (Da), *Microchloa indica* (Mi), *Panicum laetum* (Pl) and *Tribulus terrestris* (Tt). *Dactyloctenium aegyptium* (Da) and *Panicum laetum* (Pl) are characterized as high quality palatable forages while *Microchloa indica* (Mi) is less desired by the grazing animals. The legume *Alysicarpus* sp (Asp) had frequent to rare appearances in these pastures.

At the end of the grazing season, the pasture had significantly less ground cover than at the beginning of the grazing season. The legume *Zornia* sp. (Zsp), observed early in the grazing season, remained dominant at the end of the season. However, a shift in botanical composition to grasses in place of the legume *Zornia* (Zsp) was evident (see Table 15.6a and 15.6b).

The grasses *Schoenfeldia gracilis* (Sg) and *Eragrostis* sp. (Esp) were most dominant compared with the *Bracaria* sp. (Bsp) that

Table 15.4b. Final visual evaluation of ground cover and botanical composition of the ungrazed area of the Nérékoro site, November 2003, Replications 3 and 4.

			Composition (%)				Forages				
Rep[a]	Trt[b]	Pl[c]	Gr	Grass	Legume	Weed	D[d]	A[e]	F[f]	O[g]	R[h]
3	1	1	80	100	0	0	Da	Pl	Bsp	Sg	Esp
3	1	2	99	98	2	0	Pl	Da	Bsp		Isp, Cor
3	1	3	100	95	5	0	Da, Pl	Dl		Sg	Za, St
3	1	4	100	90	0	20	Da	Dl, Pl	Esp		
3	1	5	100	80	15	5	Dl, Da	Pl, Bsp	Isp,Cyp		Coc
3	1	6	100	60	20	20	Bsp	Dl, Da	Isp		
3	1	7	98	89	1	10	Da	Bsp	Cl		Cor
3	1	8	95	50	0	50	Da	Bsp	Pl		
3	2	1	100	95	5	0	Da	Pl	Bsp, Dl	Isp, Zsp	
3	2	2	99	80	10	10	Da	Pl	Isp, Dl	Esp, Gl	
3	2	3	100	82	3	15	Dl	Da, Pl	Bsp		Isp
3	2	4	95	100	0	0	Da	Bsp, Pl	Dl		
3	2	5	99	30	10	60	Da	Bsp	Dl	Ipo	
3	2	6	100	93	2	5	Da	Bsp, Dl			Cor, Isp
3	2	7	97	88	2	10	Da	Dl		Pl	Isp
3	2	8	99	98	2	0	Da	Dl	Pl	Esp	Isp
4	1	1	100	95	5	0	Da	Pl	Dl	Cor	Cl
4	1	2	100	85	5	10	Dl, Da		Pl	Isp, Gl	
4	1	3	100	99	1	0	Da	Dl	Bsp	Tp, Esp	Isp
4	1	4	20	0	15	85				Isp	
4	1	5	95	85	5	10	Bsp	Pl	Esp	Isp	Ct
4	1	6	99	99	1	0	Bsp	Dl	Da	Pl	Isp
4	1	7	100	95	5	0	Da	Pl, Bsp		Isp	Tp
4	1	8	100	95	5	0	Da	Dl, Bsp	Tp	Isp, St	
4	2	1	99	80	15	5	Pl	Da	Tp, Bsp	Isp	Z, Ct
4	2	2	98	75	10	15	Dl	Da, Pl		Isp	Cor, St
4	2	3	98	98	2	0	Da	Bsp	Pl, Isp		Sg
4	2	4	90	30	70	0	Isp	Bsp	Da	Cl, Cor	Sg
4	2	5	100	90	10	0	Ei	Da	Ipo		
4	2	6	99	94	5	1	Dl	Da	Cl, Esp	Bsp, Cor, Pl	
4	2	7	99	96	4	0	Pl, Dl	Bsp	Cl, Da	Zsp, Ct	
4	2	8	98	98	1	1	Da	Da	Gl		Isp

Notes for Tables 15.4a and 15.4b. *Grasses*: *Dactyloctenium aegyptium*: Da; *Panicum laetum*: Pl; *Eragrostis* sp.: Esp; *Bracaria* sp.: Bsp; *Pennisetum pedicellatum*: pp; *Brachiaria ramose*: Bro; *Schoenfeldia gracilis*: Sg; *Setaria* sp.: St; *Cloris* sp.: Cl; *Digitaria longitudinalis*: Dl. *Forbs*: *Zornia*: Zsp; *Ipomea* sp.: Isp; *Cassia tora*: Ct; *Corchorus tridens*: cor; *Sida alba*: Sida; *Cyperus* sp.: Cyp; grasslike weed: Gl; *Tephrosia* sp.: Tp.

was observed in abundance earlier in the season. Both *Schoenfeldia gracilis* (Sg) and *Eragrostis* sp (Esp) are highly desired, good quality forage crops. However, these grasses are late maturing compared to the other plant species adapted to the area. Also, excessive primary growth and new growth from seeds could have caused the relative abundance of these two grasses. Due to abundant available forage, the result of excessive moisture during the 2003 grazing season, forage production exceeded animal need, thereby causing under-utilization of plant species that are highly desirable at vegetative/less matured stages, such as *Schoenfeldia gracilis* (Sg) and *Eragrostis* sp. (Esp). At a mature stage (plants with seed-head), animals do not graze these two

Table 15.5. Initial visual evaluation of ground cover and botanical composition, DAFOR scale, for the Siragourou.

			Composition (%)				Forages				
Rep[a]	Trt[b]	Pl[c]	Ground cover	Grass	Legume	Weed	D[d]	A[e]	F[f]	O[g]	R[h]
1	1	1	60	40	60	0	Zsp	Da	Mi, Asp		
1	1	1	70	15	80	5	Zsp	Mi			
1	2	1	50	40	60	0	Zsp	Dsp	Dsp		
1	2	1	75	15	80	5		Zsp, Mi			Pl
2	2	1	30	10	90	0	Zsp		Bsp	Mi	Asp
2	2	1	65	30	50	20	Dsp	Bsp, Ct, Cn	Da, Zsp		Esp
2	1	1	75	5	95	0	Zsp		Bsp		Mi, Asp
2	1	1	80	20	75	5	Zsp		Mi		Bsp, Asp
3	1	1	70	20	80	70	Bsp	Zsp	Asp	Mi	
3	1	1	70	30	60	70		Zsp, Bsp			Tt, Isp
3	2	1	95	10	90	0	Bsp	Zsp			Asp, Ul
3	2	1	85	50	30	20	Bsp			Pl, Zsp	Asp
4	2	1	41	33	67	0	Bsp		Zsp		
4	2	1	40	60	35	5					Zsp, Mi
4	1	1	Bare	10	90	0	Zsp	Bsp			Mi
4	1	1	Bare	10	90	0	Zsp				Mi

Rep[a] = replication; Trt[b] = treatment; Pl[c] = paddock; D[d] = dominant; A[e] = abundant; F[f] = frequent; O[g] = occasional; R[h] = rare. *Grasses*: *Dactyloctenium aegyptium*: Da; *Panicum laetum*: Pl; *Tribulus terrestris*: Tt; *Bracaria* sp.: Bsp; *Microchloa indica*: Mi; *Digitaria* sp.: Dsp; *Digitaria longitudinalis*: Dl; *Tribulus terrestris*: Tt. *Forbs*: *Zornia*: Zsp; *Alysicarpus* sp.: Asp; *Urena lobata*: Ul; *Ipomea* sp.: Isp; *Cassia tora*: Ct; *Cassia nigricans*: Cn.

species, which explains the relative abundance of these plants at the end of the 2003 grazing season.

The cause of the botanical shift at this location could be attributed to grazing preference and forage abundance, which exceeded animal need, rather than to treatment effects (i.e. residual height). At the end of the growing season, all the grasses turned brown with the exception of the legume *Zornia* (Zsp) and a few broadleaf weeds. Regardless of soil moisture and temperature, annual grasses in this climate undergo senescence, the process of plant degeneration that generally occurs at the end of the growing season in order to produce seeds for the following season.

Visual evaluation of ungrazed areas revealed differences from the grazed areas (Tables 15.6a and 15.6b vs. 15.7a and 15.7b). Most of the grazing areas included both grass and forbs (legumes and broadleaf weeds), ranging from 15% to 95% ground cover. In most cases, regardless of treatments, both grasses and legumes appeared at similar frequencies in the ungrazed areas. In addition to grass/legume compositions, there were more plant species than in the grazed areas (Table 15.7a and 15.7b). At both locations, due to the unseasonably wet growing season, forage production (but not quality) was maintained at the highest level throughout the growing season.

Animal performance

Nérékoro and Siragourou locations

During the first week of November, the animals were removed from the pastures and weighed. The animals gained 1–3 kg over the season. For the Nérékoro location, there was only a slight difference in weight gain between treatments (see Table 15.8). According to the field technicians, the animals were heavier in August to

Table 15.6a. Final visual evaluation of ground cover and botanical composition of the grazed area for the Siragourou site, November 2003, Replications 1–3.

			Composition (%)				Forages				
Rep[a]	Trt[b]	Pl[c]	Ground cover	Grass	Legume	Weed	D[d]	A[e]	F[f]	O[g]	R[h]
1	1	1	45	100	0	0	Sg			Bsp	
1	1	2	15	40	60	0	Zsp	Sg			Esp
1	1	3	15	2	98	0	Zsp				Sg
1	1	4	25	10	90	0	Zsp			Sg	Bsp
1	1	5	20	40	60	0	Zsp			Sg	
1	1	6	70	30	70	0	Zsp		Et		Isp
1	2	1	10	100	0	0	Sg				
1	2	2	30	80	20	0	Bsp	Zsp			Ao
1	2	3	35	40	60	0	Zsp, Bsp		Sg		Ao, Ms
1	2	4	5	0	100	0	Zsp				
1	2	5	15	10	90	0	Zsp		Sg		
1	2	6	30	15	85	0	Zsp		Sg		
1	2	7	60	60	40	0	Isp, Da			Esp, Sg	Sb
2	1	1	15	80	20	0	Zsp, Sg			Esp, Ao	Da
2	1	2	10	1	99	0	Zsp				Sg, Ao
2	1	3	10	1	99	0	Zsp				Sg
2	1	4	10	10	90	0	Zsp		Bsp		
2	1	5	10	20	80	0	Sg			Sg	
2	2	1	30	40	60	0	Zsp		Sg	Bsp	Esp
2	2	2	10	0	100	0	Zsp				
2	2	3	20	5	95	0	Zsp				
2	2	4	20	50	50	0					Sg, Esp
2	2	5	20	60	40	0	Sg	Zsp			
2	2	6	40	95	5	0		Sg			Isp
3	1	1	50	85	15	0	Bsp	Pl, Zsp		Sg	Ct
3	1	2	80	80	0	20	Bsp				
3	1	3	15	15	85	0	Zsp				Sg, Esp
3	1	4	15	10	90	0	Zsp			Sg	
3	1	5	15	15	85	0	Zsp			Esp, Sg	Bsp
3	2	1	60	60	40	0	Bsp	Sg, Ao	Zsp		
3	2	2	10	50	50	0			Zsp, Sg		

mid-October, but might have lost weight due to increasingly poor forage quality as the season progressed. Although forage biomass was high, the forage quality declined with the decrease in leaf to stem ratio and increase in seed head production. Generally, such unpalatable forages are high in fibre and low in crude protein and soluble sugars. If animals were weighed on a monthly basis instead of at the end of the grazing season, the assessment of treatment effects and overall animal performance would have been better. Although the original plan was to weigh the animals at the end of each cycle (after animals had gone through cycle of eight pegs/paddocks per treatment), due to the excessive forage growth during the 2003 growing season (driven by excessive rainfall), none of the animals went through all the eight grazing areas/paddocks. By the end of the grazing season two or three pegs (grazing areas/paddock) remained ungrazed for treatment one (where animals grazed to 3 cm residual height). Animals grazing to the 3 cm

Table 15.6b. Final visual evaluation of ground cover and botanical composition of the grazed area Siragourou site, November 2003, Replications 3 and 4.

			Composition (%)				Forages				
Rep[a]	Trt[b]	Pl[c]	Ground cover	Grass	Legume	Weed	D[d]	A[e]	F[f]	O[g]	R[h]
3	2	3	5	100	0	0	Sg				
3	2	4	5	80	20	0	Sg	Sg			
3	2	5	20	60	40	0	Bsp	Sg		Sg	
4	1	1	5	10	90	0	Zsp			Sg	Bsp
4	1	2	10	100	0	0	Sg				
4	1	3	15	100	0	0	Sg				
4	1	4	20	15	85	0	Zsp Ao	Pl			Isp
4	1	5	5	10	90	0	Iv			Sg	
4	1	6	5	100	0	0	Esp				
4	1	7	15	40	60	0	Zsp	Bsp	Esp	Isp, Cb	
4	1	8	20	55	45	0	Zsp	Esp, Sg			
4	2	1	10	100	0	0	Sg				
4	2	2	15	100	0	0	Sg		Ct		
4	2	3	10	50	50	0	Zsp	Bsp, Sg			
4	2	4	20	95	5	0	Sg				Tp
4	2	5	15	98	2	0	Sg		Esp		Zsp
4	2	6	15	80	20	0	Sg	Zsp		Esp	
4	2	7	60	40	60	0	Zsp	Sg		Cb	

Notes for Tables 15.6a and 15.6b. Rep[a] = replication; Trt[b] = treatment; Pl[c] = paddock; D[d] = dominant; A[e] = abundant; F[f] = frequent; O[g] = occasional; R[h] = rare. *Grasses*: *Dactyloctenium aegyptium*: Da; *Panicum laetum*: Pl; *Eragrostis* sp.: Esp; *Bracaria* sp.: Bsp; *Cenchrus biflorus*: Cb; *Schoenfeldia gracilis*: Sg; *Agrostis* sp.: Ao. *Forbs*: *Zornia*: Zsp; *Ipomea* sp.: Isp; *Cassia tora*: Ct; *Tephrosia* sp.: Tp; *Ipomia vegans*: Iv.

residual height moved much more slowly through the cycle than those grazing to 6 cm residual height. As evidenced by the excessive forage mass left ungrazed this growing season, each pasture paddock could have carried two animals instead of one animal/peg. Also, grazing on the 3 cm residual height treatment paddocks appeared to be much more severe on most plant species (less regrowth and more bare ground) than it did on the 6 cm residual height paddocks. The effect of the 3 cm grazing height on animal performance was more evident at the Siragourou location than at the Nérékoro location, as shown by weight loss of one animal (see Table 15.9). In contrast, animals grazing to 6 cm residual height gained the most compared to the residual height of 3 cm at this location.

Summary

In areas with unpredictable rainfall and poor soil fertility (for information on soil fertility, see Chapter 3), maintaining vegetative cover and the regrowth/persistence of forages is highly critical. The growing season is 80–90 days, with an average of 41 wet days and rainfall totals varying between 203 mm and 610 mm. The lack of pasture and forage resources has been a growing problem in the region. With rapidly disappearing perennial forages, annual grasses and some legumes have become the dominant species in the region. The overall objective of this study was to improve pasture resources and increase feed production.

The research indicated that grazing intensity has a profound impact on species diversity. The response to intensive grazing

Table 15.7a. Final visual evaluation of ground cover and botanical composition of the ungrazed area of the Siragourou site, November 2003, Replications 1 and 2.

			Composition (%)				Forages				
Rep[a]	Trt[b]	Pl[c]	Ground cover	Grass	Legume	Weed	D[d]	A[e]	F[f]	O[g]	R[h]
1	1	1	80	100	0	0	Sg	Mi	Zsp		
1	1	2	85	98	2	0	Sg, Mi	Bsp	Esp		Ao, Zsp
1	1	3	25	35	65	0	Zsp	Sg			
1	1	4	45	20	80	0	Zsp		Esp	Sg	
1	1	5	55	45	55	0	Zsp	Bsp	Esp		
1	1	6	75	10	90	0	Zsp		Sg		Esp, Isp
1	1	7	80	30	70	0	Zsp	Isp	Esp	Da	Ao, Cb
1	1	8	95	40	60	0	Zsp, Da				Cb, Esp
1	2	1	60	40	59	1	Zsp, Mi	Sg	Esp	Ao	
1	2	2	45	40	60	0	Zsp	Sg		Bsp	Ind, Sp
1	2	3	70	50	50	0	Zsp	Sg, Bsp			Ao, Esp
1	2	4	65	45	55	0	Zsp	Sg		Esp	
1	2	5	90	30	70	0	Zsp, Isp	Sg	Bsp		Esp, Ao, Dl
1	2	6	85	70	30	0	Isp, Bsp, Dl, Pl	Cb	Zsp	Ao, Br	
1	2	7	60	60	40	0	Isp, Da	Zsp, Dl		Tp	
1	2	8	85	40	60	0	Isp	Dl	Esp, Sg, Da		
2	1	1	75	40	60	0	Zsp	Sg	Esp	Da	
2	1	2	45	20	80	0	Sg	Bsp	Esp, Da		
2	1	3	45	5	95	0	Zsp				Esp, Sg, Iv
2	1	4	80	20	80	0	Zsp	Bsp	Ao	Sg	Cb
2	1	5	80	60	40	0	Bsp	Zsp	Isp	Cb, Sg	Iv
2	1	6	30	55	45	0	Bsp, Zsp		Esp		
2	1	7	50	55	45	0	Zsp Pl				Esp, Sg
2	1	8	85	60	40	0	Da	Zsp, Pl	Dl		Esp, Cb, Iv
2	2	1	80	98	2	0	Mi	Bsp	Sg	Sg	Esp
2	2	2	45	50	50	0	Zsp, Sg		Bsp		Esp
2	2	3	70	35	65	0	Zsp	Sg			
2	2	4	50	20	80	0	Zsp		Bsp	Sg	Esp, Sb
2	2	5	35	85	15	0		Esp, Zsp			Cb
2	2	6	40	95	5	0		Sg			Isp
2	2	7	50	90	10	0	Sg			Cb, Bsp	Sg
2	2	8	65	45	55	0	Zsp		Esp, Sg	Sg, Da	Isp

was highly dependent on the morphological characteristics of the plant species. Plants with a horizontal growth pattern were able to regrow at a faster rate after being defoliated to 3 cm stable height than more upright types of plants. This was mainly because the growing point (the site of rapid cell division) on the horizontal plants is close to the ground, out of danger of being removed by the grazing animal. It is critical that these tissues be close to the soil surface to escape damage during defoliation and thus permit regrowth. Plants with an upright growth habit are more sensitive to defoliation, and are slower to recover than plants with a horizontal growth patterns. At the end of the growing season, the less frequent plant species found were those with upright growth habits that also happened to be highly desired (selectively grazed) by the animals.

The short- and long-term productivity and persistence of annual plants under natural condition are far more challenging than the management of perennial forbs, grasses or shrubs. The seasonal survival and

Table 15.7b. Final visual evaluation of ground cover and botanical composition of the ungrazed area of the Siragourou site, November 2003, Replications 3 and 4.

			Composition (%)				Forages				
Rep[a]	Trt[b]	Pl[c]	Ground cover	Grass	Legume	Weed	D[d]	A[e]	F[f]	O[g]	R[h]
3	1	1	65	60	40	0	Sg, Zsp	Bsp		Esp	
3	1	2	15	40	60	0	Zsp	Bsp			
3	1	3	45	20	80	0	Zsp	Sg	Bsp		Iv
3	1	4	40	5	90	0	Zsp			Sg	
3	1	5	40	15	85	0	Zsp		Et		Sg
3	1	6	40	64	35	1	Sg	Sg		Cb, Esp	Isp
3	1	7	60	20	80	0	Zsp	Sg	Cb, Esp		
3	1	8	55	35	63	2	Zsp	Esp, Bsp	Pl	Isp	Cb, Tp
3	2	1	35	50	50	0	Sg	Sg, Pl	Ind		Iv
3	2	2	40	40	60	0	Zsp	Sg	Bsp		
3	2	3	30	40	60	0	Zsp	Sg			Vi
3	2	4	60	40	60	0	Zsp	Sg	Bsp		Cb, Ao
3	2	5	50	30	70	0	Zsp	Bsp		Ct, Sb	
3	2	6	50	65	35	0	Zsp		Esp	Cn	Pl, Isp, Bsp
3	2	7	60	10	90	0	Zsp	Sg	Esp, Sg	Ao, Da	
3	2	8	45	5	95	0	Zsp		Esp, Bsp	Sb, Da	Sg
4	1	1	10	94	5	1	Sg			Sg	
4	1	2	25	98	1	1	Mi			Sg	
4	1	3	40	50	50	0	Sg, Zsp		Pl		Esp
4	1	4	85	90	10	0	Pl		Iv		Sg
4	1	5	50	95	5	0	Sg				Zsp, Iv
4	1	6	40	50	50	0	Zsp	Esp, Sg	Sb		Ao
4	1	7	15	40	60	0	Zsp	Bsp			
4	1	8	20	55	45	0	Zsp	Esp, Sg			
4	2	1	20	100	0	0	Cr	Sg		Esp	
4	2	2	40	80	20	0	Sg	Zsp			Iv
4	2	3	85	60	40	0	Pl	Da, Bsp	Zsp, Ao		
4	2	4	60	55	45	0	Bsp	Tp	Sg, Pl		
4	2	5	50	60	40	0	Esp, Sg	Sg	Bsp	Ao	
4	2	6	75	50	50	0	Zsp, Dl	Sg	Ao		Cb, Tp
4	2	7	40	0	100	0	Zsp				
4	2	8	45	20	80	0	Zsp		Sg, Da		Dl

Notes for Tables 15.7a and 15.7b. Rep[a] = replication; Trt[b] = treatment; Pl[c] = paddock; D[d] = dominant; A[e] = abundant; F[f] = frequent; O[g] = occasional; R[h] = rare. *Grasses*: *Dactyloctenium aegyptium*: Da; *Panicum laetum*: Pl; *Eragrostis* sp.: Esp.; *Bracaria* sp.: Bsp.; *Microchloa indica*: Mi; *Digitaria longitudinalis*: Dl; *Cenchrus biflorus*: Cb; *Schoenfeldia gracilis*: Sg; *Brachiaria ramosa*: Br. *Forbs*: *Zornia*: Zsp.; *Ipomea* sp.: Isp; *Tephrosia* sp.: Tp; *Agrostis* sp.: Ao; *Ipomia vegans*: Iv; *Indigofera* sp.: Ind; *Valteria indica*: Vi.

Table 15.8. The influence of tethered sheep grazing on animal performance: Nérékoro.

Animal number	Initial animal weight (kg) 31 July 2003	Final animal weight (kg) 4 November 2003	Seasonal weight gain (kg)	Average daily gain (g)
Treatment I				
2	21.0	24.0	3.0	32.0
3	18.0	19.5	1.5	16.0
4	18.0	20.5	2.5	26.9
7	18.0	19.0	1.0	11.0
Average =	18.75	20.75	2.00	
Treatment II				
8	20.0	21.0	1.0	11.0
5	19.0	22.0	3.0	32.0
6	18.0	19.5	1.5	16.0
1	18.0	20.0	2.0	22.0
Average =	18.75	20.60	1.85	

Table 15.9. The influence of tethered sheep grazing on animal performance: Siragourou.

Animal number	Initial animal weight (kg) 31 July 2003	Final animal weight (kg) 4 November 2003	Seasonal weight gain (kg)	Average daily gain (g)
Treatment I				
2	24	25.4	1.4	15.05
3	18	22.0	2.0	22.0
4	21	22.5	1.5	16.0
7	20	19.0	−1.0	–
Average =	20.75	22.10	0.98	
Treatment II				
8	24	26.0	2.0	22.0
5	18	20.0	2.0	22.0
6	20	24.0	4.0	43.0
1	19	22.0	3.0	32.0
Average =	20.50	23.00	2.75	

regeneration of these annual grasses are solely dependent on self-reseeding. Therefore, the management of these types of grass swards needs to focus on the importance of the rest period and defoliation heights. The rest period between defoliation needs to be long enough to allow sufficient regrowth for grazing as well as reseeding. This time period can be shorter when rainfall is in excess, as it was in 2003, because it allows rapid regrowth compared to under a drought condition where plant growth and persistence is severely limited. The data showed that close grazing of these annual plants will cause a shift in botanical composition that consists of redistribution and replacement of the more desirable grasses by more weedy types of plants. Based on the in-season observations, the 6 cm compared with the 3 cm grazing height would promote regrowth within the same season and possibly allow the plant to reseed for the following growing season.

Regardless of the dominant effect of the highly variable environmental condition of the Sahel regions of Africa, the basic knowledge of the response of annual grasses to defoliation can make a big difference in

Table 15.10. Soil test results obtained from sites with different vegetations in pastures at two sites (Siragourou and Nérékoro) within the Madiama Commune.

	Siragourou			Nérékoro			
Items	Areas mostly legumes	Areas mostly *Cassia tora*	Entire pasture	Areas mostly legumes	Areas mostly *Cassia tora*	Entire pasture	Typical values for Malian soils
pH (in water)	5.10	5.36	4.74	6.25	7.20	6.50	5.0–5.5
pH (KCl)	4.29	4.70	4.15	5.83	6.55	5.98	
Organic matter % C							
Nitrogen (Azote) %N	0.02	0.03	0.01	0.03	0.06	0.05	0.05
Total phosphorus, ppm	13.00	15.42	13.00	14.73	41.41*	28.24*	7.0
CEC, meq/100 g	1.76	2.67	2.02	3.71	3.58*	3.71	
Sodium (Na), ppm	0.09	0.18*	0.01	0.21*	0.09	0.17	0.01
Potassium (K), ppm	0.14	0.17	0.11	0.20	1.17	0.37	0.20
Calcium (Ca), ppm	0.59	0.97	0.51	1.43	1.97	1.34	0.70
Magnesium (Mg), ppm	0.37	0.52	0.29	0.87	1.13	0.32	0.40
Iron (Fe), ppm	30.76*	17.72	10.04	7.40	16.60	30.08*	20
Copper (Cu), ppm	0.52	0.20	0.28	–	0.12	0.48	–
Zinc (Zn), ppm	0.12	0.12	–	0.48	0.76	0.96	–
Manganese (Mn), ppm	8.80	18.40	9.40	14.36	15.52	21.56*	20

*Questionable values being re-run for accuracy.

Table 15.11. Particle size analysis of soil samples obtained from sites with different vegetations within pastures at two sites (Siragourou and Nérékoro) within the Madiama Commune.

	Siragourou			Nérékoro		
Items	Areas mostly legumes	Areas mostly *Cassia tora*	Entire pasture	Areas mostly legumes	Areas mostly *Cassia tora*	Entire pasture
Sand[a], %	87.2	83.9	86.8	89.3	75.8	78.3
Silt[b], %	7.5	11.2	9.0	7.8	20.6	16.9
Clay[c], %	5.3	4.8	4.1	2.9	3.6	4.8

[a]Sand size > 0.05 mm; [b]silt size 0.05–0.002 mm; [c]clay size < 0.002 mm.

maintaining in-season productivity and long term persistence of these grasses.

References

Abaye, A.O., Allen, V.G. and Fontenot, J.P. (1995) Influence of grazing cattle and sheep together and separately on animal performance and forage quality. *Journal of Animal Science* 72, 1013–1022.

Brodie, J. (1985) *Grassland Studies*. George Allen and Unwin, Boston, Massachusetts.

Butler, G.W., Greenwood, R.M. and Soper, K. (1959) Effects of shading and defoliation on the turnover of root and nodule tissue of plants of *Trifolium repens*, *Trifolium pratense*, and *Lotus uliginosus*. *New Zealand Journal of Agricultural Research* 2, 415–426.

Davidson, R.L. (1978) Root systems – the forgotten component of pastures. In: Wilson, J.R. (ed.) *Plant Relations in Pasture*. Commonwealth Scientific and Industrial Research Organization (CSIRO), East Melbourne, Australia, pp. 86–94.

Evans, P.S. (1977) Comparative root morphology of some pasture grasses and clovers. *New Zealand Journal of Agricultural Research* 20, 331–335.

Evans, P.S. (1978) Plant root distribution and

water use patterns of some pasture and crop species. *New Zealand Journal of Agricultural Research* 21, 261–265.

Harris, W. (1978) Plant relations in pastures. In: Wilson, J.R. (ed.) *Plant Relations in Pasture.* Commonwealth Scientific and Industrial Research Organization (CSIRO), East Melbourne, Australia, pp. 67–85.

Haynes, R.J. (1980) Competitive aspects of the grass-legume association. *Advanced Agriculture* 33, 227–261.

Schwinning, S. and Parsons, A.J. (1996) Analysis of the coexistence mechanisms for grasses and legumes in grazing systems. *Journal of Ecology* 84, 799–813.

Smith, D. (1962) Carbohydrate root reserves in lucerne, red clover birdsfoot trefoil under several management schedules. *Crop Science* 2, 75–78.

Smith, D. (1973) Physiological considerations in forage management. In: *Forages: the Science of Grassland Agriculture*, 3rd edn. Iowa State University Press, Ames, Iowa, pp. 425–436.

Thornley, J.H.M., Bergelson, J. and Parsons, A.J. (1995) Complex dynamics in a carbon-nitrogen model of a grass-legume pasture. *Annales Botanicae* 75, 79–94.

Watkin, B.R. and Clements, R.J. (1974) The effects of grazing animals on pastures. In: Wilson, J.R. (ed.) *Plant Relations in Pastures.* Commonwealth Scientific and Industrial Research Organization (CSIRO), East Melbourne, Australia, pp. 273–289.

16 Social Capital and Improved NRM

Keith M. Moore[1] and Salmana Cissé[2]

[1]*Office of International Research, Education and Development, Virginia Tech, Blacksburg, VA 24061–0378, USA;* [2]*Centre Régional de la Recherche Agronomique, Institut d'Economie Rurale, B.P. 205, Mopti, Mali*

As noted in Chapter 2, two competing systems of governance at the local level complicate the governance of natural resource management (NRM), particularly rule making and administration with respect to resource tenure. The dominant trend in African governance since independence in the 1960s has been the centralization of the state (Wunsch and Olowu, 1990). While centralization established state authority over land and other natural resources, for the most part the state did little to interfere with customary land tenure arrangements. National tenure or enclosure systems have only been invoked when external demand substantially increased land values (Woodhouse, 2003). Consequently, rural economic development has not been dependent on the willingness of local populations. Furthermore, centralization substantially reduced the civic capacity of local populations to act coherently in their own interests. In cases of rival land tenure claims, this legal pluralism led to arbitrary solutions where the relative power of the litigants was determinant. The average resource user, man, woman or youth, had little security of tenure if the value of his or her land increased significantly.

In response to this trend, the international donor community, non-government organizations (NGOs), and even national governments, having pushed for increased local participation for some time, began promoting decentralization as a necessary mechanism for the promotion of economic growth, improvement of NRM and government services, reduction in social inclusion and poverty, and development of civil society (Manor, 1999; Blair, 2000). Indeed, decentralization in various forms, whether political devolution to local government, deconcentration of government services to local level offices, or privatization of state assets and powers to private individuals, NGOs or corporations, has been increasing in West Africa. The NRM objective of this movement has been to increase local participation and thereby improve the quality of sustainable NRM decision making. In response to these local democratization efforts, customary authorities have begun to re-emerge as a political force (Ribot, 2002). The immediate consequence of this promotion of local governance has been the reinforcement of legal pluralism aggravating conflict among resource users.

In this chapter, we investigate the contribution of the concept of social capital to understanding the development of civil society and establishment of enabling conditions in which all stakeholders can have a voice in decentralized decision making. This analysis highlights the quality of social relations and their positive and negative impact on local NRM governance. The case of the Natural Resource Management Advisory Committee (NRMAC) in Madiama demonstrates the extent to which decentralization has generated the conditions for the growth of civil society's role in local decision making. The chapter begins by clarifying the concepts civil society and social capital first introduced in Chapters 2 and 7. After introducing how the concept of social capital helps to elucidate social relations in civil society, we examine various dimensions and affiliations of social capital to determine the extent to which broad-based consensus over NRM decision making can be established at the commune level. The analysis focuses on the social capital of locally relevant referents: the local State (Commune authorities), customary authorities (or the subordinate customary state), and rural civil society organizations (CSOs). An examination of the structure of rural civil society follows.

State and Civil Society in Africa – a Rural Clarification

Before going into the analysis of social capital, it is necessary to clarify the relationships between rural CSOs, customary association, and the state. Blair (2000) and Woolcock (1998) give a useful generic definition of civil society as an associational sphere intermediary between the individual/family/household and the state. Put into practice, however, this generic definition often masks quite different perspectives on its existence and qualities. In this regard, Mamdani (1996) highlighted the bifurcated state within African societies, while Docking (1999) and Guyer (1994) have noted an international dimension to civil society in Africa. The empirical reality of civil society in Africa is not the same as that which evolved out of the European experience several centuries ago.

Despite substantial growth of NGOs at the national level and sustained associational life in rural villages, it is difficult to conclude that civil society as conceived in the West has advanced very far in rural Mali. However, the work of international NGOs has led to experimentation in the development of CSOs and, consequently, a clearer definition of what constitutes them. The following contrasts rural associational life with criteria Davis (2000) specified. Civil society organizations must necessarily be autonomous from the state and customary authorities. An individual must have the right of entry into and exit from associations (i.e. membership must be voluntary). For democratic (normative) purposes it is desired that the internal procedures of civil society associations be inclusive and transparent (as demonstrated by leadership selection and decision-making processes).

According to these standards, customary village associations in rural Mali so often invoked by development agents do not stand up well. However, they are the most important organizational entities in a village and village residence is sufficient for, indeed obliges, membership, even for those of non-dominant ethnic groups who may reside in the village. Membership is not voluntary and essentially includes household heads for the primary adult association or *tonba*. Although leaders are chosen more for their personal efficacy than their clan or social origin, Kassibo (1997) notes that democratic elections have never been a part of Malian tradition. Because traditional decision making is consensual and combined with the regime of force *de jure*, minority or less powerful voices are often not heard. Civil society, such as it is at the village level, is insular and perhaps only meaningful within the context of face-to-face relations of village life and the authority structures (the customary local state) governing them. This is not to say that social capital does not reside within these relationships (more about this below).

Although there has been a tendency to romanticize past associational forms and the extent to which civil society has existed in Mali, as elsewhere in Africa, drawing on the work of Mamdani (1996), we argue that customary governance practices have been confused with the characteristics of an autonomous civil society. Customary associational structures must be distinguished from CSOs. These customary structures are characterized by village and clan solidarity, defined within the context of and in opposition to colonial and post-colonial states, and serve as an interface with local chieftaincies. Chieftaincy is a holistic concept without the European distinction between legislative, executive and judiciary roles (Alexandre, 1968), whether the particular domain be crop or pasture land or fishing waters. Indeed, 'traditional' chiefs and the apparatus of chieftaincy have both constituted a form of authority of their own, and have served to mediate relations between the state and society (van Dijk and Rouveroy van Nieuwaal, 1999).

This customary state was based on a regime of force in its external relations and provided identity and security for those living within its domain. In the past (and perhaps into the present) force has been the recognized modality on which production of surpluses for the larger society was based (Ribot, 1999). Tax collection and *corvée* labour (state required labour services) are the most obvious examples, but state-run production cooperatives have a similar reputation. On one hand, village chiefs are seen as the bulwark for the reproduction of village life, clan/ethnic solidarity, and protection from the state (van Dijk and Rouveroy van Nieuwaal, 1999). On the other, chiefs are representatives of the national state calling on the *gendarmerie* (i.e. the force furnished by the nation-state) when a situation surpasses their negotiating capacities.

Social Capital: the Quality of Associational Relations

Often linked with civil society, social capital is a complementary concept that can be applied to increase precision in the analysis of the transition to new social forms of associational life (local government, NGOs, etc.) and inform the understanding of the role of various associational forms for socioeconomic growth and development.

Social capital provides a way to understand the relationships and social forces at work in a development setting (Narayan, 2002). As Portes (1998) notes, 'the term social capital simply recaptures an insight present since the very beginnings of the discipline [sociology]' (p. 2). According to Woolcock (1998), it is an attempt to restore the role of norms and institutions lost in the evolution of economic theory. While increasingly popular among development specialists as a multidisciplinary integrating concept, it has been used in many different ways leading to considerable confusion. It has been used to describe both individual and group characteristics, as well to conflate sources and consequences and traverse disciplinary lines (Portes, 1998; Woolcock, 1998; Castle, 2002).

In order to understand the quality of social relations with which social capital is associated, it is useful to first draw on the work of Coleman (1988). In seeking to combine the economist's principle of rational action for the analysis of social systems without discarding social organization, Coleman defines social capital as the function of a set of resources within the social structure available to the actor. These resources include trustworthiness, role expectations and obligations, information channels and effective sanctions. In this conception, social capital refers to individualized resources that can be used to achieve the actor's interests. Putnam (1993) explains the collective character of the concept. It allows citizens to solve collective problems, cuts transaction costs, and improves social tolerance and public health. It is often measured by three indicative characteristics focusing on the strength of social relations themselves: association memberships, trust and the expectation of reciprocity. While intuitively reasonable, both approaches have been criticized for their tautological nature, conflating the sources and consequences of social capital, and not taking into account the negative dynamics associated with too much internal solidarity (Portes, 1998).

In his analysis of the problem of embeddedness, Granovetter (1985) provides the initial key to unravelling the confusion generated by the individual versus collective perspectives. In attempting to develop theories of action integrating economics and sociology, theorists have tended to posit 'atomized actors' that have been either under- or over-socialized into action.

Under-socialized 'atomized actors' have been placed in a context of self-regulating economic structures where social relations between them have no meaning. In over-socialized conceptions, 'atomized actors' have been restricted from acting as individuals making choices within a context of social relationships and roles. However, as Granovetter notes,

> Actors do not behave or decide as atoms outside a social context, nor do they adhere slavishly to a script written for them by the particular intersection of social categories that they happen to occupy. Their attempts at purposive action are instead embedded in concrete, ongoing systems of social relations. (p. 487)

His solution to this dichotomy was to take a middle road and 'embed' actors within structures of social relations where choice of action is an option.

For Granovetter (1985), development is a matter of a change in kind, not degree, of embeddedness that is the social ties, cultural practices and political relations which form the basis for solidarity, trust, membership, etc. Building on this insight, Woolcock (1998) argues that it is necessary to take into account the negative effects of social capital identified by Portes and Landolt (1996) and others. There are costs and benefits to social capital and they need to be measured in order to capture the multidimensionality of the concept. The issue is not a matter of more or less social capital, but of what kind of social capital. Woolcock proposes a complementary concept to embeddedness, that of autonomy (often referred to as 'bridging social capital', Warren *et al.*, 2001; or 'weak ties', Granovetter, 1973). Social relations are not only embedded in various networks in the social structure, but social relations have varying degrees of autonomy across those networks.

Autonomy refers to levels of exclusivity and universality, the capacity to reach out and access resources beyond the immediate group. In an earlier work, Granovetter (1973) had highlighted the importance of network connections to others who have few other common relationships allowing individuals and groups to draw on resources not found within the primary group. These 'weak ties' provide the bridges for coalition building and establishment of broad based alliances for community integration and development.

The model thus constructed (see Fig. 16.1) allows for measurement and analysis not only of social capital in local communities, but also of social capital at the level of the nation-state. At the micro-level, Woolcock (1998) characterizes the dimension of embeddedness as integration, referring to intra-community bonding through ties of solidarity and trust, and the dimension of autonomy as linkage, referring to bridges between primary groups forming intra- and inter-community networks. At the macro-level, these dimensions are described as synergy (see also Evans, 1996), referring to state–society relations, and organizational integrity, referring to corporate coherence and capacity at the national level, respectively. Figure 16.1 presents common characterizations of the conditions generated by the intersection of these dimensions of social capital. The subsequent analyses explore these dimensions.

At the micro level, high levels of both embeddedness (integration) and autonomy (linkage) yield conditions of social opportunity conducive to healthy community development. On the other end, low levels of integration and linkage yield conditions of amoral individualism, where each individual looks out for himself or herself and trusts no one and extra-local social relations are not developed. High levels of linkage with low levels of integration generate conditions of anomie, where extensive extra-local social relations exist, but social bonding to a primary community is minimal. Opposite this, is the condition of amoral familism where individuals are well integrated within their communities, but not well linked with the exterior world.

At the macro level, developmental states are generated by high levels of synergy and organizational integrity, whereas anarchy results from low levels of each. High levels of organizational integrity combined with low levels of synergy yield only

<table>
<tr><th colspan="2">Embeddedness</th><th colspan="2">Autonomy</th></tr>
<tr><td colspan="2" rowspan="2">Micro Level</td><td colspan="2">Linkage</td></tr>
<tr><td>Low</td><td>High</td></tr>
<tr><td rowspan="2">Integration</td><td>Low</td><td>Amoral individualism</td><td>Anomie</td></tr>
<tr><td>High</td><td>Amoral familism</td><td>Social opportunity</td></tr>
<tr><td colspan="2" rowspan="2">Macro Level</td><td colspan="2">Organizational Integrity</td></tr>
<tr><td>Low</td><td>High</td></tr>
<tr><td rowspan="2">Synergy</td><td>Low</td><td>Anarchy
(Collapsed states)</td><td>Inefficiency, ineffectiveness
(Weak states)</td></tr>
<tr><td>High</td><td>Predation, corruption
(Rogue states)</td><td>Cooperation, accountability
(Developmental states)</td></tr>
</table>

Fig. 16.1. Consequences of social capital, applying the concepts of embeddedness and autonomy at the micro and macro levels (from Woolcock, 1998).

weak ineffective conditions for state/society development. High synergy, but low organizational integrity yields conditions ripe for predation and corruption.

These dimensions of social capital provide the conceptual tools with which to analyse the evolution of civil society within rural Mali and evaluate both the national policy of decentralization, as well as the strength of local institutions. Trust and the expectation of reciprocity have been most frequently identified in the functioning of small groups, families or communities. In the case of rural Mali, trust and the expectation of reciprocity are structured ethnically within the context of village clans and organized by village elders as represented by the village chief. Embeddedness in rural Mali is manifested through family and clan solidarities, and autonomy by the density of bridges between these primary groups, as manifested by membership in CSOs.

Analysis of Social Capital Formation

The panel survey sample

Data for this analysis come from panel surveys of sample households resident in the Commune of Madiama. The initial sample design involved a two-stage process. In the first stage, five of the ten villages in the commune were selected as clusters to represent a broad cross section of the community, assuring that villages of pastoralists, sedentary farmers, and agro-pastoralists would be included. Households were randomly selected from 1996 village census lists obtained from the local administrative authorities. These households were initially interviewed in March of 1999 for an economic analysis of household resources and production activities (Brewster *et al.*, Chapter 13 this volume). A socio-institutional follow-up survey with both household heads (male) and their leading wives was conducted later that year in September. After a pre-test of the questionnaires, men and women interviewers were trained in both the content and objectives of the survey, as well as in how to elicit standardized responses. Of the initial 120 households, the first round of the socio-institutional survey participants included 118 household heads and 120 leading wives for a total of 238 interviews. In January 2003, the panel of households was re-surveyed with some minor attrition, with 115 household heads and 114 leading wives completing the interviews.

This sample represents approximately 10% of the household population in the commune. The total population of the commune was 7771 in 1999, rising to 7973 in 2001.

This panel was initially designed to analyse differences between four different production systems: those of farmers (predominantly involved in crop production with only a few animals), agro-pastoralists (substantially involved in both crop and livestock production), sedentary pastoralists (predominantly herder, but staying in the village to produce crops), and transhumant herders (technically resident, but predominantly involved in transhumant livestock management). In each village, the village chief was contacted and a meeting held with farmers and herders informing them about the survey and the information that was being sought. In the process, the villagers were asked to determine the category into which each randomly selected household fell and the consensus category thus elicited was recorded on each questionnaire.

Farmers, whether agro-pastoralist or not, are predominantly Marka. Herders, whether sedentary or transhumant, are predominantly Peul. However, over time it appears that herders are building a sedentary perspective, with more of them self-identifying simply as farmers or agro-pastoralists in the 2003 survey. The few remaining Bozo or Dogon in the sample have blended their identities within their particular village settings. One of the five official villages is composed of two hamlets settled at different times by Marka and Peul, respectively. Because these villagers see themselves as distinct from one another and they demonstrate different response sets they are treated separately in the following analysis.

Framing the analysis

The primary hypothesis of the SANREM CRSP Project in Mali has been that: When a local population is provided with a methodology for resource management and conflict resolution and an institutional vehicle for inter-village, inter-ethnic dialogue, it can become proactive in addressing major NRM issues. In this analysis we investigate factors critical to the social mobilization necessary for that endeavour.

Chapter 7 described the formation of the NRMAC as an institutional vehicle for community deliberation and management of natural resources. The NRMAC initiated and has been involved in many NRM activities. This living experiment allows for addressing a number of questions. Three issues of interest here are: (i) the relative strength of civil society in Madiama; (ii) the extent of social capital that civil society organizations can draw upon and its distribution within the social structure of the community; and (iii) the extent to which the NRMAC established confidence and credibility in the eyes of community members and thereby enhanced the strength of rural civil society.

Social capital

The survey provides a solid data set to investigate the level and quality of connectedness of social relations between and among groups in the population. Association membership (networks) and confidence in local institutions (trust) provide indicators to interpret the level of embeddedness and autonomy of social relations in order to characterize the quality of social capital in the Commune of Madiama. Autonomy is measured by the level of respondent's membership in local organizations. Integration is indicated by the degree of confidence (trust) respondents have in village and community institutions and their leaders.

Membership

There are a number of different types of formal and informal associations to which someone can belong. The most common membership for adult men is the *tonba* (or large village association). Other village-level associations include the Hunters' Association (often with occult restrictions on membership) and the Foresters' Association. Although women's membership in the *tonba* has always existed in principle, only recently has recognition of women's roles beyond the household entered into village discourse and independent women's associations have formed. At the commune or inter-village level, associations have been formed around productive activities, such as the Association of the Herders of Nérékoro and the Rice Producers Association of the *Casier du Syn*. A School Parents Association and a Health Association (linked to the government health service delivery system) also have formed over the past decade. The NRMAC at the commune level with its constituent Village NRM User Groups is the most recently formed.

Among these primary associations, men's association membership has increased minimally between 1999 and 2003 (Table 16.1). However, women's participation in associational life has increased significantly. In 1999, women did not actively manifest their membership in the village association. In 2003, 42% of women respondents claimed such membership. Membership in women's associations has also increased from 33% to 50% since 1999. Overall, 40% of the women respondents claimed membership in one association versus 33% without any membership. Men's participation is slightly higher with 45% participating in one association versus 24% without any membership. Increasing membership indicates a higher level of connectedness, but the depth of participation was not determined.

Association membership also varies by village and ethnicity. The numerically superior Marka clearly dominate organizational memberships, composing nearly the entire membership of commune level organizations such as the Association of the Casier du Syn and the NRMAC, as well as village level associations such as Hunters'Associations (culturally a Bambara/Marka tradition). Although the Herders' Association of Nérékoro is largely Peul, it still has a significant number of Marka members. Participation in village and women's associations is more balanced, reflecting the Marka/Peul composition of each village, although minority groups (e.g. Bozo, Bambara, Dogon) tend not to be members. One exception should be noted here. In the administratively twinned village of Tatia Nouna, the Marka (Tatia) clans are more likely to be association members than the Peul (Nouna) clans (more about this below).

Table 16.1. Changes in association membership by village.

Village		Men's average number of association memberships		Women's average number of association memberships		Marka in Village (%)
		1999	2003	1999	2003	
	M	1.10	1.12	0.30	1.06	
Madiama	SD	(1.03)	(0.96)	(0.47)	(0.95)	97
	n	30	33	33	34	
	M	1.20	1.90	0.50	1.60	
Tombonkan	SD	(1.32)	(1.52)	(0.53)	(1.08)	60
	n	10	10	10	10	
	M	0.83	1.13	0.39	0.96	
Promani	SD	(0.72)	(0.87)	(0.58)	(0.93)	35
	n	23	23	23	23	
	M	1.12	1.04	0.43	0.70	
Nérékoro	SD	(0.78)	(0.98)	(0.51)	(0.64)	0
	n	25	25	23	23	
	M	1.62	2.69	0.62	2.00	
Tatia	SD	(1.19)	(2.21)	(0.65)	(1.08)	92
	n	13	13	13	13	
	M	1.27	1.00	0.09	0.09	
Nouna	SD	(1.01)	(1.18)	(0.30)	(0.30)	9
	n	11	11	11	11	

Trust

In order to measure levels of trust in institutions, respondents to the 2003 survey were asked to identify their level of confidence in a list of local institutions over the past 3 years on a five point scale (1 = increased a lot; 2 = increased some; 3 = no change; 4 = decreased some; and 5 = decreased a lot). The list included the Commune Council and the Mayor, representing local government; the village chief and religious leaders, representing customary authorities; and the rice growers association (Comité de gestion de Casier de Syn), the herders' association (Association des Eleveurs de Nérékoro), and the NRMAC, representing associations of civil society. A factor analysis confirmed the separate grouping of local government, customary authorities, and CSOs.

Overall, Peul villages have lower rates of association membership. This may be due to a tradition of greater individualism among the Peul (Wilson-Fall, 2000), and consequently, they are less likely to be joiners. When the confidence of members in each of their associations is compared by ethnicity, Peul are statistically (at the 0.01 level) more likely to express confidence in village level leadership. There is relatively more confidence among Marka in inter-village associations (with the exception of the Herders' Association of Nérékoro). Nevertheless, there are no significant statistical differences in confidence found between Marka and Peul. This suggests a higher level of autonomy or linkage across networks for Marka than Peul.

Comparisons of these perceptions of trust by village reveals the extent to which the population is embedded in their primary associations. The key indicator of embeddedness is the expressed level of confidence in customary society. Difference in means tests applied to standardized village level scores for confidence in customary society highlighted a distinctive pattern of statistically significant differences (Table 16.2). For both men and women, greater confidence in customary society was expressed in the village of Promani than in the central village of Madiama and the neighbouring village of Tombonkan (at the 0.01 level). For men, this difference carries over between the village of Nérékoro and those of Madiama and Tombonkan (at the 0.01 level) as well. Women in Promani are more confident in customary authorities than those of Nérékoro (at the 0.05 level). Promani and Nérékoro are predominantly Peul villages with the lowest levels of Marka residence in the sample. However, as the analysis of the twinned village of Tatia Nouna demonstrates, confidence in customary authorities is not necessarily a function of ethnicity. This divided village exhibits extreme confidence levels, with the Marka village clans of Tatia expressing the greatest degrees of confidence. In fact, this pattern holds across all measures of confidence in local institutions (local government, customary society, and civil society at the 0.01 level). The Peul of Nouna are the least embedded group in their village.

Table 16.2. Differences in mean scores for confidence in customary society by village.

Village of residence		Confidence in customary society	
		Men	Women
Madiama	M	0.55^{ab}	0.46^{e}
	SD	(0.98)	(0.98)
	n	31	30
Tombonkan	M	0.39^{cd}	0.17^{f}
	SD	(0.36)	(0.48)
	n	10	10
Promani	M	-0.47^{ac}	-0.68^{efg}
	SD	(0.82)	(0.88)
	n	23	20
Nérékoro	M	-0.28^{bd}	0.03^{g}
	SD	(1.06)	(1.11)
	n	25	21
Tatia	M	-0.63^{h}	-0.68^{i}
	SD	(0.68)	(0.59)
	n	13	12
Nouna	M	0.66^{h}	0.57^{i}
	SD	(0.89)	(0.69)
	n	8	10

a, b, c, d, e, f, h, i All significant at the 0.01 level.
g Significant at the 0.05 level.
Positive = less trust; negative = more trust.

Figure 16.2 graphically presents the two dimensions of social capital for each of these villages. While the axis mid-points have been arbitrarily set at the standardized mean values, this presentation provides the opportunity to enrich the discussion with some qualitative insights from the typology. In the first instance, it can be noted that villages are not homogeneous with respect to social capital, despite the fact that there are few differences in association memberships between villages. Tatia is the only outlier, which places it squarely in the midst of social opportunity. This is most likely an anomaly driven by the forced integration of two distinct primary groups in one official village. Nouna is at the other extreme of the twofold characterization, amoral individualism. Such differences in self-perception do not bode well for scaling up social capital in Tatia Nouna. In response to this finding, the NRMAC recently held a conflict resolution and consensus building workshop for the combined village.

Perhaps more instructive for the region are the other villages. The village of Madiama is the most populous and houses the seat of Commune governance. Factionalism divides the different *quartiers* of the village. One *quartier*, where recent in-migrants reside, is clearly being marginalized by the *grand* families among the original settlers who clearly dominated village life, and now are expanding their dominance to the commune. Relations between the NRMAC President and the Commune Mayor, who are from different *quartiers* of Madiama, were originally tenuous. However, they have found that they can work together for mutual benefit. Nevertheless, this improvement of relations does not appear to have filtered down to others in their respective *quartiers*. Furthermore, the proximity to local government resources and almost urbanized conditions appears to lead to low levels of confidence in customary authorities, but not necessarily significant network building through associational memberships. Indeed,

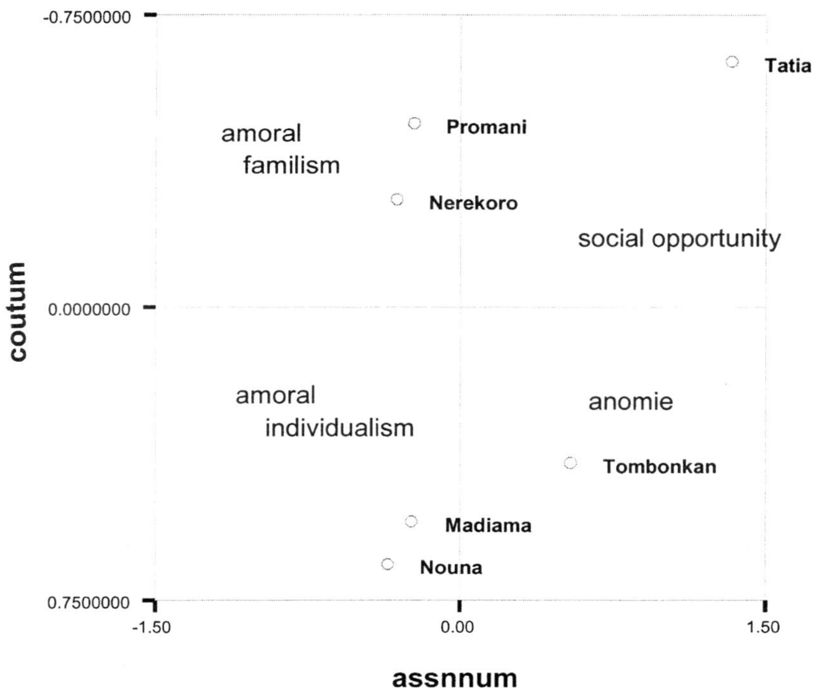

Fig. 16.2. Men's social capital characterization of villages according to indicators (2003) of embeddedness (confidence in customary society) and autonomy (association membership).

the characterization of amoral individualism is consistent with local gossip. Everyone is out for his or her own good and cooperation/collaboration for the mutual achievement of goals is difficult at best.

Tombonkan provides an interesting contrast. Although a small village with a predominantly Marka and farmer population whose lands are encroached on by the neighbouring village of Madiama, it has benefited from straddling a paved road, thereby gaining the attention of passing NGOs, and has built up a relatively higher level of associational memberships. However, the same individual holds key positions in each association so that the benefits of building bridges to others are not reinforced by a dense network. There have been several occasions on which he could not participate in some activities, because he was busy on another association's business. Also, linkages outside of the village are primarily limited to him.

The predominantly Peul villages of Promani and Nérékoro, although both characterized in terms of amoral familism, show high promise in that only a little more participation in secondary associations could generate a successful spiral into social opportunity. However, the strong, but insular, solidarity of these communities and the distrust among the dominant Marka population may inhibit such a shift. Promani was an old administrative centre of the area. Its central position has clearly shifted as Madiama has grown with its weekly market and siting of various services (school, health centre, rice association warehouse, IER outpost, other offices). However, the more inclusive membership of the Nérékoro Herders' Association may have a countervailing effect and with its multiethnic membership is clearly the type of network building that can strengthen social capital within the Commune as a whole.

Analysis of NRM Conflict and Decision Making

Although the preceding analysis demonstrates variability in the quality of social capital between villages (the level of primary group loyalty), network building has increased, and for the most part, the strength of primary social ties has remained stable if not improved as well. The next stage of the analysis investigates the impact and implications of this positive movement in social capital. The working hypothesis is that improved social relations along both dimensions of social capital should have a direct and immediate impact on conflict and cooperation in the commune.

Conflict and cooperation

Overall, from 1999 to 2003 the seriousness with which a list of 23 socioeconomic and environmental problems was perceived has decreased, even though the year 2002 had been one of the worst in the past decade with both low rainfall for rainfed agriculture and a very weak inundation of the rice fields and *bourgoutière*. Both men and women perceived four problems that had significantly increased and shared three of them: the drought, poor soils, and the lack of delta flooding. Men also noted the insufficiency of pastureland, and women the lack of training opportunities. Those aspects perceived as improving for men included: reduced conflict between farmers and herders, improved inter-community cooperation, education opportunities, means of communication and labour availability, and reduced soil erosion. Perceived improvements for women included improved inter-community cooperation and access to credit and firewood.

The interest here is in the improvement in perceptions of farmer–herder conflict and inter-community cooperation (Table 16.3). The pattern of decreased seriousness of farmer–herder conflict and improved inter-community cooperation was consistent across all villages between 1999 and 2003 with the following exceptions: the seriousness of farmer–herder conflict increased for men in Nérékoro, and for women in Madiama and Nouna; the problem of inter-community cooperation increased for men in Tatia and Nouna, and for women, in Promani and Tatia. None of these differences

Table 16.3. Community perceptions concerning the seriousness of farmer–herder conflict and inter-community collaboration.

Perception of problems regarding		1999	2003
Conflict between farmers and herders	Men[a]	3.50	3.24
	Women	3.19	3.06
Inter-community cooperation	Men[a]	2.31	1.95
	Women[b]	2.33	1.83

[a] Difference in means significant at the 0.05 level.
[b] Difference in means significant at the 0.01 level.

was statistically significant. Significant differences in perceived levels between villages, however, are notable. Overall, men and women in Madiama and Tombonkan perceived that conflict between farmers and herders and inter-community cooperation are less serious problems than their counterparts in the herder village of Nérékoro perceived them to be, despite the perception of improvements by all.

Consensus over decision making responsibilities

Given that there has been an increase in social capital in the commune and it appears to have reduced inter-community tensions, the question shifts to how this translates into the community's perceptions of how to deal with specific natural resource conflicts and management issues. Here the investigation focuses on how the social capital associated with different local institutions is related to perspectives on NRM decision-making responsibility. The working hypotheses are: (i) for embeddedness, those who believed that customary authorities should decide natural resource issues will express high levels of confidence in those same authorities; and (ii) for autonomy, the linkage of respondents to multiple local organizations (networks) will be positively related to civil society control over natural resource decision making. This comparative analysis uses the same indicators as above to measure trust/confidence in civil society organizations and in local government.

Respondents were asked who should be the responsible authority for local NRM conflict resolution and decision making (Tables 16.4 and 16.5). Their responses were collapsed into four broad categories to facilitate the following analyses. The plurality of men and women believe that village institutions (particularly the village chief) are best for conflict resolution and decision making, but many are beginning to see a role for local government. Only rarely did respondents suggest that the state should still be responsible, and a few more felt that civil society was best situated to influence such decision making and arbitration.

To assess community support for these institutions as makers of these decisions, the levels of institutional trust were compared. Among men, it is those who believe that local government should be responsible for NRM issues who consistently expressed high confidence in all types of local institutions. Those men favouring village institutions expressed less confidence in each set of institutions. These differences are statistically significant for levels of trust in local government and in civil society. For women, however, it is those few who believe that civil society should be responsible who tended to express the highest levels of confidence overall. Only those women who believe that local government should be responsible tended to express higher levels of confidence in local government as an institution. Although a clear plurality of both men and women believe that village institutions should decide on NRM matters, their expression of confidence in village institutions is weakest.

Only one difference was noted with respect to the measures of autonomy. Men preferring that local government be responsible for NRM conflict resolution and decision making were more likely to be members of their village associations (at the 0.01 level). There were no differences in levels of association membership for either men or women. The expectation was that participation in civil society organizations would

Table 16.4. Men's trust in local institutions according to who should be responsible for conflict and NRM.

		Who should be responsible for conflict and NRM decisions			
Level of trust in		The state	Local government	Civil society	Village institutions
Local government	M	1.08[a]	−0.39[ab]	−0.09	0.26[b]
	SD	(1.35)	(0.99)	(0.99)	(0.87)
	n	4	35	14	53
Civil society	M	0.16	−0.36[c]	−0.04	0.21[c]
	SD	(0.80)	(1.04)	(0.88)	(0.97)
	n	4	35	14	53
Customary society	M	−0.70[def]	−0.08[f]	0.23[d]	0.06[e]
	SD	(0.32)	(0.98)	(0.99)	(1.04)
	n	4	35	14	53

[a, b, c, d, e] *t*-test for difference in means significant at the 0.01 level.
[f] *t*-test for difference in means significant at the 0.05 level.
Positive = less trust; negative = more trust.

Table 16.5. Women's trust in local institutions according to who should be responsible for conflict and NRM

		Who should be responsible for conflict and NRM decisions			
Level of trust in		The state	Local government	Civil society	Village institutions
Local government	M	0.81[abc]	−0.25[b]	−0.08[a]	0.04[c]
	SD	(0.49)	(0.99)	(0.58)	(1.00)
	n	5	35	11	46
Civil society	M	0.66[d]	0.02	−0.58[d]	0.06
	SD	(0.63)	(0.97)	(0.87)	(1.09)
	n	5	35	11	46
Customary society	M	0.00	0.07	−0.42	−0.02
	SD	(0.84)	(0.96)	(0.99)	(0.97)
	n	5	35	11	46

[a] *t*-test for difference in means significant at the 0.01 level.
[b, c, d] *t*-test for difference in means significant at the 0.05 level.
Positive = less trust; negative = more trust.

have differentiated confidence levels, but it did not. This could be because associational membership might be qualitatively different for some members, particularly in the case of non-voluntary membership in village associations.

Perceptions of the NRMAC

The NRMAC has played a central role in diminishing inter-group conflict and enhancing conditions for improved community (multi-village) NRM decision making. Tables 16.6 and 16.7 present measures of the two dimensions of social capital for men and women, respectively. Essentially, the same observations can be made for both men and women. Indicators of autonomy, that is, linkages external to the primary group (number of associations an individual is a member of and confidence in civil society organizations), are significantly higher for those having participated in NRMAC activities. This is also so for the measure of confidence in local government for women.

Table 16.8 shows that while the confidence inspired by the village chief and reli-

Table 16.6. Measures of social capital according to men's participation in NRMAC activities.

		Number of associations	Level of trust in: Local government	Customary society	Civil society
No knowledge of NRMAC	M	0.95[a]	0.10	−0.051	0.33[c]
	SD	(0.77)	(1.05)	(1.10)	(1.03)
	n	41	37	37	37
Awareness of NRMAC	M	0.97[b]	−0.00	−0.04	−0.07
	SD	(0.81)	(1.05)	(1.01)	(0.99)
	n	39	38	38	38
Participation in NRMAC	M	2.20[ab]	−0.11	0.10	−0.27[c]
	SD	(1.80)	(0.91)	(0.89)	(0.91)
	n	35	35	35	35

[a, b, c] Difference in means significant at the 0.01 level.

Table 16.7. Measures of social capital according to women's participation in NRMAC activities.

		Number of associations	Level of trust in: Local government	Customary society	Civil society
No knowledge of NRMAC	M	0.77[a]	0.19[c]	0.11	0.27[b]
	SD	(0.76)	(1.00)	(1.09)	(0.96)
	n	71	64	64	64
Awareness and participation in NRMAC	M	1.48[a]	−0.30[c]	−0.18	−0.44[b]
	SD	(1.17)	(0.94)	(0.81)	(0.91)
	n	42	39	39	39

[a, b] Difference in means significant at the 0.01 level.
[c] Difference in means significant at the 0.05 level.

gious leaders is significantly superior to that of the NRMAC, the NRMAC inspires greater confidence than other local institutions of civil society or local government. This is clearly so for men. A parallel survey of village chiefs, elected officials, development agents and CSO leaders confirms this finding (Touré, 2003).

Discussion

Although social capital exists to some degree in all communities, its strength and breadth can vary. In addition, it can be either viewed positively or negatively for poverty alleviation, NRM or economic development according to how the community fits into the greater socioeconomic context. Through collective action, poor groups can increase their capacity to take advantage of opportunities to enhance their well-being. People who trust each other and cooperate for a specific purpose may have a general resource available for other cooperative endeavours. However, unlike money, social capital is not a universal resource, anonymous and fungible; it is tied to specific organizational forms for specific purposes (Warren *et al.*, 2001; Evans, 1996). However, societal well-being is dependent on a shift from exclusive loyalty to primary groups to networks of secondary associations (Narayan, 2002). The problem of social capital for development purposes is in 'scaling up', that is, how to combine the strength of social solidarities at the village level across a broad set of villages.

One of the constraints to this transition can be ethnolinguistic fragmentation (Rodrik,

Table 16.8. Comparison of levels of confidence in various local institutions.

	Men	Women
Religious leaders	1.89[a]	2.20
	(0.78)	(1.00)
Village chief	1.93[b]	2.00[g]
	(0.98)	(1.02)
NRMAC	2.17[abcdef]	2.38[ghi]
	(0.84)	(0.82)
Mayor	2.44[c]	2.39
	(1.08)	(1.02)
Rice Growers'	2.54[d]	2.83[h]
Association	(0.85)	(0.71)
Herders'Association	2.56[e]	2.78[i]
	(0.87)	(0.79)
Commune council	2.61[f]	2.49
	(1.06)	(0.98)
N	110	103

[a, d, e, f, g, h, i] Difference in means significant at the 0.01 level.
[b,c] Differences in means significant at the 0.05 level.

1997, cited in Narayan, 2002) that can also lead to income inequalities. Brewster *et al.* (Chapter 13 this volume) document the disparities between household production systems leading to such social differentiation. Under these conditions, if the institutions of conflict management are weak, exogenous shocks can easily trigger conflicts. The analysis of social capital suggests that the relatively low level of confidence expressed in customary authorities is indicative of the need for new mechanisms for conflict resolution to emerge.

Most of the Madiama population still perceives NRM decision-making as necessarily resting at the village level. However, it is these same people who express less confidence in customary authorities than those who would have the local government regulate NRM issues. Although it is not possible to determine from such cross-sectional data, there appears to be increased confidence in local government, a product of decentralization. However, it is possible that increased inter-village networking has begun a shift towards increased confidence in local government. Certainly, overall confidence is higher for those seeking NRM leadership from local government or civil society than for those seeking it from customary authorities. This shift in confidence in local government and civil society appears to be associated with the NRMAC.

This suggests that the population is divided, or possibly simply confused in their primary loyalties. Civil society is emerging built on strong ties embedded at the village level and new bridges linking villages. But not all villagers are equal participants, and some villages have more positive social capital than others. Whether this new rural civil society will be more inclusive of minority interests is unclear. Women are still tied to the village level loyalties, because these are most accessible to them. Herders, represented by the villages of Nérékoro and Promani, appear to have a strong solidarity base at the village level as well, but could easily be isolated by the farmer majority at the commune level.

Although we have taken a bottom-up approach in this analysis and in practice, it is also worthwhile to discuss top-down issues for a moment. Without State support for innovation in social relationships across civil society, many of the types of changes beginning to occur would not be possible. Discussing relations between public and private actors, Evans (1996) notes that 'effective states deliver rule-governed environments which strengthen and increase efficiency of local organizations and institutions' (p. 1120). The State can contribute to social capital formation by providing complementary resources of rules, information and infrastructure. It can most effectively do so when (at least some) State actors are embedded in the local community in which they work and actively assist in the co-production of collective goods. He goes on to cite numerous examples of such synergetic relations and how social capital may inhere in state-private sector relations. Interestingly, the only negative example is the single African case cited by Ostrom (1996), who describes the over-centralization of the Nigerian state bureaucracy as inhibiting the potential for public–private sector synergy. Although the rigidity of over-centralization is problematic for African development, the present analysis highlights the countervailing force which, in part incites it, but most certainly complicates the situation. To be successful, decentralization must come to terms with the local customary state whose insular environment left on its own will not transcend local loyalties.

Can social capital be constructed or reinvigorated? We have demonstrated the potential to do so. Evans (1996) suggests three elements for consideration in this regard: people's perceptions of themselves shape the social structure and these are malleable; the soft technologies of organizational design can have a major influence; and collective difficulties can be addressed through redefining the problem in ways conducive to collective action (see Chapter 8). The experience of the NRMAC (Chapter 7) shows that it indeed constitutes a new 'soft technology'. Shifting from exclusive loyalty to primary groups to participation in secondary associations requires the development of new skills and the learning of new behaviours. It has allowed villagers to adopt new identities at the Commune level, thereby reshaping their self-perceptions. Training in conflict resolution has provided the committee with self-confidence, a means to collectively reframe NRM problems, and the capacity to share this message with village groups.

Conclusion

Has a broad-based local consensus over resource use been established? No. However, the building blocks for an autonomous civil society are beginning to emerge. On this basis an inclusive approach can be developed locally to resolve the impasse of legal pluralism. Through a sector-specific (NRM) initiative, disparate groups are coalescing around broad-based, but narrowly defined organizational themes.

Previously social capital was never explicitly mobilized due to the ignorance or contempt of the administration or the rigidity of its rules and procedures. It can now be seen to have possibilities. For example, all the village chiefs of the Commune of Madiama questioned concerning their perception of the creation of the rural communes (decentralization) noted that it had led to the breaking of relations with the Commandant (*sous-prefet*), that is, with the administration (Touré, 2003). Those same village chiefs also noted the positive contributions of the NRMAC to commune life.

This analysis of social capital demonstrates the importance of building on traditionally valued social relationships and combining these with linkages across groups for the management of various conflict situations (involving land tenure or not). In this way, viable negotiated solutions can be achieved and a new social contract realized, thus scaling up social capital.

We conclude that donor/NGO emphasis on building village level associations, while successfully mobilizing local resources for development has three drawbacks: (i) the scale is too small for the costs of extended replication; (ii) village social capital is too insular for these associations to have a transformative effect on rural social structures and dynamics; and consequently, (iii) these associations can easily revert to neo-neocolonial mechanisms of divide and control and for the extraction of surpluses from the rural population. For rural civil society to grow, linkages between villages must be developed, and citizen networks established. In particular, we recommend the reinforcement of all commune-wide associations which multiply the ties between agriculturalists and pastoralists. We must qualify this in regard to the development of women's role in rural society. Often constrained to remain in the village by tradition, women's village associations are serving to mobilize women in their own struggles for improved quality of life. These opportunities for women to formally associate should be encouraged.

Changes at the national level supportive of independent CSOs are necessary in order that rural civil society can prosper and they are beginning to occur. Decentralization has created the opportunity for the NRMAC to build traditional bonds at the village level into a network of relations creating a modern tool at the commune level. The association is the manifestation of a public good whose bridging social capital allows it the autonomy to function successfully, drawing the strengths of customary associations into rural civil society.

References

Alexandre, P. (1968) The problem of chieftaincies in French Africa. In: *West African Chiefs: Their Changing Status Under Colonial Rule and Independence* (Institute of African Studies, No. 7). University of Ife, Ife, Nigeria, pp. 1–42.

Blair, H. (2000) Participation and accountability at the periphery: democratic local governance in six countries. *World Development* 28(1), 21–39.

Castle, E.N. (2002) Social capital: an interdisciplinary concept. *Rural Sociology* 67(3), 331–349.

Coleman, J.S. (1988) Social capital in the creation of human capital. *American Journal of Sociology* 94(Suppl.), 95–120.

Davis, J.U. (2000) Classrooms of democracy? The educational prospects of Malian civil society. In: Bingen, J.R., Robinson, D. and Staatz, J.M. (eds) *Democracy and Development in Mali.* Michigan State University Press, East Lansing, Michigan, pp. 289–320.

Docking, T.W. (1999) International influence on civil society in Mali: the case of the cotton farmers' union, SYCOV. Ph.D. dissertation, Boston University, Boston, Massachusetts.

Evans, P. (1996) Government action, social capital and development: reviewing the evidence on synergy. *World Development* 24(6), 1119–1132.

Granovetter, M. (1973) The strength of weak ties. *The American Journal of Sociology* 78(6), 1360–1380.

Granovetter, M. (1985) Economic action and social structure: the problem of embeddedness. *American Journal of Sociology* 91(3), 481–510.

Guyer, J. (1994) The spatial dimensions of civil society in Africa: an anthropologist looks at Nigeria. In: Harbeson, J.W., Rothchild, D. and Chazan, N. (eds), *Civil Society and the State in Africa.* Lynne Reinner Publishers, Boulder, Colorado, pp. 215–229.

Kassibo, B. (ed.) (1997) *La Decentralisation au Mali: État des Lieux* (Bulletin No. 14). LitVerlag, Hamburg, Germany.

Mamdani, M. (1996) *Citizen and Subject: Contemporary Africa and the Legacy of Late Colonialism.* Princeton University Press, Princeton, New Jersey.

Manor, J. (1999) *The Political Economy of Democratic Decentralization.* World Bank. Washington, DC.

Narayan, D. (2002) Bonds and bridges: social capital and poverty. In: Isham, J., Kelly, T. and Ramaswamy, S. (eds), *Social Capital and Economic Development.* Edward Elgar Publishing Limited, Northampton, Massachusetts, pp. 58–81.

Ostrom, E. (1996) Crossing the great divide: coproduction, synergy and development. *World Development* 24(6), 1073–1087.

Portes, A. (1998) Social capital: its origins and applications in modern sociology. *Annual Review of Sociology* 24, 1–24.

Portes, A. and Landolt, P. (1996) The downside of social capital. *The American Prospect* 26, 18–22.

Putnam, R.D. (1993) *Making Democracy Work: Civic Traditions in Modern Italy.* Princeton University Press, Princeton, New Jersey.

Ribot, J.C. (1999) Decentralization, participation and accountability in Sahelian forestry: legal instruments of political–administrative control. *Africa* 69(1), 23–65.

Ribot, J. (2002) *African Decentralization: Local Actors, Powers and Accountability* (Programme on Democracy, Governance and Human Rights, Paper No. 8). United Nations Research Institute for Social Development, Geneva, Switzerland.

Touré, B. (2003) *Enquêtes Leadership Local.* Sustainable Agriculture and Natural Resource Management Collaborative Resource Support Program, Institut de l'Economie Rurale, Centre Régional de la Recherche Agronomique, Mopti, Mali.

van Dijk, R. and van Rouveroy van Nieuwaal, E.A.B. (1999) Introduction: the domestication of chieftaincy: the imposed and the imagined. In: van Rouveroy van Nieuwaal, E.A.B. and van Rijk, D. (eds) *African Chieftaincy in a New Socio-Political Landscape.* Lit Verlag/Transaction Publishers, Hamburg, Germany.

Warren, M.R., Thompson, J.P. and Saegert, S. (2001) The role of social capital in combating poverty. In: Saegert, S., Thompson, J.P. and Warren, M.R. (eds) *Social Capital and Poor Communities.* Russell Sage Foundation, New York, New York, pp. 1–28.

Wilson-Fall, W. (2000) Conflict prevention and resolution among the Fulbe. In: Zartman, I.W.

(ed.) *Traditional Cures for Modern Conflicts: African Conflict 'Medicine'*. Lynne Rienner Publishers, Boulder, Colorado, pp. 49–66.

Woodhouse, P. (2003) African enclosures: a default mode of development. *World Development* 31(10), 1705–1720.

Woolcock, M. (1998) Social capital and economic development: toward a theoretical synthesis and policy framework. *Theory and Society* 27(2), 151–208.

Wunsch, J.S. and Olowu, D. (1990) The failure of the centralized African State – Introductory chapter. In: Wunsch, J.S. and Olowu, D. (eds) *The Failure of the Centralized African State.* Westview Special Studies on Africa, Boulder, Colorado, pp. 1–22.

17 What We Have Learned

Keith M. Moore,[1] Charlene Brewster[1] and Michael Bertelsen[2]

[1]Office of International Research, Education and Development, Virginia Tech, Blacksburg, VA 24061–0378, USA; and [2]College of Agriculture and Life Sciences, Virginia Tech, Blacksburg, VA 24061–0334, USA

We have seen how resource allocations have shifted from a largely extensive pastoral system to an expanding, but still extensive agricultural system. The expansion of crop production in the wake of the introduction of animal traction has had a degrading effect on the natural resource base of Madiama. Continuous cropping with little or no fallowing has depleted soil fertility and forced over-grazing of fragile range resources. Furthermore, legal pluralism with respect to land tenure has thwarted the transition to more intensified production systems, except in a few cases where powerful actors could easily realize marketable returns. For the majority, a new consensus concerning the rules of the game has to be achieved for sustainable development to occur. The concatenation of production systems, once complementary, has become contentious with increasingly frequent conflicts. Meanwhile, social capital has remained strong at the primary group level (villages and clans), but weak at the inter-community scale of rural civil society where new rules need to be established.

The introduction of the SANREM project in Mali coincided with the initiation of major steps in government decentralization in the late 1990s. These steps involved establishment of Rural Communes, election of Commune Councils, and the devolution of many natural resource management (NRM) responsibilities. However, the newly elected officials have been poorly prepared for these new responsibilities. Recognizing this weakness, the SANREM project adapted its programme to closely articulate with this process by establishing a multi-village committee at the commune level, the Natural Resource Management Advisory Committee (NRMAC). The initial mandate of the NRMAC was to act as a local interlocutor for NRM research and development services at the Commune level. Having matured and become an officially recognized civil society organization (CSO), the NRMAC now: (i) bridges the rural knowledge divide linking rural civil society with local government and deconcentrated technical services; (ii) provides advice and support to the Commune Council on sustainable NRM issues; and (iii) builds social relations and consensus between and among communities, researchers, and other relevant stakeholders.

Social infrastructure alone, however, cannot resolve the crisis of sustainable growth; intensification of crop and livestock production will also be required to assure continued livelihoods for the rural population. Examples of this intensification can be seen in the cash cropping of watermelons and in cattle fattening operations.

Intensification is also evident in: selected soil fertility enhancements; greater efforts to develop manure exchange relations; the transformation of *Cassia tora* into a valued forage crop through ensiling; and the development of more intensively managed grazing regimes. Although the transition will be challenging to all, it will be particularly hard on those pastoral populations who have lost access to the primary resource on which their management strategy was based – free movement through extensive pasturelands.

During the 1999 Participatory Landscape/Lifescape Appraisal (PLLA) farmers identified soil fertility and water retention capacity as primary constraints to sustainable livelihoods. Through extensive literature review and consultation with farmers, SANREM researchers identified a wide range of low-cost soil fertility-enhancing technologies that have been tested and refined on farmers' fields. Various combinations of crop rotations and soil amendments, incorporating farmers' perceptions and preferences (for example, including the use of local, rather than research-generated cultivars, and inclusion of small ruminant manure which farmers believe to be the most potent) were examined. Farm trials and computer modelling evaluated the yield effect and long-term sustainability of different applications of organic manure, natural phosphate, and chemical fertilizers, including a locally adapted International Crops Research Institute for the Semi-Arid Tropics (ICRISAT) protocol for micro-dosing inorganic fertilizer that minimizes farmer input expenses. One consequence of these interventions is that the traditional manure-fodder exchange has been revitalized through overnight paddocking of animals on fields.

The SANREM project also established data and information products and decision support tools, including a household level socioeconomic panel survey database and a multi-year, geo-referenced database of weather, soil, cropping system, land use and water point data. This latter database has been used for constructing maps and modelling at the commune level. The calibrated CropSyst Model simulating the long-run viability of potential new technologies now provides information for agroeconomic decision-making from the commune to the regional level. By integrating agronomic research and economic analysis, Institut de l'Ecomomie Rurale (IER) researchers are also able to more quickly identify promising new technologies and farmers are able to understand the economic implications of new cropping practices. The Social Accounting Matrix (SAM) developed for Madiama Commune provides guidance to researchers and commune leadership about strengths and weaknesses in the local economy, potential impact of new technologies, and how these impacts would be distributed between different groups within the community. The socioeconomic database has also provided data to evaluate the extent of social capital formation and the impact of the NRMAC.

Adaptive Learning through the SANREM Approach

The four pillars of SANREM, participation, landscape/lifescape scale, multiple stakeholders and interdisciplinarity, provide the frame of reference structuring this work in Madiama. The working hypothesis was that when a local population is provided with: (i) methods for natural resource and conflict management, and (ii) an institutional vehicle for inter-village, inter-ethnic dialogue, the population can become proactive in addressing major agricultural productivity and natural resource management issues. There were three points of intervention in this multi-pronged approach: (i) addressing the manifestation of NRM problems through directly developing the capacities to confront and resolve conflicts and the underlying factors which shape the conditions generating those conflicts; (ii) introducing new technologies and strategies to increase the efficiency and sustainability of the natural resource base; and (iii) developing social networks to increase local power in the governance of natural resources.

In addition to being a symptom of underlying causes, conflict has an independent impact on both biophysical resources and the social relations governing resource use. In the first instance, resources are over-exploited in a competition to gain some benefits before or instead of others leading to rapid degradation. More importantly, latent conflict inhibits groups from coming together to resolve their collective problems. By providing tools and capacities to manage conflict, we hoped to increase the confidence of the population in itself and in its neighbours.

The same tools used to resolve conflicts also helped to develop organizational skills for consensus building (human capital investment) and have led to the establishment of a new form of social infrastructure bridging relationships between local communities (social capital investment). Through inter-village networking, sharing of knowledge, tree planting and developing co-management accords, social capital was built and this led to further consensus building and the trust necessary for wider discussions throughout rural civil society. In addition, we linked the communities with scientific research services to identify and introduce technologies and strategies that have the potential to increase the productive capacity of the natural resource base.

The process of empowering local people to take charge of their own economic and social development is slow and complex. Participation, the core of the SANREM approach, is not only inherent in many of the tools introduced to assist local populations in realizing their goals, but is also a development outcome in its own right. Let us consider what we learned over the short duration of the SANREM project in Mali.

Participation: towards a new vision

The first pillar addresses an inherited problem, participation. Traditionally, development projects have approached local populations in a mode similar to that of national governing structure:

> A centralized and vertical approach . . . left no room for local people to participate in the decision-making process. . . On the one hand, the technicians acted in accordance with a conventional technology-transfer model, whereby they presented themselves as providers of know-how to local people. On the other, local people, constantly seeing their own frame of reference, knowledge and know-how disregarded, came to suffer from a lack of self-esteem, and adopted in some places what might be referred to as a 'welfare mentality'. In other words, they tended to turn more and more to outside agencies for solutions to local problems.
>
> (Guèye, n.d., p.1)

This paternalistic approach placed the local population in the role of passive receivers of a project's largesse. To the extent that participation existed, local people simply took on the role of middlemen for the capture and/or distribution of project benefits. The idea that an independent local leadership could be capable of deciding priorities was never considered.

Into this local social, economic, and political context the SANREM CRSP–West Africa Project introduced its approach to locally controlled NRM. SANREM sought to be inclusive of everyone in the Commune, as individuals regardless of occupation, gender, age or other characteristics. SANREM also included the associations that had been formed to represent various socio-professional groups, such as rice cultivators, hunters, and pastoralists. Representatives of local government, technical services and NGOs were all present during the creation of the new civil society organization, the NRMAC. Through real participation in the decision making structures of local governance, local people would learn to behave in new ways that would empower them to take control over the management of their natural resources as provided for in the new decentralization laws.

Social learning

Learning new roles does not unfold in a simple linear fashion. While villages enthusiastically received the project into their communities, their previous experiences with development projects shaped their expectations. In other words, they expected to become traditional project beneficiaries. But this kind of passivity was counter to project objectives. During the relatively long training process that followed, new skills were developed to strengthen local leadership capacities, but local enthusiasm waned. While the cohort of local leaders was being taught new skills, the necessary reinforcement for the new roles they were expected to fill was slow in coming.

The process of learning new roles is separate from the process of learning the skills to accomplish those roles. New roles are only truly learned when the practice of those skills is reinforced through validation by others and through the realization of material benefits. In rural societies where agricultural seasons shape life, the value and meaning of these new roles cannot materialize until at least a full annual cycle has been completed. Inter-village discussions and networking to manage herder–farmer conflicts can only be tested with the annual harvest season livestock passage through the commune.

The learning of new roles is also not a one-sided affair. Roles are defined by relationships. SANREM Project agents and their partners in the CRRA/Mopti, CARE/Djenné, Groupe de Recherche Actions pour le Développement (GRAD) and other deconcentrated technical services have also been learning the new skills and roles necessary to empower local populations. Another full cycle of seasons must pass before these new relationships can be consolidated into accepted behavioural practices where role representatives properly respond to their counterpart's role. Understanding each other's new roles and building mutual trust has been a central project task. At stake is the transformation away from the 'welfare mentality'. This transformation is difficult, because it depends not only on development agent behaviour, but also on the behaviour among and between segments of the population. The new roles that the various groups in the population should play must be reinforced by the successes of their actions and the manner in which their partners respond to them in their activities.

Scale Issues

The second pillar of SANREM is landscape/lifescape scale. Scaling-up has two dimensions: one social and one biophysical. The landscape/lifescape scale requires going beyond the village based *gestion de terroir* approach. As scale increases, more stakeholders become involved and ecosystem limits expand as well. *Gestion de terroir* succeeded on the village level where face-to-face relations between actors occur on a daily basis. The decisions to be made at the landscape scale not only concern individual adoption of production or conservation technologies, but also collective choices about land/resource use, whether on common property or private lands. All the actors using the same space must be taken into account. Here, accountability is an important issue. Including stakeholders on a lifescape scale higher than the village involves gathering together those who do not have routine daily contacts, even though they are involved in social relations through economic exchanges or sharing the same resources.

The NRMAC was designed with downward accountability in mind. Each village chief designates members who are expected to report back to village user groups on a routine basis. Indeed, as committee members themselves have recognized, this reporting has posed serious difficulties. Village chiefs appear to receive reports, but broader dissemination throughout the village is often limited to informal and chance occasions, with the exception of meetings held for specific purposes, such as discussing and negotiating a multi-village *bourgoutière* management agreement.

For villagers, commune and regional decision makers to think and act responsibly, they need knowledge and information about those outside their village *terroir* who share the same resource base and how these resources can best be exploited. The biophysical, geographic and socio-economic modelling tools developed by SANREM researchers have produced results that can be extrapolated beyond the boundaries of the commune. This is as true for technologies adapted to the Madiama landscape types found throughout the Inland Delta as it is for intra-village dependencies within commune economies.

Multiple stakeholders/networking

The SANREM pillar of involving multiple stakeholders invokes another dimension of participation, social networking. Encouraging participation of multiple stakeholders in community activities without precedent models that the population and researchers have experienced is time consuming and difficult. Scheduling and other obligations hamper collaborative action, because it takes time to coordinate the activity schedules of different partners. Conflicting agendas and obligations of the various partners also inhibit the consensus building necessary for collaborative action. Behaviour patterns of different partners need to coalesce for coherent programme implementation, and their schedules need to synchronize so that the fundamental communication necessary for collaborative participation can occur. Each aspect of the participatory process requires negotiation among all participants, and where possible structural impediments, differential power relations and latent conflicts must be removed.

In the SANREM research and development programme, we experimented with various methods of building a new network of relationships within rural civil society between neighbouring villages, various ethnic groups, production systems, the population, technical services, local government and other partners. Providing a community with social infrastructure creates positive conditions for participation. To assure that all key stakeholders have the same level of information, training was first provided to a wide range of development agents and local officials before focusing efforts on training commune level stakeholders. Communication of commune level activities at the village level has been encouraged and has required vigilance in monitoring and follow-up efforts. The object of this social infrastructure is to multiply connections and networking between all the different groups.

In recent years, two approaches to collective NRM have been taken: one, targeting processes for establishing local agreements for co-management; and the other, targeting the development of human capital and social infrastructure to create the social capital necessary for successful co-management. While one cannot be done without the other, SANREM activities favoured the second approach. Examples of the first approach are mentioned in the next section.

We began with the assumption that social infrastructure can be developed to increase the chance that various ideas and people can intensify open debate, increase the circulation of information, especially technical information, and bring about more considered decisions (Flora, 1998). However, skills in developing/negotiating stakeholder relations and organizational networks do not guarantee success in achieving inter-community management agreements. The SANREM experience clearly illustrates the difficulties of achieving substantial NRM achievements in the short term, when the early project focus is on developing the skills to master the processes of decentralized and participatory management.

The question becomes what social learning proceeds in such a situation. The SANREM experience validates observations of Ribot (1999) who noted that the processes of decentralization and participation can become perverted in the act of managing natural resources. Instrumentalist approaches to NRM inhibit empowerment. Although often characteristic of government and project agents, the instrumentalist approach may also be replicated by NGOs as they become project implementers. However, NGO facilitation has proven to be crucial to the success of civil society efforts. NGOs can serve an important role by providing the 'guide on the side', instigating committee members to step outside of their preconceived roles and become the community leaders and spokespersons for rural civil society, facilitating and monitoring committee activities, and coordinating human capacity development among leaders and the community. However, differentiating facilitation for social empowerment from administration of a population is often difficult. Government service providers have always had great difficulty maintaining an empowering stance as their formally defined role is as administrators of the population.

This raises the question of the proper roles and behaviours for NGO, administration and technical service agents. The Malian government has taken an initial step by opening up the terrain for rural CSOs and the decentralized administration of services. It is important for local officials and technical service agents to cultivate relations with these CSOs. Following the conclusions of Evans (1996), this synergy (the quality of the interaction between state and civil society) should develop in two dimensions. State agents should focus on building relationships through providing goods and services complementary to those provided by the private sector. Further, state agents at the local level need to become embedded in the community, building trust and the networks for increased social capital in rural areas.

Lessons Learned

By learning new empowered roles, the SANREM-trained local community leaders have brought about behavioural changes in community relations leading to attitudinal changes in the commune population as a whole. Future successes will be built on this cohort of local leaders, and their mobilization of social capital. These new leaders have learned to act on behalf of their communities in the development of sustainable and profitable practices for decentralized NRM.

While SANREM has been working in Madiama, similar inter-village level experiments in decentralized NRM have taken place at the local level in Mali. In two cases, they have been assisted by locally based NGOs: SOS Sahel in Bankass and the Near East Foundation (NEF) in Douentza (Bocoum *et al.*, 2003). A third case, *Siwaa*, was supported by CRRA/Sikasso researchers in collaboration with extension personnel (Hilhorst and Aarnink, 1999). In all three cases, the NGO developed relations between the local population and government service agencies in support of co-management of natural resources. However, each evolved a different social infrastructure. In Bankass, the Projet d'Appui à la gestion de l'environnement (PAGE) formed no permanent inter-village association; they simply facilitated the linkage between government services and individual village associations. In the zone of Kelka (near Douentza), the NEF coordinated with government services to support a traditional multi-village association called *Waldé Kelka*. The CRRA/Sikasso in collaboration with the forestry and extension services helped develop a local forest co-management agreement among three villages. These cases confirm and reinforce the following lessons learned from the NRMAC experience in Madiama.

Lesson 1: Including all stakeholders is a necessary but problematic task. All four projects experienced difficulty maintaining sustained communications and assuring the full participation of women and pastoralists (particularly non-resident pastoralists) in their activities and decision making structures. Furthermore, participation is not limited to natural resource end users; customary authorities, government officials, technical service agents and local NGOs should also be included. Accountability and sustained communication are necessary reinforcements for stakeholder inclusion.

Lesson 2: There is no single model for building social capital and developing local management agreements. The process must be iterative, allowing for the adaptive learning that takes a considerable amount of time (years) to bring all stakeholders together in an informed and voluntary manner, and to negotiate consensus on each component of an accord. As we have documented, the process needs time to evolve and mature, because building trust between groups is a time-consuming process.

Lesson 3: Project and partner personnel need to be well trained to encourage open debate, foster consensus building and guide without leading. Respect for and recognition of local know-how is essential. This is something which we, as development agents can control, because respect and recognition depend on our own behaviour regarding the population. This is as true for local leaders as it is for government service agents. Mistakes will be made and should be expected. They can be used as lessons. Training programmes in conflict resolution and consensus building have made major contributions in this regard.

Lesson 4: Power relations and stakeholder interests need to be carefully taken into account. If the interests of any stakeholder, whether socially powerful or not, are ignored, full implementation of collective endeavours will be compromised. Training in conflict management can assist in bringing divisive issues into the open so that they can be confronted and managed.

Lesson 5: Development agents must focus interventions on developing synergy between the public and private sectors and on reinforcing the networks of local associations, cultivating horizontal bonds across communities, thereby facilitating the mobilization of viable, locally defined initiatives. This is the new agenda for research and development of community life in West Africa.

The research/development process: interdisciplinarity

The final SANREM pillar, interdisciplinarity, is a form of participation peculiar to the philosophers and planners of development. In this respect, the SANREM CRSP Landscape/Lifescape approach (amplified by Holistic Management) has integrated local perspectives and allowed researchers to address real world problems in ways that are meaningful to all natural resource users. Further, we have seen that listening to farmer perspectives improves the quality of decision making tools (models) for higher-level decision makers. The landscape types Badini and Dioni introduced in Chapter 3 and applied in structuring the analysis and modelling in Chapter 11, build on farmer insights noted by Earl *et al.* in Chapter 6 and Crane and Traoré in Chapter 10. The tethered grazing research conducted by Abaye *et al.* in Chapter 15 was shaped by interaction with the model of Holistic Management (HM) introduced in Chapter 9. The evaluation of technologies by Wyeth *et al.* in Chapter 12 relies on layers of interdisciplinary work from the field through several different laboratories.

Interdisciplinarity assures dialogue between multiple perspectives in the process of holistic research. As researchers, we are obliged to choose an aspect, to isolate and examine it, and finally to apply the findings to improve livelihoods. This reductionist approach is the way modern science advances, but it has also produced monoculture cropping and the weaknesses of an unbalanced landscape. Through interdisciplinarity, we have attempted to re-establish the balance in our recommendations. However, it should be noted that there are tensions between disciplinary and holistic perspectives. It has not always been an easy path to follow, particularly across the divide between the social and biophysical sciences. Some of these issues were highlighted in Bingham's (Chapter 9) discussion of HM which advocated a unified perspective for SANREM that conceptually linked the biophysical and social sciences.

Although it has provided a framework for discussion among researchers and development agents and with local natural resource users, HM has been very difficult to implement in practice. Scientists from divergent disciplinary perspectives began to think about and understand some basic concepts across disciplines. However, these discussions often broke down into a multi-disciplinary parallelism, leaving researchers to integrate individual insights into their own narrowly focused work. The forming of joint projects did not really occur until individual scientists had made disciplinary advances of their own.

Ultimately, HM concepts did help drive research directions, such as manure exchange, through demonstration of field stabling of livestock and the identification of grazing management indicators for holistic pasture management. Also, by providing a framework for community discussion and decision-making concerning NRM alternatives, these concepts helped lay to rest agriculture versus environment debates and highlight complementarities and integration.

This work may be criticized for missing or ignoring important issues or not going deeply enough into specific themes. Indeed, we have recognized these shortfalls and expect criticism, because our particular disciplinary and professional perspectives often focus attention on certain issues and blind us to others. However, multidisciplinarity (the foundation for interdisciplinarity) is in itself difficult to achieve given the resources available to such research projects. Inter- or multi-disciplinarity poses serious financial constraints on a programme because of the necessity to fund individuals representing the number of disciplines required for adequate problem identification and sufficient cross fertilization of ideas and perspectives. Furthermore, there is a need for greater flexibility in financial management of projects working to empower local populations. The iterative process of bringing all stakeholders to consensus does not follow a set schedule.

Conclusion

The SANREM CRSP–West Africa Research Project has been a participatory programme building science-based solutions while promoting a holistic approach for sustainable agriculture and natural resource management. Our work in Madiama is replicable in other communes of Mali and throughout West Africa. Indeed, we have already found considerable interest among local leaders in neighbouring communes. This approach has consisted of institution building activities supported by agronomic and livestock management interventions, databases, and decision-support systems. The process reinforces decentralized NRM by applying conflict resolution, leadership development and empowerment, and by using science-based technical expertise in agricultural research and social infrastructure development to create technological options and build social capital for local decision making.

Despite its short history, SANREM can point to significant project impacts, the most important of which have been social and institutional. Building capacity through conflict resolution has been empowering, as has the networking of villagers created by the NRMAC. Together these two interventions have led to increased social capital, building greater trust and ultimately lowering transaction costs for community development activities. The NRMAC is now widely perceived as an important actor in environmental protection and economic development for the Commune. Its role in sustaining and developing gains made through the project is evidenced by their support from within the community (collection of membership fees). The mandate of the current NRMAC was renewed with their re-election to a new term and the support of the mayor and village chiefs. The NRMAC is building social capital within the community and region by bridging relationships between villages within and outside the Commune, and with government services and local authorities. In one of its first activities, the committee organized and implemented the Mayor's campaign for protection of the *Acacia albida*, a nitrogen-fixing tree native to the Sahel, thereby directly impacting soil fertility and productivity. The NRMAC is also orchestrating the establishment of inter-village accords for management of wetlands (*bourgoutières*) and for open-range rotational grazing. Training in HM has built the NRMAC's capacity to act as the primary mechanism for channelling farmers' and herders' input in assessing soil fertility and pasture management technologies being experimented with by the project.

More importantly, the NRMAC has proved to be an effective catalyst for improvements in social relations between villages and ethnic groups within the commune. The leadership and socio-institutional surveys indicate that there has been a marked decrease in the occurrence of NRM conflicts and in people's perception of conflict as a major problem in the commune since the committee has been established. On at least two occasions the NRMAC has intervened to avert escalation of conflict by applying concepts and skills learned in the conflict management training workshops. The *Sous-Prefet*, responsible for several communes in the *Cercle*, has testified that the project's institutional innovations and capacity building interventions have enabled the successful uptake of decentralized NRM responsibilities by the Rural Commune of Madiama to a much greater degree than elsewhere.

NRMAC members have been trained in livestock and soil management decision making through the introduction of Holistic Management as a diagnostic and management tool for NRM interventions. On the basis of this training, the NRMAC has defined a common holistic vision for developing their community and its members now function as trainers and facilitators building local capacity and commitment to attain their collective goal. In addition to the documented advances in local institutional capacity, the database and modelling efforts have improved resource managers and planners' capacity to make NRM decisions that result in sustainable and equitable outcomes. The CropSyst Model and the SAM are providing decision-making information to researchers as well as planners at various administrative levels by enabling them to analyse the long-term implications of various technology and policy options.

References

Bocoum, A., Cochrane, K., Diakité, M. and Kane, O. (2003) L'inclusion sociale pour une gestion équitable et durable des ressources naturelles: deux expériences au Mali. Securing the commons, No. 7. SOS-Sahel International Institute for Environment and Development, London, UK, pp. 1–45.

Evans, P. (1996) Government action, social capital and development: reviewing the evidence on synergy. *World Development* 24(6), 1119–1132.

Flora, J. (1998) Social capital and communities of place. *Rural Sociology* 63(4), 481–506.

Guèye, Bara (no date) *Whither Participation? Experience from francophone West Africa.* MARP Sahel Programme, Dakar, Sénégal.

Hilhorst, T. and Aarnink, N. (1999) Co-managing the commons: setting the stage in Mali and Zambia. *Bulletin 346.* Royal Tropical Institute, Amsterdam, The Netherlands, pp. 1–77.

Ribot, J.C. (1999) Decentralization, participation and accountability in Sahelian forestry: Legal instruments of political-administrative control. *Africa* 69(1), 23–65.

Index

Page numbers in **bold** refer to illustrations and tables

Crop Pollination by Bees
Keith S. Delaplane and Daniel F. Mayer

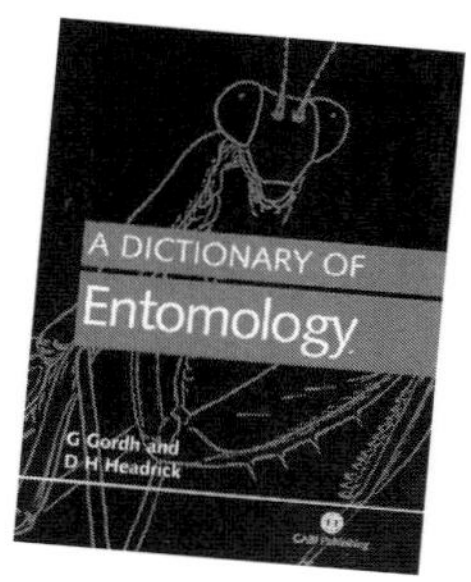
A DICTIONARY OF
Entomology
G Gordh and
D H Headrick

Principles of
CATTLE PRODUCTION
C.J.C. Phillips

Seeds
THE ECOLOGY OF REGENERATION IN PLANT COMMUNITIES
2nd EDITION
Edited by
Michael Fenner